# Electrical Engineering and Intelligent Systems

**Lecture Notes in Electrical Engineering**

Volume 130

For further volumes:
http://www.springer.com/series/7818

Sio-Iong Ao • Len Gelman

Editors

# Electrical Engineering
# and Intelligent Systems

 Springer

*Editors*
Sio-Iong Ao
International Association of Engineers
Hong Kong, China

Len Gelman
School of Engineering
Applied Mathematics and Computing
Cranfield University
Cranfield, Bedford MK43 0AL, UK

ISSN 1876-1100          e-ISSN 1876-1119
ISBN 978-1-4899-9526-1    ISBN 978-1-4614-2317-1 (eBook)
DOI 10.1007/978-1-4614-2317-1
Springer New York Dordrecht Heidelberg London

Printed on acid-free paper

Springer is part of Springer Science+Business Media (www.springer.com)

# Preface

A large international conference in electrical engineering and intelligent systems was held in London, UK, 6–8 July, 2011, under the World Congress on Engineering (WCE 2011). The WCE 2011 was organized by the International Association of Engineers (IAENG); the Congress details are available at: http://www.iaeng.org/WCE2011. IAENG is a non-profit international association for engineers and computer scientists, which was founded originally in 1968. The World Congress on engineering serves as good platforms for the engineering community to meet with each other and exchange ideas. The conferences have also struck a balance between theoretical and application development. The conference committees have been formed with over 200 members who are mainly research center heads, faculty deans, department heads, professors, and research scientists from over 30 countries. The conferences are truly international meetings with a high level of participation from many countries. The response to the Congress has been excellent. There have been more than 1,300 manuscript submissions for the WCE 2011. All submitted papers have gone through the peer review process, and the overall acceptance rate is 56.93%.

This volume contains 33 revised and extended research articles written by prominent researchers participating in the conference. Topics covered include computational intelligence, control engineering, network management, wireless networks, signal processing, internet computing, high performance computing, and industrial applications. The book offers the state of the art of tremendous advances in electrical engineering and intelligent systems and also serves as an excellent reference work for researchers and graduate students working on electrical engineering and intelligent systems.

Hong Kong, China
Cranfield, Bedford MK43 0AL, UK

Sio-Iong Ao
Len Gelman

# Contents

**1 Partial Exploration of State Spaces and Hypothesis Test for Unsuccessful Search** ............................................... 1
Eleazar Jimenez Serrano

**2 Duality in Complex Stochastic Boolean Systems** ..................... 15
Luis González

**3 Reconfigurable Preexecution in Data Parallel Applications on Multicore Systems** .................................................. 29
Ákos Dudás and Sándor Juhász

**4 Efficient Implementation of Computationally Complex Algorithms: Custom Instruction Approach** .......................... 39
Waqar Ahmed, Hasan Mahmood, and Umair Siddique

**5 Learning of Type-2 Fuzzy Logic Systems by Simulated Annealing with Adaptive Step Size** .................................. 53
Majid Almaraashi, Robert John, and Samad Ahmadi

**6 Genetic Algorithms for Integrated Optimisation of Precedence-Constrained Production Sequencing and Scheduling** ........................................... 65
Son Duy Dao and Romeo Marin Marian

**7 A Closed-Loop Bid Adjustment Method of Dynamic Task Allocation of Robots** ........................................... 81
S.H. Choi and W.K. Zhu

**8 Reasoning on OWL2 Ontologies with Rules Using Metalogic** ...... 95
Visit Hirankitti and Trang Xuan Mai

**9 Bayesian Approach to Robot Group Control** ....................... 109
Tomislav Stipancic, Bojan Jerbic, and Petar Curkovic

**10**  **A Palmprint Recognition System Based on Spatial Features** ....... 121
Madasu Hanmandlu, Neha Mittal, and Ritu Vijay

**11**  **An Interactive Colour Video Segmentation:**
**A Granular Computing Approach** .................................... 135
Abhijeet Vijay Nandedkar

**12**  **A Novel Approach for Heart Murmurs Detection**
**and Classification** ...................................................... 147
Maamar Ahfir and Izzet Kale

**13**  **A Fuzzy Logic Approach to Indoor Location**
**Using Fingerprinting** .................................................. 155
Carlos Serodio, Luis Coutinho, Hugo Pinto, Joao Matias,
and Pedro Mestre

**14**  **Wi-Fi Point-to-Point Links: Extended Performance Studies**
**of IEEE 802.11 b,g Laboratory Links Under**
**Security Encryption** ..................................................... 171
J.A.R. Pacheco de Carvalho, H. Veiga, N. Marques,
C.F. Ribeiro Pacheco, and A.D. Reis

**15**  **Wi-Fi IEEE 802.11 B,G Wep Links: Performance Studies**
**of Laboratory Point-to-Point Links** ................................. 183
J.A.R. Pacheco de Carvalho, H. Veiga, N. Marques,
C.F. Ribeiro Pacheco, and A.D. Reis

**16**  **Integrating Mobility, Quality-of-service and Security**
**in Future Mobile Networks** ........................................... 195
Mahdi Aiash, Glenford Mapp, Aboubaker Lasebae,
Raphael Phan, and Jonathan Loo

**17**  **Decision Under Uncertainties of Online Phishing** ................... 207
Ping An Wang

**18**  **Using Conceptual Graph Matching Methods**
**to Semantically Mediate Ontologies** ................................. 219
Gopinath Ganapathy and Ravi Lourdusamy

**19**  **Asymmetric Optimal Hedge Ratio with an Application** ............. 231
Youssef El-Khatib and Abdulnasser Hatemi-J

**20**  **Computations of Price Sensitivities After**
**a Financial Market Crash** ............................................. 239
Youssef El-Khatib and Abdulnasser Hatemi-J

**21**  **TRP Ratio and the Black–Litterman Portfolio**
**Optimisation Method** .................................................. 249
Gal Munda and Sebastjan Strasek

22  An Overview of Contemporary and Individualised
    Approaches to Production Maintenance............................  263
    James Hogan, Frances Hardiman, and Michael Daragh Naughton

23  Weibull Prediction Limits for a Future Number
    of Failures Under Parametric Uncertainty.........................  273
    Nicholas A. Nechval, Konstantin N. Nechval, and Maris Purgailis

24  Vendors' Design Capabilities Enabler Towards Proton
    Internationalization Strategy........................................  285
    Ana Sakura Zainal Abidin, Rosnah Mohd. Yusuff,
    Nooh Abu Bakar, Mohd. Azni Awi, Norzima Zulkifli,
    and Rasli Muslimen

25  Effects of Leaked Exhaust System on Fuel Consumption
    Rate of an Automobile................................................  301
    Peter Kayode Oke and Buliaminu Kareem

26  Designing a Robust Post-Sales Reverse Logistics Network..........  313
    Ehsan Nikbakhsh, Majid Eskandarpour,
    and Seyed Hessameddin Zegordi

27  A Case Study of Lean Manufacturing Implementation
    Approach in Malaysian Automotive
    Components Manufacturer............................................  327
    Rasli Muslimen, Sha'ri Mohd. Yusof,
    and Ana Sakura Zainal Abidin

28  Investing in the Sheep Farming Industry: A Study Case
    Based on Genetic Algorithms........................................  337
    Iracema del Pilar Angulo-Fernandez, Alberto Alfonso
    Aguilar-Lasserre, Magno Angel Gonzalez-Huerta,
    and Constantino Gerardo Moras-Sanchez

29  High-Precision Machining by Measuring
    and Compensating the Error Motion of Spindle's
    Axis of Rotation in Radial Direction...............................  347
    Ahmed A.D. Sarhan and Atsushi Matsubara

30  Modeling a Complex Production Line Using Virtual Cells.........  361
    Luís Pinto Ferreira, Enrique Ares Gómez,
    Gustavo Peláez Lourido, and Benny Tjahjono

31  Optimal Quantity Discount Strategy for an Inventory
    Model with Deteriorating Items ....................................  375
    Hidefumi Kawakatsu

**32  Effect of Cutting Parameters on Surface Finish
for Three Different Materials in Dry Turning** ...................... 389
Mohammad Nazrul Islam and Brian Boswell

**33  Impact of Radio Frequency Identification
on Life Cycle Engineering** ............................................. 403
Can Saygin and Burcu Guleryuz

**Index** ................................................................. 415

# Contributors

**Ana Sakura Zainal Abidin** Department of Mechanical and Manufacturing Engineering, Faculty of Engineering, Universiti Malaysia Sarawak, Sarawak, Malaysia

**Alberto Alfonso Aguilar-Lasserre** Division de Estudios de Posgrado e Investigacion, Instituto Tecnologico de Orizaba, Orizaba, Veracruz, Mexico

**Maamar Ahfir** Department of Informatics, University of Laghouat, Laghouat, Algeria

**Samad Ahmadi** The Centre for Computational Intelligence, De Montfort University, Leicester, UK

**Waqar Ahmed** Department of Electronics, Quaid-i-Azam University, Islamabad, Pakistan

**Mahdi Aiash** School of Engineering and Information Sciences, Middlesex University, Middlesex, UK

**Majid Almaraashi** The Centre for Computational Intelligence, De Montfort University, Leicester, UK

**Iracema del Pilar Angulo-Fernandez** Division de Estudios de Posgrado e Investigacion, Instituto Tecnologico de Orizaba, Orizaba, Veracruz, Mexico

**Mohd. Azni Awi** Group Procurement, Perusahaan Otomobil Nasional Sdn. Bhd., Selangor, Malaysia

**Nooh Abu Bakar** School of Graduate Studies, UTM International Campus, Kuala Lumpur, Malaysia

**Brian Boswell** Department of Mechanical Engineering, Curtin University, Perth, WA, Australia

**S.H. Choi** Department of Industrial and Manufacturing Systems Engineering, The University of Hong Kong, Hong Kong, China

**Luis Coutinho** UTAD, Vila Real, Portugal

**Petar Curkovic** Faculty of Mechanical Engineering and Naval Architecture, University of Zagreb, Zagreb, Croatia

**J.A.R. Pacheco de Carvalho** Unidade de Detecção Remota, Universidade da Beira Interior, Covilhã, Portugal

**Son Duy Dao** School of Advanced Manufacturing and Mechanical Engineering, University of South Australia, Mawson Lakes, SA, Australia

**Ákos Dudás** Department of Automation and Applied Informatics, Budapest University of Technology and Economics, Budapest, Hungary

**Youssef El-Khatib** Department of Mathematical Sciences, UAE University, Al-Ain, United Arab Emirates

**Majid Eskandarpour** Department of Industrial Engineering, Faculty of Engineering, Tarbiat Modares University, Tehran, Iran

**Luís Pinto Ferreira** Escola Superior de Estudos Industriais e de Gestão (Technical Scientific Unity of Industrial Engineering and Production), Instituto Politécnico do Porto, Porto, Portugal

**Gopinath Ganapathy** Department of Computer Science, Bharathidasan University, Tiruchirappalli, India

**Enrique Ares Gómez** Área Ingeniería de los Procesos de Fabricación, Universidad de Vigo, Vigo, Spain

**Luis González** Research Institute IUSIANI, Department of Mathematics, University of Las Palmas de Gran Canaria, Las Palmas de Gran Canaria, Spain

**Magno Angel Gonzalez-Huerta** Division de Estudios de Posgrado e Investigacion, Instituto Tecnologico de Orizaba, Orizaba, Veracruz, Mexico

**Burcu Guleryuz** Mechanical Engineering Department, The University of Texas at San Antonio, San Antonio, TX, USA

**Madasu Hanmandlu** Electrical Engineering Department, Indian Institute of Technology, New Delhi, India

**Frances Hardiman** Department of Mechanical and Automobile Technology, Limerick Institute of Technology, Limerick, Ireland

**Abdulnasser Hatemi-J** Department of Economics and Finance, UAE University, Al-Ain, United Arab Emirates

**Visit Hirankitti** School of Computer Engineering, King Mongkut's Institute of Technology Ladkrabang, Ladkrabang, Bangkok, Thailand

**James Hogan** Department of Mechanical and Automobile Technology, Limerick Institute of Technology, Limerick, Ireland

**Mohammad Nazrul Islam** Department of Mechanical Engineering,
Curtin University, Perth, WA, Australia

**Bojan Jerbic** Faculty of Mechanical Engineering and Naval Architecture,
University of Zagreb, Zagreb, Croatia

**Robert John** The Centre for Computational Intelligence,
De Montfort University, Leicester, UK

**Sándor Juhász** Department of Automation and Applied Informatics,
Budapest University of Technology and Economics, Budapest, Hungary

**Izzet Kale** Applied DSP and VLSI Research Group, Department of
Electronic Systems, University of Westminster, London, UK

**Buliaminu Kareem** Department of Mechanical Engineering,
The Federal University of Technology, Akure, Ondo State, Nigeria

**Hidefumi Kawakatsu** Department of Economics and Information Science,
Onomichi University, Onomichi, Japan

**Aboubaker Lasebae** School of Engineering and Information Sciences,
Middlesex University, Middlesex, UK

**Jonathan Loo** School of Engineering and Information Sciences,
Middlesex University, Middlesex, UK

**Ravi Lourdusamy** Department of Computer Science,
Sacred Heart College, Tirupattur, India

**Gustavo Peláez Lourido** Área Ingeniería de los Procesos de Fabricación,
Universidad de Vigo, Vigo, Spain

**Hasan Mahmood** Department of Electronics, Quaid-i-Azam University,
Islamabad, Pakistan

**Trang Xuan Mai** International College, King Mongkut's Institute of Technology
Ladkrabang, Ladkrabang, Bangkok, Thailand

**Glenford Mapp** School of Engineering and Information Sciences,
Middlesex University, Middlesex, UK

**Romeo Marin Marian** School of Advanced Manufacturing and Mechanical
Engineering, University of South Australia, Mawson Lakes, SA, Australia

**N. Marques** Centro de Informática, Universidade da Beira Interior,
Covilhã, Portugal

**Joao Matias** Centre for Mathematics - UTAD, Vila Real, Portugal

**Atsushi Matsubara** Department of Micro Engineering,
Graduate School of Engineering, Kyoto University, Kyoto, Japan

**Pedro Mestre** CITAB-UTAD, Vila Real, Portugal

**Neha Mittal** Electronics & Communication Engineering Department,
J.P. Institute of Engineering & Technology, Meerut, India

**Constantino Gerardo Moras-Sanchez** Division de Estudios de Posgrado
e Investigacion, Instituto Tecnologico de Orizaba, Orizaba, Veracruz, Mexico

**Gal Munda** PricewaterhouseCoopers, Times Valley, Uxbridge, UK

**Rasli Muslimen** Department of Mechanical and Manufacturing Engineering,
Faculty of Engineering, Universiti Malaysia Sarawak, Sarawak, Malaysia

**Abhijeet Vijay Nandedkar** Department of Electronics and Telecommunication
Engineering, S.G.G.S. Institute of Engineering & Technology,
Vishnupuri, Nanded, Maharashtra, India

**Michael Daragh Naughton** Department of Mechanical and Automobile
Technology, Limerick Institute of Technology, Limerick, Ireland

**Nicholas A. Nechval** Department of Statistics, EVF Research Institute,
University of Latvia, Riga, Latvia

**Konstantin N. Nechval** Department of Applied Mathematics,
Transport and Telecommunication Institute, Riga, Latvia

**Ehsan Nikbakhsh** Department of Industrial Engineering, Faculty of Engineering,
Tarbiat Modares University, Tehran, Iran

**Peter Kayode Oke** Department of Mechanical Engineering,
The Federal University of Technology, Akure, Ondo State, Nigeria

**C.F. Ribeiro Pacheco** Unidade de Detecção Remota,
Universidade da Beira Interior, Covilhã, Portugal

**Raphael Phan** Department of Electrical and Electronic Engineering,
Loughborough University, Loughborough, UK

**Hugo Pinto** UTAD, Vila Real, Portugal

**Maris Purgailis** Department of Cybernetics, University of Latvia, Riga, Latvia

**A.D. Reis** Unidade de Detecção Remota, Universidade da Beira Interior,
Covilhã, Portugal

**Ahmed A.D. Sarhan** Center of Advanced Manufacturing and Material Processing,
Department of Engineering Design and Manufacture, Faculty of Engineering,
University of Malaya, Kuala Lumpur, Malaysia

**Can Saygin** Mechanical Engineering Department, The University of Texas
at San Antonio, San Antonio, TX, USA

**Carlos Serodio** CITAB-UTAD, Vila Real, Portugal

**Eleazar Jimenez Serrano** Department of Automotive Science, Graduate School
of Integrated Frontier Science, Kyushu University, Fukuoka, Japan

**Umair Siddique** Research Center for Modeling and Simulation,
National University of Sciences and Technology (NUST), Islamabad, Pakistan

**Tomislav Stipancic** Faculty of Mechanical Engineering and Naval Architecture,
University of Zagreb, Zagreb, Croatia

**Sebastjan Strasek** The University of Maribor, Maribor, Slovenia

**Benny Tjahjono** Manufacturing Department, Cranfield University,
Cranfield, UK

**H. Veiga** Centro de Informática, Universidade da Beira Interior,
Covilhã, Portugal

**Ritu Vijay** Electronics and Communication Engineering Department,
Banasthali Vidyapeeth, Banasthali University, Rajasthan, India

**Ping An Wang** Graduate School Cybersecurity Department,
University of Maryland University College, Adelphi, MD, USA

**Sha'ri Mohd. Yusof** Department of Manufacturing and Industrial Engineering,
Faculty of Mechanical Engineering, Universiti Teknologi Malaysia,
UTM Skudai, Johor, Malaysia

**Rosnah Mohd. Yusuff** Department of Mechanical and Manufacturing Engineering,
Faculty of Engineering, Universiti Putra Malaysia, Selangor,
UPM Serdang, Malaysia

**Seyed Hessameddin Zegordi** Department of Industrial Engineering,
Faculty of Engineering, Tarbiat Modares University, Tehran, Iran

**W.K. Zhu** Department of Industrial and Manufacturing Systems Engineering,
The University of Hong Kong, Hong Kong, China

**Norzima Zulkifli** Department of Mechanical and Manufacturing Engineering,
Faculty of Engineering, Universiti Putra Malaysia, Selangor,
UPM Serdang, Malaysia

# Chapter 1
# Partial Exploration of State Spaces and Hypothesis Test for Unsuccessful Search

Eleazar Jimenez Serrano

## 1 Introduction

Validating the correctness of systems by generating the corresponding state space (SSp) is a common method limited by the possible exponential growth of its size, known as the SSp explosion problem. Probabilistic methods focus on analyzing just a fraction of the SSp by means of partial exploration, but the probability of omitting states is greater than zero, sometimes incorrect evaluations are bound to happen and conclusiveness is not achievable all the time.

Our research is to determine the existence of certain states without having to explore the entire SSp. This is done with a reachability analysis, by using a partial exploration algorithm.

The lack of conclusiveness in the analysis upon unsuccessful search is confronted here. We present how to treat the results with a statistical hypothesis test and decide with a level of confidence if certain state exists or not.

In the confines of reachability analysis in SSp generated from Petri net (PN) systems, there is a large list of works doing partial exploration [1, 5, 6, 11]. Although good results exist for some specific cases, others lack conclusiveness due to unsuccessful search [7, 8, 10].

The output of any partial exploration can be seen as a probabilistic sample. Upon unsuccessful search, a hypothesis test can be conducted for completeness using the sample. In general, random walk is used in the partial exploration algorithm, but random walk cannot be used as a sampling method because the probability that a given state is visited is far from being uniform [13]. Therefore, the sample is biased and any posterior hypothesis test will be incorrect.

E.J. Serrano (✉)
Department of Automotive Science, Graduate School of Integrated Frontier Science,
Kyushu University, 744 Motooka, Nishi-ku, Fukuoka 819-0395, Japan
e-mail: eleazar.jimenez.serrano@kyudai.jp

S.-I. Ao and L. Gelman (eds.), *Electrical Engineering and Intelligent Systems*,
Lecture Notes in Electrical Engineering 130, DOI 10.1007/978-1-4614-2317-1_1,
© Springer Science+Business Media, LLC 2013

In this chapter first we show how to obtain a sample during the partial exploration of the SSp which is utilizable in any posterior hypothesis test, and second we explain how to give conclusiveness to the reachability analysis in the case of unsuccessful search by means of a hypothesis test.

This chapter presents a partial exploration algorithm guided by a quota sampling. It is basically a table of quotas to fill in during the exploration. The sampling method is neither totally unbiased nor valid variance estimate, but it goes in the direction toward obtaining a representative sample. Then, we show a proper treatment to the sample by means of a hypothesis test [10] to provide conclusiveness to the analysis.

## 2 Petri Nets and State Spaces

One analysis strategy for the state space (SSp) is exploration. The exploration can be complete and partial. A major difficulty in the partial exploration is to conduct a random search with proper length and simultaneously traversing it all.

To verify that a random sample of the SSp has been obtained from the corresponding exploration, we need the probability distribution of the variable $Y$, a metric of every state in the set $\Psi$ of the SSp. Since it is difficult if not impossible to know in advance the distribution of $Y$, by looking at a variable $X$, same metric of every state of the easily known superset $\Omega$ of the SSp, some parameters of $Y$ can be estimated.

To explain this, a model of system behavior like Petri nets will be used to graphically identify the superset $\Omega$ of the SSp. This section will introduce briefly the necessary theoretical concepts required to continue.

### 2.1 Petri Nets

A Petri net is a tuple $(P, T, A, B, Q)$, where $P$ is a finite nonempty set of $i$ places, $T$ a finite nonempty set of $j$ transitions, $A$ the set of directed arcs connecting places to transitions, $B$ the set of directed arcs connecting transitions to places, and $Q$ is a capacity function for the places mapping $P \rightarrow N$.

A Petri net is mathematically represented with the pre- and post-incident matrices $\vec{A}$ and $\vec{B}$, respectively, having both $i$ rows and $j$ columns, with values of $[A(p_i, t_j)]$ and $[B(p_i, t_j)]$, respectively.

The way to describe a state of the system is by putting tokens in the corresponding places. Tokens are black dots that exist only in the places. The function $m$ called marking maps $P \rightarrow N$, and $m_0$ is the initial marking. A PN with initial marking is called a PN system. We say that a place $p$ is marked when $m(p) > 0$. A marking $m$ is represented with a vector $\vec{m} \in N^i$.

The finite set of all possible markings (i.e., the reachability space) of a PN system is denoted by $\Psi(m_0)$, and $r_{m_0}$ denotes the number of elements in the set $\Psi(m_0)$.

The number of states generated from a PN system depends on the input and output arcs, the initial marking, and the way how the occurrence of transitions is specified. Occurrence of single transition is carried out for complete exploration of the SSp of a PN system. Occurrence of solely concurrent transitions could lead to incomplete exploration of the SSp.

In the general case, the only way to know all the markings that could result from one initial marking $m_0$ is throughout complete exploration of the SSp. Each marking can be generated with the state transition function $\vec{m}_n = \vec{m}_c + (\vec{B} - \vec{A}) \times \vec{\sigma}$. In the formula $\vec{m}_c$ is the current marking, $\vec{m}_n$ is the next marking, and $\vec{\sigma}$ is a firing count vector representing the number of times every transition has fired. For additional details about PN the reader is addressed to [12].

Due to the SSp explosion problem many times complete exploration is neither practical nor feasible. However, a raw estimation of the number $r_{m_0}$ can be calculated from the superset $\Omega$ of the SSp of the PN system as:

$$\hat{r}_{m_0} = \prod_{n=1}^{i} (q_n + 1).$$

In the formula $q_n$ is the token capacity of the $i$ place.

Also, it is not easy to calculate the maximal number of tokens $k_{m_0}$ that could exist among all the configurations of the SSp without complete exploration, but a raw estimate using the superset $\Omega$ is calculated as

$$\hat{k}_{m_0} = \sum_{n=1}^{i} (q_n).$$

## 2.2  Marked Graph Petri Net

The pre-conditions of a transition $t$ are in the set of input places $\bullet t$ and the post-conditions in $t\bullet$. The pre-events of a place $p$ are in the set $\bullet p$ and the post-events in the set $p\bullet$.

A marked graph is a PN such that $|\bullet p| = |p\bullet| = 1 \ \forall p \in P$, i.e., single arc as input and output for each place.

For the case of the SSp of an Marked Graph Petri Net (MGPN) system having places with infinite token capacity, by assigning a fixed value to $q_n$ we can estimate $r_{m_0}$ and $k_{m_0}$. We will not explain how to obtain the estimators $\hat{r}_{m_0}$ and $\hat{k}_{m_0}$; nevertheless for an understanding in these matters the reader is addressed to the theory and results from [3, 4, 9, 16–18].

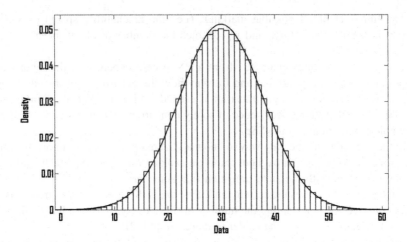

**Fig. 1.1** A histogram of an arbitrary $X$

In the rest of this section we will assume $\hat{r}_{m_0}$ and $\hat{k}_{m_0}$ obtained from the superset $\Omega$ are refined estimations of the SSp of an MGPN system with initial marking $m_0$ such that the differences $\hat{r}_{m_0} - r_{m_0} \geq 0$ and $\hat{k}_{m_0} - k_{m_0} \geq 0$.

## 2.3    Probability Distribution of the Variable X of the Superset

For any marking $m \in \Omega$ let us define the variable $X$ as the number of tokens in $m$ (the cardinality of $m$). Independently of the number of places and the different but finite token capacity of the places of the PN, if we create a histogram of $X$ we get a distribution of frequencies which looks like the one in Fig. 1.1. The appearance of $X$ is of a normal probability distribution function, i.e., $X \sim N(\bar{x}, s^2)$ and its values the integers from 0 to $\hat{k}_{m_0}$.

For any PN, with any number of places and finite tokens capacities, any distribution of $X$ appears like a normal probability distribution. Formally, only a test of goodness of fit can confirm this statement, but in this paper the previous statement is assumed to be true.

The mean of $X$ can be estimated as $\hat{\bar{x}} = \hat{k}_{m_0}/2$. On the other hand, the variance represents a small challenge. Taken from the statistical "68–95–99.7" rule, stating that for a normal distribution of a large population nearly all the values lie within $3\sigma$ (three standard deviations from the mean), an estimator of the standard deviations is $\hat{s} = \hat{\bar{x}}/3 = \hat{k}_{m_0}/6$ (see Fig. 1.2).

A better estimator of the standard deviations involves having a factor of proportional error $\tau$ between the estimation and the real value with the form $\hat{k}_{m_0}/6\tau$. Figure 1.3 shows in ascending order the proportional error $\tau$ for 1, 10, 100, and 1,000 token capacity in the places, respectively.

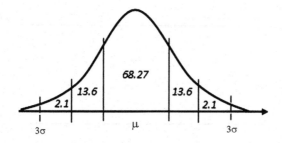

**Fig. 1.2**  A segmented histogram of an arbitrary $X$

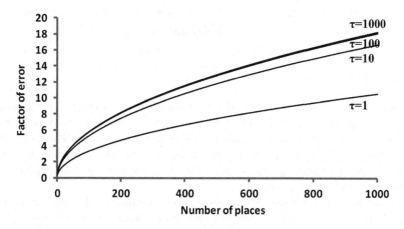

**Fig. 1.3**  Proportional error between estimated and real variance

The error $\tau$ can be estimated with a function $\varphi \times i^{1/2}$, where $\varphi$ is a constant associated with the token capacity of the places and $i$ the number of places. The values of $\varphi$ for the 1, 10, 100, and 1,000 token capacity can be seen in Fig. 1.4.

Finally, in the rest of this chapter we are assuming for the case when $\hat{r}_{m_0} - r_{m_0} \cong 0$ and $\hat{k}_{m_0} - k_{m_0} \cong 0$, the probability distribution of $Y \cong X$ and the mean $\bar{x}_{m_0}$ and variance $s^2_{m_0}$ of $X$ are estimations of the mean $\mu$ and variance $\sigma^2$ of $Y$, respectively.

# 3   Partial Exploration by Quota Sampling

A probability sample is one for which every unit in a finite population has a positive probability of selection, but not necessarily equal to that of other units, like random (or uniform probability) samples [15].

Exploration methods, called sampling methods when there is no place for confusion, may be probabilistic or not. Random walk is the main algorithm in many probabilistic methods of SSp exploration, but so far there is no universal

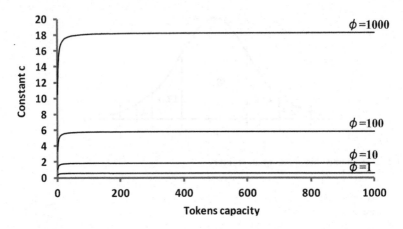

**Fig. 1.4** Values of the constant φ

solution in the framework of random walk based partial exploration methods. Moreover, the random walk cannot be considered to be random because the probability that a given state is visited is far from being uniform. Random walk as a method of SSp exploration is only useful for small models, having not practical interest in model checking, and so it is that the choice of the best exploration method is model dependent [13].

Nevertheless, non-probability samples are useful for quick and cheap studies, for case studies, for qualitative research, for pilot studies, and for developing hypotheses for future research. While random sampling allows use of statistics and tests hypotheses, non-probability (nonrandom) sampling are for exploratory research and generate hypotheses, where population parameters are not of interest (and cannot be estimated), and adequacy of the sample cannot be known due to the bias (non-probability sampling introduces unknown biases). In general, any generalization obtained from a non-probability sample must be filtered through the analyst knowledge of the topic being studied.

Non-probabilistic sampling methods appeal more to notions of representativity [14]. In this section we explain how to get a representative sample of the SSp according to the variable $Y$ which is assumed to be normally distributed. The sample is obtained with a partial exploration algorithm guided by a quota sampling [14].

## 3.1   Exploration Algorithm

The result of our analysis is to decide whether a target marking exists or not in the SSp of a PN system. The method uses an algorithm primarily directing the exploration toward fulfilling the quotas of a table of samples while searching for the target marking $m_t$ under analysis.

First we calculate the number of states to explore (the number of samples to collect). We use the proportion $\rho$ of markings with the same cardinality as the target marking $m_t$ taken from the assumed normal probability distribution of the SSp with respect to $X$. A classical estimator to calculate the number of samples in a large population for a proportion is in the next formula, where $\upsilon$ is the confidence level (e.g., $\upsilon = 1.29$ for 99%) and $e$ is the acceptable error (e.g., $e = 0.05$ means we accept a difference up to 5% in the proportion) [2].

$$ n = \frac{\upsilon^2 \rho(1 - \rho)}{e^2}. $$

In our exploration, each state might be a valid sample. All samples are segmented into six mutually exclusive groups. The groups are represented in a table of quotas with six slots ($sl_g$ with $g = 1, \ldots, 6$). The table of quotas accounts only for $n$ valid samples. Each slot is used as a counter of only valid samples.

The proportion of states in a group will be applied to the quota of a slot. The proportions are taken from the assumed normal probability distribution of the SSp with respect to $Y$ with estimated mean and variance $\hat{\bar{x}}_{m_0}$ and $\hat{s}^2_{m_0}$, respectively.

The first slot $sl_1$ will account for $n \times N(\hat{\bar{x}}_{m_0}, \hat{s}^2_{m_0}; \hat{s}_{m_0})$ samples. Those are markings with cardinality less than or equal to $\hat{s}_{m_0}$.

From the second to the sixth slot ($g = 2, \ldots, 6$), they will account for $n \times [N(\hat{\bar{x}}_{m_0}, \hat{s}^2_{m_0}; g \times \hat{s}_{m_0}) - N(\hat{\bar{x}}_{m_0}, \hat{s}^2_{m_0}; (g-1) \times \hat{s}_{m_0})]$ samples. The markings in each slot have cardinality greater than $(g-1) \times \hat{s}_{m_0}$ and less than or equal to $g \times \hat{s}_{m_0}$.

The algorithm directs the exploration toward markings for slots with the lowest ratio of completion of their quota.

At each step of the exploration, with each sample, the algorithm takes into consideration the proportions of completeness in the previous step because it must ensure that the sample is representative of the SSp.

All unique explored markings are registered in a record of states, but only markings which are also entries in the table of quotas are valid samples. A marking with cardinality not in the range of any slot of the table is a type-1 not valid sample. A marking with cardinality of a slot with complete quota is a type-2 not valid sample. Repeated markings are discarded as samples.

The number of type-2 not valid samples is reduced by one every time a valid sample corresponding to an incomplete quota of a slot is found.

The exploration stops if the target marking is found or when all quotas are complete. In addition, the exploration stops following the same statistical "68-95-99.7" rule for detecting outliers, once the number of type-1 not valid samples is greater than or equal to 0.27% with respect to the $\hat{r}_{m_0}$.

Also, the exploration stops if the number of valid samples plus the number of type-2 not valid samples are greater than $n$. This number is artificially imposed to put a limit to the amount of collected samples (visited states) in the range of the assumed distribution of the SSp.

Finally, it should be noted that a major restriction in our analysis is the relation between $\hat{s}_{m_0}$ and the target marking. The cardinality of the target marking must be within the limits of the assumed normally distribution of $Y$.

$$\hat{\bar{x}}_{m_0} - 3\hat{s}_{m_0} \leq card(m_t) \leq \hat{\bar{x}}_{m_0} + 3\hat{s}_{m_0}$$

Although the restriction limits the analysis scope of our method, it gives robustness to the outcome given the assumed probability distribution of the variable $Y$ of the SSp.

## 3.2 Sampling Result

In theory, any random sample taken from a normally distributed population should give a normally distributed sample. In practice, we cannot take a random sample from the exploration of the SSp of a PN system. However, by using a quota sampling strategy, we can direct our exploration toward getting the result of a random sampling, a pseudo-normally distributed sample.

The sample taken with the table of quotas can be seen as a set $\Psi$ of markings. The distribution of a variable $W$, cardinality of the markings in the set $\Psi$, appears like a pseudo-normal probability distribution, i.e., $W \sim N(\bar{x}_w, s_w^2)$.

For each marking $m$ in the set $\Psi$ let us define the binomial variable $Q$ as follows:

$$Q = \begin{cases} 1 & if \ card(m) = c \\ 0 & if \ card(m) \neq c \end{cases}$$

For a target marking $m_t$ with cardinality of $c$, $q$ is the proportion of markings in the sample with the same cardinality as $m_t$.

## 4  Standard Hypothesis Test

In our reachability analysis, in searching a target marking $m_t$ with cardinality of $c$, the partial exploration algorithm fills in the quota of samples for a table of $n$ markings from the SSp.

Just for the case of unsuccessful search of the marking $m_t$, the explored states filling the quotas of the six slots belong to a set $\Psi$ of samples. Those samples can be treated with a hypothesis test.

## 4.1  State of Nature

Suppose we have a refined estimation for $r_{m_0}$ and $k_{m_0}$ taken from the superset $\Omega$ such that $\hat{r}_{m_0} - r_{m_0} \cong 0$ and $\hat{k}_{m_0} - k_{m_0} \cong 0$ and positive, where the estimators of the parameters of the mean $\mu$ and standard deviation $\sigma$ of the variable $Y$, cardinality of the markings and normal probability distribution from the set $\Psi$ of the SSp, are $\bar{x}$ and $s^2$, respectively, of the variable $X$.

The parameter $\theta$ is the proportion of markings with cardinality of $c$ in $\Psi$. An estimation of $\theta$ is calculated with the probability $N(\bar{x}_{m_0}, s^2_{m_0}; c)$.

If the estimation $\hat{\theta}$ is such that $\hat{\theta} - \theta \cong 0$ and positive, then the maximal number of markings with cardinality of $c$ in $\Psi$ is $\hat{\theta} \times n$.

## 4.2   Hypothesis Test for Proportions

We want to decide if the proportion of markings in the SSp with the same cardinality of $m_t$ is greater than the estimated proportion $\hat{\theta}$. For that we conduct the following hypothesis test for proportions:

$$H_0 : p = \hat{\theta}$$
$$H_1 : p > \hat{\theta}$$

The testatistic computed for the hypothesis test of proportions is

$$\pi = \frac{q - \hat{\theta}}{\sqrt{q(1 - q)/n}}.$$

The decision obtained with this hypothesis test will be: if we reject $H_0$ it means with a $1 - \alpha$ level of significance that other markings with cardinality $c$ exist in the SSp. But if we cannot reject $H_0$ it means there might not be other markings with the same cardinality as the target marking in the SSp, therefore upon unsuccessful search the target marking might not be in the SSp of the PN system.

An additional advantage of our analysis method is that, even in the case of stopping the exploration before completing the quota sampling, sometimes is still possible to conduct a hypothesis test with the partial results upon confirmation with a test of goodness of fit that the sample represents a normal distribution.

## 5   Evaluation

We have observed correct results of the hypothesis test with the following types of PNs [10], like the MGPN in the Fig. 1.5.

Let us take a larger step toward formal verification. In what is next we will extend the results to larger networks by combining the three networks presented before. We prepared some simulations designed in order to cope with some possible scenarios for some target markings with different cardinality.

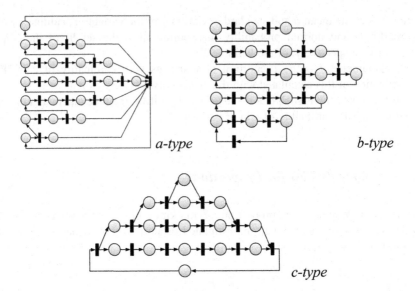

**Fig. 1.5** Marked graph Petri nets

We search for the sample requirements rejecting $H_0$. We look through the smallest acceptable error $e$ in the proportion less than or equal to 5%. Our target was to obtain a representative sample with traversal coverage, meaning in our case reaching an SSp exploration of less than or equal to 90% of the entire SSp and a sample with maximal size equal to 20% of the SSp.

We used good estimations of the mean and the standard deviation of the variable $Y$, where the dimension in the error of the estimations is smaller than 0.1 for the mean and the standard deviation.

Two samples were obtained for each reachability problem by conducting a breadth-first and depth-first search exploration. We compared the results of both algorithms to know which one conducts a better exploration toward markings with same cardinality as the target marking.

A hypothesis test rejecting $H_0$ means the proportion of markings in the SSp with same cardinality as the target marking is larger, indicating in the case of unsuccessful search that such marking might exist in the unexplored portion of the SSp. It also means that the algorithm does not visit enough markings with cardinality of $c$.

Tables 1.1 and 1.2 have the results of some explorations of the SSp for markings with certain cardinality, using the breadth-first (upper result) and depth-first (lower result) algorithms.

We use a 95% level of significance in our hypothesis test. The standard normal $z$-table gives a value of 1.645. A result from the testatistic greater than this value rejects $H_0$.

**Table 1.1** Results for card $(m_0) = 1$ with estimations of $\hat{\bar{x}}_{m_0} = 7.2$ and $\hat{s}_{m_0} = 1.6$

| Card $(m_t)$ | $e$ | Explored (%) | Sampled (%) | Testatistic |
|---|---|---|---|---|
| 5 | 0.04 | 79 | 11 | −1.35 |
|   | 0.04 | 67 | 11 | 0.42 |
| 6 | 0.05 | 85 | 12 | **2.54** |
|   | 0.05 | 87 | 12 | −3.31 |
| 7 | 0.05 | 85 | 14 | −3.67 |
|   | 0.05 | 87 | 14 | 0.54 |
| 8 | 0.05 | 85 | 13 | **2.90** |
|   | 0.05 | 87 | 13 | **2.92** |
| 9 | 0.05 | 79 | 9 | −0.02 |
|   | 0.04 | 86 | 14 | −2.37 |

**Table 1.2** Results for card $(m_0) = 4$ with estimations of $\hat{\bar{x}}_{m_0} = 7.2$ and $\hat{s}_{m_0} = 1.6$

| Card $(m_t)$ | $e$ | Explored (%) | Sampled (%) | Testatistic |
|---|---|---|---|---|
| 5 | 0.04 | 71 | 11 | 0.42 |
|   | 0.04 | 72 | 14 | 0.49 |
| 6 | 0.05 | 92 | 11 | **2.03** |
|   | 0.05 | 72 | 12 | −2.29 |
| 7 | 0.05 | 92 | 14 | −4.25 |
|   | 0.05 | 72 | 14 | −0.09 |
| 8 | 0.05 | 92 | 13 | **2.92** |
|   | 0.05 | 72 | 13 | **2.92** |
| 9 | 0.05 | 70 | 8 | −0.02 |
|   | 0.05 | 36 | 8 | −1.38 |

# 6  Conclusion and Future Work

A common problem for statisticians is calculating the sample size required to yield a certain power for a test given a predetermined level of significance. Compared with other sampling method, ours is somehow more intricate because in addition we must find how not to explore the entire SSp while getting the sample.

Obtaining a representative sample of markings from an MGPN system with the desired SSp coverage is possible but highly dependent on the exploration algorithm. Only simple depth and breadth search algorithms were used in our experiments guided by fulfilling the sampling quota. After the results we understood that a more sophisticated guidance in the exploration could largely reduce the portion of the exploration, especially since the cardinality of the initial markings also plays a major role in the result of any algorithm, as shown in the results of the two tables.

Despite the SSp exploration is not a process of randomly generated independent markings, the approach by quota sampling allows us to confirm the highly possible appropriateness of the hypothesis test in order to add completeness to unsuccessful searches, but just in MGPN systems with the three forms presented here. The results belong to rather small MGPN systems and the simulations do not exhaustively cover all possible scenarios, but the usability of the test in larger networks and other scenarios is straightforward.

The usage of the hypothesis test presented here is important because it statistically shows if it is significant to continue the search of the target marking. In addition, with our method, we can identify the best exploration algorithm by their effectiveness on rejecting $H_0$. From the results we can observe the breadth-first algorithm rejects $H_0$ more times than depth-first by one, meaning the second one reached more markings with the cardinality of the target marking.

Finally, since reachability has been completely characterized, the future success of our research relies in the transformation of more general PN into MGPN like the ones used here. With this, our exploration method could be used with any type of PN.

# References

1. Ciardo G (2004) Reachability set generation for Petri nets: can brute force be smart? In: Proceedings of 25th international conference on applications and theory of petri nets, 2004. LNCS 3099. Springer, Berlin, pp 17–34
2. Cochran WG (1963) Sampling techniques, 2nd edn. Wiley, New York
3. Commoner F, Holt AW (1971) Marked directed graphs. J Comput Syst Sci 5:511–523
4. Deo N (1974) Graph theory with applications to engineering and computer science. Prentice-Hall, Englewood Cliffs, NJ
5. Holzmann GJ (1988) Algorithms for automated protocol verification. AT&T Tech J 69(2):32–44
6. Jensen K, Kristensen LM (2009) Coloured Petri nets – Modeling and validation of concurrent systems. Springer, Berlin
7. Jiménez SE (2010) A probability-based state space analysis of Petri nets. Technical Report, vol 110, no.283, Institute of Electronics, Information and Communication Engineers (IEICE), pp 7–12
8. Jiménez SE (2010) Using metaheuristics and SPC in the analysis of state spaces of Petri nets. Medwell J Eng Appl Sci 5(6):413–419
9. Jiménez SE(2011) Improved estimation of the size of the state space of Petri nets for the analysis of reachability problems. In: Proceedings of the 5th european computing conference (WSEAS 2011), pp 92–99.
10. Jimenez SE (2011) Hypothesis test for unsuccessful partial explorations of state spaces of Petri nets. In: Proceedings of world congress on engineering (WCE 2011). Lecture Notes in Engineering and Computer Science. London, UK, 6–8 July 2011, pp 1254–1259
11. Kuntz M, Lampka K (2004) Probabilistic methods in state space analysis. Validation of stochastic systems. LNCS 2925. Springer, Berlin, pp 339–383
12. Murata T (1989) Petri nets: properties, analysis and applications. Proc IEEE 77:541–580
13. Pelanek R, Hanzl T, Cerna I, Brim L (2005) Enhancing random walk state space exploration. In: Proceedings of the 10th international workshop on formal methods for industrial critical systems (FMICS 05). ACM Press, New York, NY, USA, pp 98–105
14. Särndal C, Swensson B, Wretman J (1992) Model assisted survey sampling. Springer, New York
15. Schreuder HT, Gregoire TG, Weyer JP (2001) For what applications can probability and non-probability sampling be used? Environ Monitor Assess J 66(3):281–291

16. Watson III JF, Desrochers AA (1992) Methods for estimating state-space size of Petri nets. In: Proceedings of the IEEE international conference on robotics and automation, pp 1031–1036
17. Watson III JF, Desrochers AA (1992) State-space size estimation of conservative Petri nets. In: Proceedings of the IEEE international conference on robotics and automation, pp 1037–1042
18. Watson JF III, Desrochers AA (1994) State-space size estimation of Petri nets: a bottom–up perspective. IEEE Trans Robot Automat 10(4):555–561

# Chapter 2
# Duality in Complex Stochastic Boolean Systems

Luis González

## 1 Introduction

The study of complex systems is at present one of the most relevant research areas in Computer Science and Engineering. In this paper, we focus our attention on the complex stochastic Boolean systems (CSBSs), that is, those complex systems which depend on a certain number $n$ of random Boolean variables. These systems can appear in any knowledge area, since the assumption "random Boolean variables" is satisfied very often in practice.

Using the statistical terminology, a CSBS can be modeled by the $n$-dimensional Bernoulli distribution. As is well known (see, e.g., (10)), this distribution consists of $n$ random variables $x_1, \ldots, x_n$, which only take two possible values, 0 or 1, with probabilities

$$\Pr\{x_i = 1\} = p_i, \quad \Pr\{x_i = 0\} = 1 - p_i \quad (1 \leq i \leq n).$$

In the following, we assume that the marginal Bernoulli variables $x_1, \ldots, x_n$ are mutually independent, so that the probability of occurrence of each binary $n$-tuple, $u = (u_1, \ldots, u_n) \in \{0, 1\}^n$, can be computed as the product

$$\Pr\{(u_1, \ldots, u_n)\} = \prod_{i=1}^{n} \Pr\{x_i = u_i\} = \prod_{i=1}^{n} p_i^{u_i}(1 - p_i)^{1-u_i}, \tag{2.1}$$

that is, $\Pr\{(u_1, \ldots, u_n)\}$ is the product of factors $p_i$ if $u_i = 1$, $1 - p_i$ if $u_i = 0$. Throughout this paper, the binary $n$-tuples $(u_1, \ldots, u_n)$ of 0s and 1s will be

L. González (✉)
Research Institute IUSIANI, Department of Mathematics, University of Las Palmas de Gran Canaria, Campus Universitario de Tafira, 35017 Las Palmas de Gran Canaria, Spain
e-mail: luisglez@dma.ulpgc.es

S.-I. Ao and L. Gelman (eds.), *Electrical Engineering and Intelligent Systems*,
Lecture Notes in Electrical Engineering 130, DOI 10.1007/978-1-4614-2317-1_2,
© Springer Science+Business Media, LLC 2013

also called binary strings or bitstrings, and the parameters $p_1, \ldots, p_n$ of the $n$-dimensional Bernoulli distribution will be also called basic probabilities.

*Example 1.1* Let $n = 3$ and $u = (1, 0, 1) \in \{0, 1\}^3$. Let $p_1 = 0.1$, $p_2 = 0.2$, and $p_3 = 0.3$. Then, using (2.1), we have

$$\Pr\{(1, 0, 1)\} = p_1(1 - p_2)p_3 = 0.024.$$

One of the most relevant questions in the analysis of CSBSs consists of ordering the binary strings $(u_1, \ldots, u_n)$ according to their occurrence probabilities. Of course, the theoretical and practical interest of this question is obvious. For instance, in (2; 8) the authors justify the convenience of using binary $n$-tuples with occurrence probabilities as large as possible, in order to solve, with a low computational cost, some classical problems in Reliability Theory and Risk Analysis.

Of course, computing and ordering all the $2^n$ binary $n$-tuple probabilities (in decreasing or increasing order) are only feasible for small values of $n$. For large values of the number $n$ of basic Boolean variables (the usual situation in practice), we need an alternative strategy. For this purpose, in (2) we have established a simple, positional criterion that allows one to compare two given binary $n$-tuple probabilities, $\Pr\{u\}, \Pr\{v\}$, without computing them, simply looking at the positions of the 0s and 1s in the $n$-tuples $u$, $v$. We have called it the *intrinsic order criterion*, because it is independent of the basic probabilities $p_i$ and it *intrinsically* depends on the positions of the 0s and 1s in the binary strings.

The intrinsic order, denoted by " $\preceq$ ", is a partial order relation on the set $\{0, 1\}^n$ of all binary $n$-tuples. The usual representation of this kind of binary relations is the Hasse diagram (9). In particular, the Hasse diagram of the partially ordered set $(\{0, 1\}^n, \preceq)$ is referred to as the *intrinsic order graph* for $n$ variables.

In this context, the main goal of this paper is to state and rigorously prove some properties of the intrinsic order graph. Some of these properties can be found in (7). In particular, we focus our attention on several duality properties of this graph. For this purpose, this paper has been organized as follows. In Sect. 2, we present some previous results on the intrinsic order relation and the intrinsic order graph, enabling nonspecialists to follow the paper without difficulty and making the presentation self-contained. Section 3 is devoted to provide different duality properties of the intrinsic order graph. Finally, conclusions are presented in Sect. 4.

# 2   The Intrinsic Order Relation and Its Graph

## 2.1   The Intrinsic Order Relation

In the context of the CSBSs defined in Sect. 1, the following simple question arises: Given a certain $n$-dimensional Bernoulli distribution, how can we order two given binary $n$-tuples, $u, v \in \{0, 1\}^n$, by their occurrence probabilities, without computing them? Of course, the ordering between $\Pr(u)$ and $\Pr(v)$ depends, in general, on the parameters $p_i$ of the Bernoulli distribution, as the following simple example shows.

*Example* 2.1 Le $n = 3$, $u = (0, 1, 1)$ and $v = (1, 0, 0)$. Using (2.1) for $p_1 = 0.1$, $p_2 = 0.2$, and $p_3 = 0.3$, we have:

$$\Pr\{(0, 1, 1)\} = 0.054 < \Pr\{(1, 0, 0)\} = 0.056,$$

while for $p_1 = 0.2$, $p_2 = 0.3$, and $p_3 = 0.4$, we have:

$$\Pr\{(0, 1, 1)\} = 0.096 > \Pr\{(1, 0, 0)\} = 0.084.$$

However, for some pairs of binary strings, the ordering between their occurrence probabilities is independent of the basic probabilities $p_i$, and it only depends on the relative positions of their 0s and 1s. More precisely, the following theorem (2; 3) provides us with an intrinsic order criterion—denoted from now on by the acronym IOC—to compare the occurrence probabilities of two given $n$-tuples of 0s and 1s without computing them.

**Theorem 2.2** *Let $n \geq 1$. Let $x_1, \ldots, x_n$ be $n$ mutually independent Bernoulli variables whose parameters $p_i = \Pr\{x_i = 1\}$ satisfy*

$$0 < p_1 \leq p_2 \leq \cdots \leq p_n \leq \frac{1}{2}. \tag{2.2}$$

*Then, the probability of the n-tuple $v = (v_1, \ldots, v_n) \in \{0, 1\}^n$ is intrinsically less than or equal to the probability of the n-tuple $u = (u_1, \ldots, u_n) \in \{0, 1\}^n$ (that is, for all set $\{p_i\}_{i=1}^n$ satisfying (2.2)) if and only if the matrix*

$$M_v^u := \begin{pmatrix} u_1 & \cdots & u_n \\ v_1 & \cdots & v_n \end{pmatrix}$$

*either has no $\binom{1}{0}$ columns or for each $\binom{1}{0}$ column in $M_v^u$ there exists (at least) one corresponding preceding $\binom{0}{1}$ column (IOC).*

*Remark 2.3* In the following, we assume that the parameters $p_i$ always satisfy condition (2.2). Note that this hypothesis is not restrictive for practical applications because, if for some $i : p_i > 0.5$, then we only need to consider the variable $\bar{x}_i = 1 - x_i$, instead of $x_i$. Next, we order the n Bernoulli variables by increasing order of their probabilities.

*Remark 2.4* The $\binom{0}{1}$ column preceding to each $\binom{1}{0}$ column is not required to be necessarily placed at the immediately previous position, but just at previous position.

*Remark 2.5* The term corresponding, used in Theorem 2.2, has the following meaning: For each two $\binom{1}{0}$ columns in matrix $M_v^u$, there must exist (at least) two different $\binom{0}{1}$ columns preceding to each other. In other words: For each $\binom{1}{0}$ column in matrix $M_v^u$, the number of preceding $\binom{0}{1}$ columns must be strictly greater than the number of preceding $\binom{1}{0}$ columns.

*Remark 2.6* IOC can be equivalently reformulated in the following way, involving only the 1-bits of $u$ and $v$ (with no need to use their 0-bits). Matrix $M_v^u$ satisfies IOC if and only if either $u$ has no 1-bits (i.e., $u$ is the zero n-tuple) or for each 1-bit in $u$ there exists (at least) one corresponding 1-bit in $v$ placed at the same or at a previous position. In other words, either $u$ has no 1-bits or for each 1-bit in $u$, say $u_i = 1$, the number of 1-bits in $(v_1, \ldots, v_i)$ must be greater than or equal to the number of 1-bits in $(u_1, \ldots, u_i)$.

The matrix condition IOC, stated by Theorem 2.2 or by Remark 2.6, is called the intrinsic order criterion, because it is independent of the basic probabilities $p_i$ and it only depends on the relative positions of the 0s and 1s in the binary $n$-tuples $u$, $v$. Theorem 2.2 naturally leads to the following partial order relation on the set $\{0, 1\}^n$ (3). The so-called intrinsic order will be denoted by " $\preceq$ ," and we shall write $u \succeq v$ ($u \preceq v$) to indicate that $u$ is intrinsically greater (less) than or equal to $v$. The partially ordered set (from now on, poset, for short) $(\{0, 1\}^n, \preceq)$ on $n$ Boolean variables will be denoted by $I_n$.

**Definition 2.7** For all $u, v \in \{0, 1\}^n$

$$v \preceq u \quad \text{iff} \quad \Pr\{v\} \le \Pr\{u\} \quad \text{for all set } \{p_i\}_{i=1}^n \quad \text{s.t. } (2.2)$$

$$\text{iff } M_v^u \text{ satisfies IOC.}$$

*Example 2.8* Neither $(0, 1, 1) \preceq (1, 0, 0)$, nor $(1, 0, 0) \preceq (0, 1, 1)$ because the matrices

$$\begin{pmatrix} 1 & 0 & 0 \\ 0 & 1 & 1 \end{pmatrix} \text{ and } \begin{pmatrix} 0 & 1 & 1 \\ 1 & 0 & 0 \end{pmatrix}$$

do not satisfy IOC (Remark 2.5). Therefore, $(0, 1, 1)$ and $(1, 0, 0)$ are incomparable by intrinsic order, i.e., the ordering between $\Pr\{(0, 1, 1)\}$ and $\Pr\{(1, 0, 0)\}$ depends on the basic probabilities $p_i$, as Example 2.1 has shown.

*Example 2.9* $(1, 1, 0, 1, 0, 0) \preceq (0, 0, 1, 1, 0, 1)$ because matrix

$$\begin{pmatrix} 0 & 0 & 1 & 1 & 0 & 1 \\ 1 & 1 & 0 & 1 & 0 & 0 \end{pmatrix}$$

satisfies IOC (Remark 2.4). Thus, for all $\{p_i\}_{i=1}^6$ s.t. (2.2)

$$\Pr\{(1, 1, 0, 1, 0, 0)\} \le \Pr\{(0, 0, 1, 1, 0, 1)\}.$$

*Example 2.10* For all $n \ge 1$, the binary $n$-tuples

$$\left(0, \overset{n}{\ldots}, 0\right) \equiv 0 \quad \text{and} \quad \left(1, \overset{n}{\ldots}, 1\right) \equiv 2^n - 1$$

are the maximum and minimum elements, respectively, in the poset $I_n$. Indeed, both matrices

$$\begin{pmatrix} 0 & \cdots & 0 \\ u_1 & \cdots & u_n \end{pmatrix} \quad \text{and} \quad \begin{pmatrix} u_1 & \cdots & u_n \\ 1 & \cdots & 1 \end{pmatrix}$$

satisfy IOC, since they have no $\binom{1}{0}$ columns!

Thus, for all $u \in \{0,1\}^n$ and for all $\{p_i\}_{i=1}^n$ s.t. (2.2)

$$\Pr\left\{ \left(1, \overset{n}{\ldots}, 1\right) \right\} \leq \Pr\{(u_1, \ldots, u_n)\} \leq \Pr\left\{ \left(0, \overset{n}{\ldots}, 0\right) \right\}.$$

Many different properties of the intrinsic order can be immediately derived from its simple matrix description IOC (2; 3; 5). For instance, denoting by $w_H(u)$ the Hamming weight—or weight, simply—of $u$ (i.e., the number of 1-bits in $u$), by $u_{(10}$ the decimal representation of $u$, and by $\leq_{lex}$ the usual lexicographic (truth-table) order on $\{0, 1\}^n$, i.e.,

$$w_H(u) := \sum_{i=1}^n u_i, \quad u_{(10} := \sum_{i=1}^n 2^{n-i} u_i, \quad u \leq_{lex} v \text{ iff } u_{(10} \leq v_{(10}$$

then we have the following two necessary (but not sufficient) conditions for intrinsic order (see (3) for the proof).

**Corollary 2.11** *For all $n \geq 1$ and for all $u, v \in \{0, 1\}^n$*

$$u \succeq v \Rightarrow w_H(u) \leq w_H(v),$$

$$u \succeq v \Rightarrow u_{(10} \leq v_{(10}.$$

## 2.2   A Hasse Diagram: The Intrinsic Order Graph

Now, the graphical representation of the poset $I_n = (\{0, 1\}^n, \preceq)$ is presented. The usual representation of a poset is its Hasse diagram (see (9) for more details about these diagrams). Specifically, for our poset $I_n$, its Hasse diagram is a directed graph (digraph, for short) whose vertices are the $2^n$ binary $n$-tuples of 0s and 1s, and whose edges go upward from $v$ to $u$ whenever $u$ covers $v$, denoted by $u \triangleright v$. This means that $u$ is intrinsically greater than $v$ with no other elements between them, i.e.,

$$u \triangleright v \quad \Leftrightarrow \quad u \succ v \text{ and } \not\exists \, w \in \{0, 1\}^n \text{ s.t. } u \succ w \succ v.$$

A simple matrix characterization of the covering relation for the intrinsic order is given in the next theorem; see (4) for the proof.

**Theorem 2.12 (Covering relation in $I_n$)** *Let $n \geq 1$ and let $u, v \in \{0,1\}^n$. Then, $u \triangleright v$ if and only if the only columns of matrix $M_v^u$ different from $\binom{0}{0}$ and $\binom{1}{1}$ are either its last column $\binom{0}{1}$ or just two columns, namely one $\binom{1}{0}$ column immediately preceded by one $\binom{0}{1}$ column, i.e., either*

$$M_v^u = \begin{pmatrix} u_1 \ldots u_{n-1} & 0 \\ u_1 \ldots u_{n-1} & 1 \end{pmatrix} \tag{2.3}$$

*or there exists $i$ $(2 \leq i \leq n)$ s.t.*

$$M_v^u = \begin{pmatrix} u_1 \ldots u_{i-2} & 0 & 1 & u_{i+1} \ldots u_n \\ u_1 \ldots u_{i-2} & 1 & 0 & u_{i+1} \ldots u_n \end{pmatrix}. \tag{2.4}$$

*Example 2.13* For $n = 4$, we have

$$6 \triangleright 7 \quad \text{since } M_7^6 = \begin{pmatrix} 0 & 1 & 1 & 0 \\ 0 & 1 & 1 & 1 \end{pmatrix} \quad \text{has the pattern (2.3),}$$

$$10 \triangleright 12 \quad \text{since } M_{12}^{10} = \begin{pmatrix} 1 & 0 & 1 & 0 \\ 1 & 1 & 0 & 0 \end{pmatrix} \quad \text{has the pattern (2.4).}$$

The Hasse diagram of the poset $I_n$ will be also called the *intrinsic order graph* for $n$ variables, denoted as well by $I_n$.

For small values of $n$, the intrinsic order graph $I_n$ can be directly constructed by using either Theorem 2.2 (matrix description of the intrinsic order) or Theorem 2.12 (matrix description of the covering relation for the intrinsic order). For instance, for $n = 1$: $I_1 = (\{0, 1\}, \preceq)$, and its Hasse diagram is shown in Fig. 2.1.

Indeed $I_1$ contains a downward edge from 0 to 1 because (see Theorem 2.2) $0 \succ 1$, since matrix $\binom{0}{1}$ has no $\binom{1}{0}$ columns! Alternatively, using Theorem 2.12, we have that $0 \triangleright 1$, since matrix $\binom{0}{1}$ has the pattern (2.3)! Moreover, this is in accordance with the obvious fact that

$$\Pr\{0\} = 1 - p_1 \geq p_1 = \Pr\{1\}, \quad \text{since} \quad p_1 \leq 1/2 \text{ due to (2.2)!}$$

However, for large values of $n$, a more efficient method is needed. For this purpose, in (4) the following algorithm for iteratively building up $I_n$ (for all $n \geq 2$) from $I_1$ (depicted in Fig. 2.1) has been developed.

**Fig. 2.1** The intrinsic order
graph for $n = 1$

```
0
|
1
```

**Fig. 2.2** The intrinsic order graphs for $n = 1, 2, 3, 4$

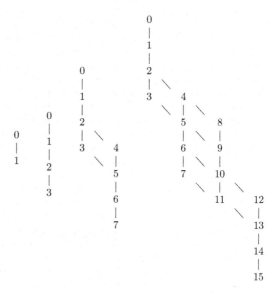

**Theorem 2.14 (Building up $I_n$ from $I_1$)** *Let $n \geq 2$. The graph of the poset $I_n = \{0, \dots, 2^n - 1\}$ (on $2^n$ nodes) can be drawn simply by adding to the graph of the poset $I_{n-1} = \{0, \dots, 2^{n-1} - 1\}$ (on $2^{n-1}$ nodes) its isomorphic copy $2^{n-1} + I_{n-1} = \{2^{n-1}, \dots, 2^n - 1\}$ (on $2^{n-1}$ nodes). This addition must be performed placing the powers of 2 at consecutive levels of the Hasse diagram of $I_n$. Finally, the edges connecting one vertex $u$ of $I_{n-1}$ with the other vertex $v$ of $2^{n-1} + I_{n-1}$ are given by the set of $2^{n-2}$ vertex pairs*

$$\left\{ (u, v) \equiv \left( u_{(10}, 2^{n-2} + u_{(10)} \right) \mid 2^{n-2} \leq u_{(10} \leq 2^{n-1} - 1 \right\}.$$

Figure 2.2 illustrates the above iterative process for the first few values of $n$, denoting all the binary $n$-tuples by their decimal equivalents. Basically, we first add to $I_{n-1}$ its isomorphic copy $2^{n-1} + I_{n-1}$. This addition must be performed by placing the powers of two, $2^{n-2}$ and $2^{n-1}$, at consecutive levels in the intrinsic order graph. The reason is simply that

$$2^{n-2} \triangleright 2^{n-1} \quad \text{since matrix } M_{2^{n-1}}^{2^{n-2}} \text{ has the pattern (2.4).}$$

Then, we connect one-to-one the nodes of "the second half of the first half" to the nodes of "the first half of the second half": A nice fractal property of $I_n$!

Each pair $(u, v)$ of vertices connected in $I_n$ either by one edge or by a longer path, descending from $u$ to $v$, means that $u$ is intrinsically greater than $v$, i.e., $u \succ v$. On the contrary, each pair $(u, v)$ of non-connected vertices in $I_n$ either by one edge or by a longer descending path means that $u$ and $v$ are incomparable by intrinsic order, i.e., $u \not\succ v$ and $v \not\succ u$.

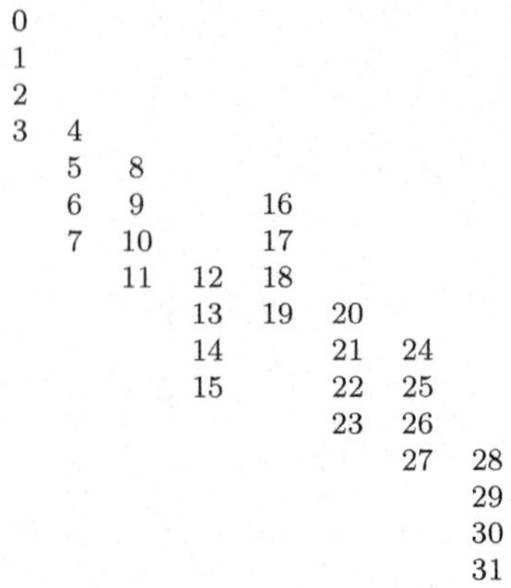

```
0
1
2
3   4
    5   8
    6   9        16
    7   10       17
        11  12  18
            13  19  20
            14      21  24
            15      22  25
                    23  26
                        27  28
                            29
                            30
                            31
```

**Fig. 2.3** The edgeless intrinsic order graph for $n = 5$

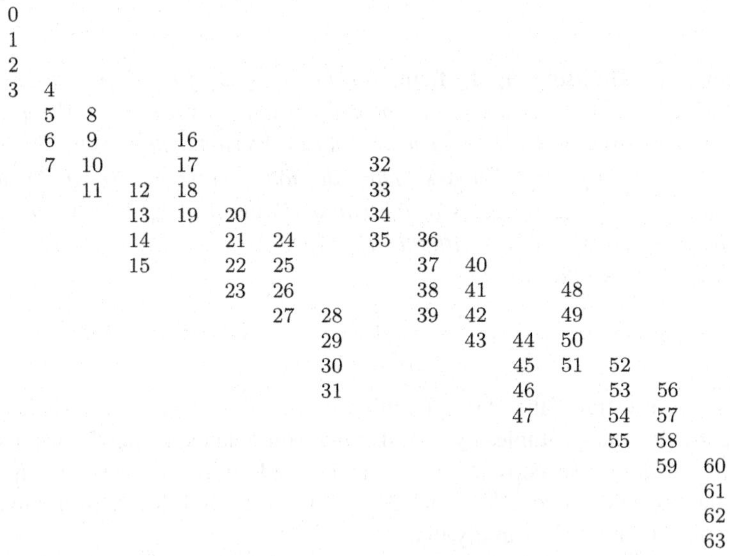

```
0
1
2
3   4
    5   8
    6   9       16
    7   10      17                  32
        11  12  18                  33
            13  19  20              34
            14      21  24      35  36
            15      22  25          37  40
                    23  26          38  41          48
                        27  28      39  42          49
                            29          43  44  50
                            30          45  51  52
                            31          46      53  56
                                        47      54  57
                                                55  58
                                                    59  60
                                                        61
                                                        62
                                                        63
```

**Fig. 2.4** The edgeless intrinsic order graph for $n = 6$

The edgeless graph for a given graph is obtained by removing all its edges, keeping its nodes at the same positions. In Figs. 2.3 and 2.4, the edgeless intrinsic order graphs of $I_5$&$I_6$, respectively, are depicted.

For further theoretical properties and practical applications of the intrinsic order and the intrinsic order graph, we refer the reader to, e.g., (2; 4–8).

# 3  Duality Properties in the Intrinsic Order Graph

First, we need to set the following nomenclature and notation.

**Definition 3.1** The complementary $n$-tuple $u^c$ of a given binary $n$-tuple $u = (u_1, \ldots, u_n) \in \{0,1\}^n$ is obtained by changing its 0s by 1s and its 1s by 0s

$$u^c = (u_1, \ldots, u_n)^c = (1 - u_1, \ldots, 1 - u_n).$$

The complementary set $S^c$ of a given subset $S \subseteq \{0,1\}^n$ of binary $n$-tuples is the set of the complementary $n$-tuples of all the $n$-tuples of $S$

$$S^c = \{u^c \mid u \in S\}.$$

*Remark 3.2* Note that for all $u = (u_1, \ldots, u_n) \in \{0,1\}^n$ and for all $S, T \subseteq \{0,1\}^n$, we obviously have

$$\text{(i) } (u^c)^c = u, \quad \text{(ii) } (S^c)^c = S, \quad \text{(iii) } u \in S \Leftrightarrow u^c \in S^c,$$

$$\text{(iv) } (S \cup T)^c = S^c \cup T^c, \quad \text{(v) } w_H(u) + w_H(u^c) = n.$$

The following proposition states a duality property of the intrinsic order, which explains the symmetric structure of the intrinsic order graph.

**Proposition 3.3** *Let $n \geq 1$ and $u, v \in \{0,1\}^n$. Then*

$$u \triangleright v \Leftrightarrow v^c \triangleright u^c,$$
$$u \succeq v \Leftrightarrow v^c \succeq u^c.$$

*Proof.* Clearly, the $\binom{0}{0}, \binom{1}{1}, \binom{0}{1}$, and $\binom{1}{0}$ columns in matrix $M_v^u$, respectively, become $\binom{1}{1}, \binom{0}{0}, \binom{0}{1}$, and $\binom{1}{0}$ columns in matrix $M_{u^c}^{v^c}$.

Hence, on one hand, using Theorem 2.12, we have that $u \triangleright v$ iff matrix $M_v^u$ has either the pattern (2.3) or the pattern (2.4) iff matrix $M_{u^c}^{v^c}$ has either the pattern (2.3) or the pattern (2.4), respectively, iff $v^c \triangleright u^c$.

On the other hand, using Theorem 2.2, we have that $u \succeq v$ iff matrix $M_v^u$ satisfies IOC iff matrix $M_{u^c}^{v^c}$ satisfies IOC iff $v^c \succeq u^c$.  $\square$

Next corollary provides us with two easy criteria for rapidly identifying pairs of complementary binary strings in the intrinsic order graph.

**Corollary 3.4** *Let $n \geq 1$ and $u, v \in \{0,1\}^n$. Then, $u$ and $v$ are complementary $n$-tuples if and only if their decimal equivalents sum up to $2^n - 1$ if and only if they are placed at symmetric positions (with respect to the central point) in the (edgeless) graph $I_n$.*

*Proof.* Using Definition 3.1, we have that $u$ and $v$ are complementary $n$-tuples if and only if

$$u + v = \left(1, \overset{n}{\ldots}, 1\right) \equiv 2^n - 1.$$

Using Proposition 3.3, we have that $u$ and $v$ are complementary $n$-tuples if and only if they are placed at symmetric positions (with respect to the central point) in the (edgeless) graph $I_n$.                                                    □

Hence, the simplest way to verify that two binary $n$-tuples are complementary, when we use their decimal representations, is to check that they sum up to $2^n - 1$.

*Example 3.5* The binary 5-tuples $6 \equiv (0, 0, 1, 1, 0)$ & $25 \equiv (1, 1, 0, 0, 1)$ are complementary, since $6 + 25 = 31 = 2^5 - 1$. Alternatively, we can see that 6 and 25 are placed at symmetric positions (with respect to the central point) in the edgeless graph $I_5$, depicted in Fig. 2.3.

*Example 3.6* The complementary 6-tuple of the binary 6-tuple $50 \equiv (1, 1, 0, 0, 1, 0)$ is $13 \equiv (0, 0, 1, 1, 0, 1)$, since $(2^6 - 1) - 50 = 63 - 50 = 13$. Alternatively, we can see that 13 is the symmetric node (with respect to the central point) of 50 in the edgeless graph $I_6$, depicted in Fig. 2.4.

Many different consequences can be derived from Proposition 3.3. Some of them are presented in the following corollaries. Before each of them we give some definitions required to understand the statements of the corollaries.

**Definition 3.7** For every binary $n$-tuple $u \in \{0, 1\}^n$, the set $C^u$(the set $C_u$, respectively) is the set of all binary $n$-tuples $v$ whose occurrence probabilities $\Pr\{v\}$ are always less (greater, respectively) than or equal to $\Pr\{u\}$, i.e., according to Definition 2.7, those $n$-tuples $v$ intrinsically less (greater, respectively) than or equal to $u$, i.e.,

$$C^u = \{v \in \{0, 1\}^n \mid u \succeq v\},$$

$$C_u = \{v \in \{0, 1\}^n \mid v \succeq u\}.$$

**Definition 3.8** For every binary n-tuple $u \in \{0, 1\}^n$, $Inc(u)$ is the set of all binary n-tuples v intrinsically incomparable with u, i.e.,

$$Inc(u) = \{v \in \{0, 1\}^n \mid u \not\succeq v, \ u \not\preceq v\} = \{0, 1\}^n - (C^u \cup C_u).$$

**Corollary 3.9** *For all $n \geq 1$ and for all $u \in \{0, 1\}^n$, we have*

(i) $(C^u)^c = C_{u^c}$,      (ii) $(C_u)^c = C^{u^c}$,      (iii) $(Inc(u))^c = Inc(u^c)$.

*Proof.* To prove (i) it suffices to use Remark 3.2, Proposition 3.3, and Definition 3.7. Indeed

$$v \in (C^u)^c \Leftrightarrow v^c \in C^u \Leftrightarrow u \succeq v^c \Leftrightarrow v \succeq u^c \Leftrightarrow v \in C_{u^c}.$$

Clearly, (ii) is equivalent to (i); see Remark 3.2. Finally, to prove (iii), we use (i), (ii), Remark 3.2, and Definition 3.8

$$v \in (Inc(u))^c \Leftrightarrow v^c \in Inc(u) \Leftrightarrow v^c \notin (C^u \cup C_u)$$
$$\Leftrightarrow v \notin (C^u \cup C_u)^c \Leftrightarrow v \notin C_{u^c} \cup C^{u^c} \Leftrightarrow v \in Inc(u^c),$$

as was to be shown. □

The following definition (see (9)) deals with the general theory of posets.

**Definition 3.10** Let $(P, \leq)$ be a poset and $u \in P$. Then,
(i) The lower shadow of u is the set

$$\Delta(u) = \{v \in P \mid v \text{ is covered by } u\} = \{v \in P \mid u \triangleright v\}$$

(ii) The upper shadow of u is the set

$$\nabla(u) = \{v \in P \mid v \text{ covers } u\} = \{v \in P \mid v \triangleright u\}$$

Particularly, for our poset $P = I_n$, on one hand, regarding the lower shadow of $u \in \{0, 1\}^n$, using Theorem 2.12, we have

$$\Delta(u) = \{v \in \{0, 1\}^n \mid u \triangleright v\}$$
$$= \{v \in \{0, 1\}^n \mid M_v^u \text{ has either the pattern (2.3) or (2.4)}\}.$$

and, on the other hand, regarding the upper shadow of $u \in \{0, 1\}^n$, using again Theorem 2.12, we have

$$\nabla(u) = \{v \in \{0, 1\}^n \mid v \triangleright u\}$$
$$= \{v \in \{0, 1\}^n \mid M_u^v \text{ has either the pattern (2.3) or (2.4)}\}.$$

**Corollary 3.11** *For all $n \geq 1$ and for all $u \in \{0, 1\}^n$, we have*

$$\text{(i) } \Delta(u) = \nabla^c(u^c), \qquad \text{(ii) } \nabla(u) = \Delta^c(u^c).$$

*Proof.* To prove (i), we use Remark 3.2, Proposition 3.3, and Definition 3.10. Indeed,

$$v \in \Delta(u) \Leftrightarrow u \triangleright v \Leftrightarrow v^c \triangleright u^c \Leftrightarrow v^c \in \nabla(u^c) \Leftrightarrow v \in \nabla^c(u^c).$$

Clearly, (ii) is equivalent to (i); see Remark 3.2. □

The following definition (see (1)) deals with the general theory of graphs.

**Definition 3.12** The neighbors of a given vertex $u$ in a graph are all those nodes adjacent to $u$ (i.e., connected by one edge to $u$). The degree of a given vertex $u$—denoted by $\delta(u)$—in a graph is the number of neighbors of $u$.

In particular, for (the cover graph of) a Hasse diagram, the neighbors of vertex $u$ either cover $u$ or are covered by $u$. In other words, the set $N(u)$ of neighbors of a vertex $u$ is the union of its lower and upper shadows, i.e.,

$$N(u) = \Delta(u) \cup \nabla(u), \quad \delta(u) = |N(u)| = |\Delta(u)| + |\nabla(u)|. \tag{2.5}$$

**Corollary 3.13** *For all $n \geq 1$ and for all $u \in \{0,1\}^n$, the sets of neighbors of $u$ and $u^c$ are complementary. In particular, any two complementary $n$-tuples $u$ and $u^c$ have the same degree.*

*Proof.* Using Remark 3.2, Corollary 3.11, Definition 3.12, and (2.5), we have

$$\begin{aligned}
N^c(u) &= [\Delta(u) \cup \nabla(u)]^c = \Delta^c(u) \cup \nabla^c(u) \\
&= \nabla(u^c) \cup \Delta(u^c) = N(u^c)
\end{aligned}$$

and, consequently,

$$\delta(u) = |N(u)| = |N^c(u)| = |N(u^c)| = \delta(u^c),$$

as was to be shown.                                                                    □

# 4   Conclusions

The analysis of CSBSs can be performed by using the intrinsic ordering between binary $n$-tuples of 0s and 1s. The duality property of the intrinsic order relation for complementary $n$-tuples (obtained by changing 0s into 1s and 1s into 0s) implies many different properties of CSBSs. Some of these properties have been rigorously proved and illustrated by the intrinsic order graph.

**Acknowledgments** This work was partially supported by the Spanish Government, "Ministerio de Economía y Competitividad", and FEDER, through Grant contract: CGL2011-29396-C03-01.

# References

1. Diestel R (2005) Graph theory, 3rd edn. Springer, Heidelberg
2. González L (2002) A new method for ordering binary states probabilities in reliability and risk analysis. Lect Notes Comput Sci 2329:137–146

3. González L (2003) *N*-tuples of 0s and 1s: necessary and sufficient conditions for intrinsic order. Lect Notes Comput Sci 2667:937–946

4. González L (2006) A picture for complex stochastic Boolean systems: the intrinsic order graph. Lect Notes Comput Sci 3993:305–312

5. González L (2007) Algorithm comparing binary string probabilities in complex stochastic Boolean systems using intrinsic order graph. Adv Complex Syst 10(Suppl.1):111–143

6. González L (2010) Ranking intervals in complex stochastic Boolean systems using intrinsic ordering. In: Rieger BB, Amouzegar MA, Ao S-I (eds) Machine learning and systems engineering, Lecture notes in electrical engineering, vol 68. Springer, Heidelberg, pp 397–410

7. González L (2011) Complex stochastic Boolean systems: new properties of the intrinsic order graph. Lecture notes in engineering and computer science: proceedings of the world congress on engineering 2011, WCE 2011, London, pp 1194–1199, 6–8 July 2011

8. González L, García D, Galván B (2004) An intrinsic order criterion to evaluate large, complex fault trees. IEEE Trans Reliab 53:297–305

9. Stanley RP (1997) Enumerative combinatorics, vol 1. Cambridge University Press, Cambridge

10. Stuart A, Ord JK (1998) Kendall's advanced theory of statistics, vol 1. Oxford University Press, New York

# Chapter 3
# Reconfigurable Preexecution in Data Parallel Applications on Multicore Systems

Ákos Dudás and Sándor Juhász

## 1 Introduction

The term data intensive applications or data intensive computing [1–3] is used to name a specific class of applications. Examples are web search providers, medical imaging, astronomical data analysis, weather simulation, or analyzing data collected during physical experiments. These problems all involve processing large amounts of data. The performance of such applications is usually limited by data access times rather than by pure computational power.

The speed gap between modern CPUs and their attached memory systems is an ever increasing problem in such applications as the performance of CPUs has been growing faster than the speed of memories. At the same time simultaneous multithreading and multicore CPUs put even more strain on the memory system by every executed thread requiring data in parallel.

There have been numerous approaches proposed to fight this memory wall. Cache memories are fast associative memories integrated into the CPU (L1 and L2 caches) or onto the same die (last level caches). They are small in size, but are very fast. They speed up memory accesses in case of temporally and spatially local requests. Branch prediction and speculative (out-of-order) execution target latency by not waiting for a jump condition to evaluate but work ahead by loading and decoding instructions and issuing data prefetches along a chosen branch. These techniques are available in current CPUs and require no knowledge or effort on behalf of the programmer.

Another promising technique is data prefetching. Predicting the memory addresses that will be accessed in the near future they are brought closer to the CPU before they are actually asked for. Than at the actual request they are returned

Á. Dudás (✉) • S. Juhász
Department of Automation and Applied Informatics, Budapest University of Technology and Economics, Magyar Tudósok körútja 2, 1117 Budapest, Hungary
e-mail: akos.dudas@aut.bme.hu; juhasz.sandor@aut.bme.hu

S.-I. Ao and L. Gelman (eds.), *Electrical Engineering and Intelligent Systems,*
Lecture Notes in Electrical Engineering 130, DOI 10.1007/978-1-4614-2317-1_3,
© Springer Science+Business Media, LLC 2013

by the fast cache. The two key components are prediction and timing. The prediction should correctly foresee the memory addresses; and the data should be brought into the cache before needed, but not too early (so that they are not purged by the time they are accessed).

The prediction can be based on discovering patterns. Automatic hardware prefetch methods in modern CPUs bring entire cache lines into the last level cache triggered by cache misses. Based on the recorded history of requests they search for and discover stride patterns, which makes them perfectly suited for linear array accesses; but they do not work with random memory access patterns.

Luk in [3] argues that irregular memory access patterns (responsible for increased memory latencies and inefficient hardware prefetch) cannot be predicted only through executing the code. This is the basic idea of preexecution [3–10] (terminology also includes assisted execution, prefetching helper threads, microthreading, and speculative precomputation). The algorithm itself is able to calculate the exact memory addresses it needs; but it is just too late by that time and the execution has to suffer latency due to the cache miss. If the calculation had happened earlier, and the correct address had been available, it could have been be preloaded in time.

In this paper we propose a novel preexecution model for data parallel applications [11]. We present a purely software-based adaptive preexecution approach. It is lightweight, easy to implement, and requires no special hardware other than a multicore CPU. With run-time parameter tuning the algorithm configures itself for maximum performance and adapts itself to the execution environment and the memory access characteristics of the application. We apply this method in a real-life data-intensive application of data coding and show that it achieves an average speedup of 10–30%.

The rest of the paper is organized as follows. Section 2 presents the literature of preexecution. In Sect. 3 we present our model of preexecution which is evaluated in Sect. 4 by applying it to data transformation. We conclude in Sect. 5 with a summary of our findings.

## 2 Related Works

This section presents the related literature including some of the practical applications of preexecution.

The most important component of preexecution is the prediction engine, which is derived from the original algorithm. Most helper threads execute speculative slices [10, 12], which are extracts of the original algorithm. Since some instructions are removed from the original algorithm, there is no guarantee that the calculated result will be correct; this is accepted in exchange for faster execution. These slices can be created manually [3, 10] or using profiling and compiler support [4, 6, 13–15]. Another approach is the use of heuristics. For example, when traversing binary trees Roth et al. [8] suggests that an automatic prefetch is issued for the child nodes when a particular node is visited.

The trigger of preexecution is either an instruction in the software code or a hardware event. Software triggers must be placed in the source code in compile time either manually [10] or automatically [6, 13]. Hardware triggers can be for example load operations during the instruction decoding stage (used by Ro and Gaudiot [16]) or cache miss (used by Dundas and Mudge [17]).

Mutlu et al. showed that preexecution, or as they refer to it "runahead execution" [18], is an alternative to large instruction windows. They claim that runahead execution combined with a 128-entry window approximately has the same performance as a significantly larger 384-entry wide instruction window (without preexecution). From this point of view software preexecution reduces hardware complexity by allowing smaller instruction windows.

Malhotra and Kozyrakis implemented preexecution for several data structures in the C++ STL library [7] and achieved an average speedup of 26% for various applications using the library. Zhou et al. in [9] have also shown that preexecution can bring 30–70% speedup in a relational database. These examples support our theory that preexecution has a good chance of improving performance in data-intensive applications at a relatively low cost.

# 3  Software Preexecution for Data Parallel Applications

This section presents our preexecution strategy for data parallel applications.

In data parallel applications [11] the same operation is repeated for a set of items. This part of the application is an ideal candidate for preexecution. We advocate a purely software solution, which is lightweight, has low complexity, targets data intensive applications at a high level without the need to intensively profile the application, has low overhead on the execution, and applies adaptive run-time reconfiguration.

Software approach is our choice for it is fast and easy to prototype requiring no special hardware. It is also guaranteed that a software-only solution can be evaluated by execution on physical hardware and the performance gain can be measured in terms of wall-clock execution time.

We also want to note that data parallel applications are often easy to parallelize; in this paper we propose preexecution not as an alternative to parallelization but as a technique which can be used as an accessory to single threaded and multithreaded algorithms as well.

In most works the preexecution helper thread is created and disposed by the main thread explicitly. During its lifetime it prefetches a single item, and then terminates. Instead of trigger points and short-lived helper threads, we propose that the helper thread be a continuous thread. In order for the preexecution engine to be of use,

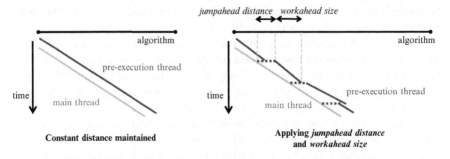

**Fig. 3.1** *Jumpahead distance* and *workahead size* parameters

it needs to know what the main thread is doing (i.e., which item in the data set is being processed). This information is required so that the preexecution thread works ahead, and prefetches data for an item to be processed in the near future.

The point where the algorithm is altered is the loop iterating over the items. In this loop the main thread publishes its loop counter via a shared variable. The helper thread, by reading this variable (which, along with writing it, is an atomic operation on x86 architectures), decides what item to preload. In order for the preexecution to be useful, we propose that a safe "distance" is always maintained between the threads. We call this distance *jumpahead distance* [19]. The value, however, cannot be a simple constant. It must reflect the characteristics of the system.

There should be frequent synchronization between the threads (i.e., when the shared loop counter is updated); however, we do not require the preexecution engine to strictly keep the jumpahead distance at all times. We can significantly reduce the overhead if the progress is not monitored so tightly. We can apply a batch-like execution and intervene only after a given number of steps have been completed by the threads. In order to regulate the frequency of intervention we introduce a second parameter called *workahead size* [20]. See Fig. 3.1 for an illustration of the meaning of the parameters.

The optimal values for these two parameters depend on a number of factors. It is affected by the memory characteristics of the application, the size of the caches, the speed of the CPU, and many other things. In order to achieve maximum performance we can use runtime reconfiguration.

In a previous work [19] we showed through empirical tests that there are optimal configurations regarding these two variables. In order to find these configurations we can apply an optimization technique, such as the Nelder–Mead method [20] to approximate the optimal values of the two parameters *jumpahead distance* and *workahead size* given the objective function is the performance of the application. The performance we measure in terms of average execution time between two samplings.

The complete pseudocode with the Nelder–Mead approximation of the parameters is listed below (see Algorithm 3.1).

**Algorithm 3.1** Pseudocode of the proposed preexecution strategy

shared volatile int *idx*
shared volatile bool *end* = false
shared volatile int *jumpAheadDist*
shared volatile int *workAheadSize*

**procedure** main(*dataset[1..n]*)
  startThread(preExecutionEngine)
  for *i* = *0* to *n/workAheadSize*
   *idx* = *i*
   for *j* = *0* to *workAheadSize*
    process(*dataset[i*workAheadSize* + *j]*)
  *end* = *true*
  return

**procedure** preExecutionEngine(*dataset[1..n]*)
  int samples = 0
  time *start* = getTime()
  while *end* == *false*
   for *i* = *0* to *workAheadSize*
    if *haveEnoughSamples(samples)*
     *{jumpAheadDist,workAheadSize}* = evaluate(getTime() − *start*)
     *start* = getTime()
     *samples* = 0
    process(*dataset[min(idx + jumpAheadDist + i,n)])*
    *samples* = *samples* + 1
  return

# 4  Evaluation

This section presents a real-life application which we used for evaluating and testing the proposed preexecution strategy.

## 4.1  *Implementation Details*

We implemented the proposed preexecution model manually. There have been numerous works published on compiler support for the creation of the helper thread [4, 6, 13–15]. This allowed us to focus our attention to the preexecution model rather than the code behind it.

Executing two threads simultaneously comes at some cost. However, having a multicore CPU at our disposal, we can execute multiple threads on a single CPU without any performance penalty. Although the cores compete for some of the system resources (including the main memory and the last level of shared cache), the level of instruction level parallelism increases: while one of the executing threads stalls on a memory read operation the other thread can still execute, making better use of the CPU and reducing the overall execution time.

Cache pollution is a concern. If the shared variables are changed by one thread executed on one of the physical cores the entire cache line must be immediately written to system memory and removed from the cache of every other core. We must assure that these writes do not pollute other variables which may reside in the same cache line as the shared variable. For this reason, the shared integer variables are each implemented as cache line-long byte arrays and are aligned on cache lines.

## 4.2 Test Environment

Preexecution was implemented for a real-life application where a set of items (billions of records) are pushed through a pipeline of transformation steps, one of which is an encoding with the use of a dictionary. This problem (data transformation) is data intensive: the transformation is a series of simple computations and a lookup/insert in the dictionary.

The dictionary is large in size (two orders of magnitude larger than the cache). For best performance a hash table is used to realize the dictionary. Three different hash tables were used: *STL hash_map*, *Google's sparsehash*, and a custom bucket hash table [21]. All three are implemented in C++ and are carefully optimized. The preexecution algorithm is the same for all three hash tables; only the hash table invocation differs in the test cases.

The applications are compiled with Microsoft Visual Studio 2010 (using default release build optimization settings) and are executed on two different computers, both running Microsoft Windows. The first machine is a desktop computer with Intel Core i7-2600 CPU (4 cores, 2 × 32 kB L1 cache and 256 kB L2 cache per core, 8 MB shared L3 cache) with 8 GB RAM. The second one is a laptop with Intel Core 2 Duo L7100 CPU (dual core, 2 × 32 kB L1 cache per core and a shared 2 MB L2 cache) with 3 GB RAM. The various hardwares are expected to yield different memory system performance and, thus, different optimal parameter configurations.

## 4.3 Performance Characteristics

For fair evaluation five executions are recorded in each test scenario and their average is reported below. The preexecuted application (*adaptive*) is tested against the baseline single threaded version (*baseline*).

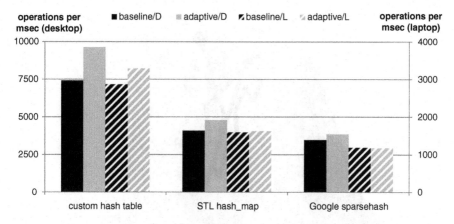

**Fig. 3.2**  Performance evaluation of adaptive preexecution

**Table 3.1**  Final *jumpahead distance and workahead size* values

| Configuration | Jumpahead distance | Workahead size |
|---|---|---|
| Custom hash table/desktop | 4,250 | 700 |
| Custom hash table/laptop | 3,046 | 1,458 |
| STL hash_map/desktop | 937 | 50 |
| STL hash_map | 734 | 61 |
| Google sparsehash/desktop | 3,003 | 500 |
| Google sparsehash/laptop | 500 | 100 |
| .NET dictionary/desktop | 4,031 | 312 |
| .NET dictionary/laptop | 4,967 | 3,845 |

Figure 3.2 plots the number of operations (insert into the hash table or lookup in the hash table) executed per millisecond for the three different hash tables executed on two different hardwares, and Table 3.1 lists the final *jumpahead distance* and *workahead size* values estimated by the algorithm. The filled bars correspond to the desktop machine (*D*) plotted on the left vertical axes and the striped ones to the laptop (*L*) on the right axis (they are of different scales).

All of the hash tables have different internal structures, hence the different performance of the baselines. The adaptive preexecution model we presented achieves 2–30% speedups in all but one of the test cases (Google sparsehash executed on the laptop) on both machines. It is also notable that this speedup is also present for the third-party STL hash_map and Google sparsehash.

## 4.4  Parameter Estimation

The Nelder–Mead method is prone to finding local optima. To illustrate the capabilities of the adaptive algorithm Fig. 3.3 plots the execution times as a 3D surface for selected test cases using various manually specified *jumpahead distance*

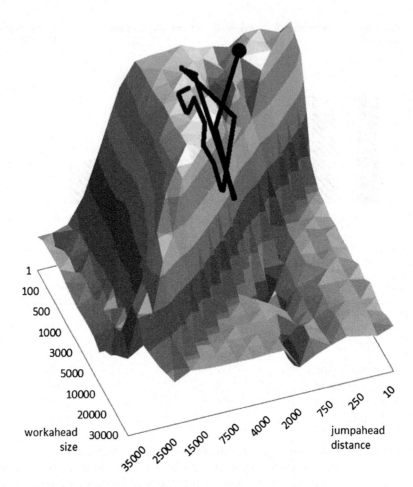

**Fig. 3.3** Performance for various jumpahead distance and workahead size values and the convergence path of the adaptive algorithm

and *workahead size* values. The higher the point is on the surface the better the performance is. The line drew on the surface shows the *jumpahead distance–workahead size* estimates chosen by the adaptive algorithm during its configuration phase. The starting point (manually specified) is marked with the circle. The fact that the starting point is near optimal is a coincidence; the algorithm does not require this.

The two machines used for the evaluation have different performance characteristics and different cache size. It is expected that the optimal parameter estimate for both will reflect these differences. Examining the values, for example, for the test case using the Google sparsehash the desktop machine calculates *jumpahead distance* as 3,003 and *workahead size* as 500 while the estimates of the laptop are 500 and 100, respectively (see Table 3.1). The *jumpahead distance* value, we remind, is the distance between the main thread and the helper thread and so it controls the amount of data resident in the cache. The smaller the cache is, the smaller this value should be; this is verified by the numbers above: the laptop has smaller cache and the resulting *jumpahead distance* is smaller indeed.

## 5  Conclusion and Discussion

This paper was motivated by data prefetch mechanisms for data-intensive applications suffering from memory to CPU latency. Preexecution is a method having promising results.

We presented a novel software-only, lightweight, adaptive preexecution algorithm, which—when applied to a data parallel, data-intensive application—masks, memory latency by bringing data into the cache of the CPU at the right time. The model we advocate attacks memory latency at a high level and does not require detailed knowledge and profiling of the applications.

The algorithm applies two parameters which greatly affect its performance. The optimal values of these parameters depend on the execution environment and the algorithm itself. We proposed an adaptive, runtime reconfiguring solution to find the optimal values of these parameters.

Through practical evaluation we have reached 2–30% speedup and have shown that the algorithm can successfully adapt to the execution environment by reconfiguring itself.

**Acknowledgment**  This project is supported by the New Hungary Development Plan (Project ID: TÁMOP-4.2.1/B-09/1/KMR-2010-0002).

## References

1. Bryant RE (2007) Data-intensive supercomputing: the case for DISC. Technical report CMU-CS-07-128. School of Computer Science, Carnegie Mellon University
2. Perkins LS, Andrews P, Panda D, Morton D, Bonica R, Werstiuk N, Kreiser R (2006) Data intensive computing. In: Proceedings of the 2006 ACM/IEEE conference on supercomputing SC'06, New York, NY, USA, p 69
3. Luk C-K (2001) Tolerating memory latency through software-controlled pre-execution in simultaneous multithreading processors. ACM SIGARCH Comput Architect News 29(2): 40–51
4. Chappell RS, Stark J, Kim SP, Reinhardt SK, Patt YN (1999) Simultaneous subordinate microthreading (SSMT). Int Symp Comput Architect 27(2):186–195
5. Dubois M (2004) Fighting the memory wall with assisted execution. In: Proceedings of the 1st conference on computing frontiers. Ischia, Italy, pp 168–180
6. Kim D et al (2004) Physical experimentation with prefetching helper threads on Intel's hyper-threaded processors. In: Proceedings of the international symposium on code generation and optimization, pp 27–38

7. Malhotra V, Kozyrakis C (2006) Library-based prefetching for pointer-intensive applications. Technical Report. Available online. http://csl.stanford.edu/~christos/publications/2006. library_prefetch.manuscript.pdf

8. Roth A, Moshovos A, Sohi GS (1998) Dependence based prefetching for linked data structures. ACM SIGPLAN Notices 33(11):115–126

9. Zhou J, Cieslewicz J, Ross KA, Shah M (2005) Improving database performance on simultaneous multithreading processors. In: Proceedings of the 31st international conference on very large data bases. Trondheim, Norway, pp 49–60

10. Zilles C, Sohi G (2001) Execution-based prediction using speculative slices. ACM SIGARCH Comput Architect News 29(2):2–13

11. Hillis WD, Steele GL (1986) Data parallel algorithms. Commun ACM 29(12):1170–1183

12. Cintra M, Llanos D (2003) Toward efficient and robust software speculative parallelization on multiprocessors. In: Proceedings of the ninth ACM SIGPLAN symposium on principles and practice of parallel programming, pp 13–24

13. Song Y, Kalogeropulos S, Tirumalai P (2005) Design and implementation of a compiler framework for helper threading on multi-core processors. In: Proceedings of 14th international conference on parallel architectures and compilation techniques (PACT'05), pp 99–109

14. Kim D, Yeung D (2002) Design and evaluation of compiler algorithms for pre-execution. ACM SIGPLAN Notices 37(10):159

15. Kim D, Yeung D (2004) A study of source-level compiler algorithms for automatic construction of pre-execution code. ACM Trans Comput Syst 22(3):326–379

16. Ro WW, Gaudiot J-L (2004) SPEAR: a hybrid model for speculative pre-execution. In: Proceedings of the 18th international parallel and distributed processing symposium, pp 75–84

17. Dundas J, Mudge T (1997) Improving data cache performance by pre-executing instructions under a cache miss. In: Proceedings of the 11th international conference on supercomputing (ICS'97). New York, NY, USA, pp 68–75

18. Mutlu O, Stark J, Wilkerson C, Patt YN (2003) Runahead execution: an alternative to very large instruction windows for out-of-order processors. In: Proceedings of the 9th international symposium on high-performance computer architecture, pp 129–140

19. Dudás Á, Juhász S (2011) Using pre-execution and helper threads for speeding up data intensive applications. In: Proceedings of the world congress on engineering 2011 (WCE 2011). Lecture Notes in Engineering and Computer Science, London, UK, 6–8 July 2011, pp 1288–1293

20. Nelder JA, Mead R (1965) A simplex method for function minimization. The Comput J 7(4): 308–313

21. Juhász S, Dudás Á (2009) Adapting hash table design to real-life datasets. In: Proceedings of the IADIS european conference on informatics 2009 (Part of the IADIS multiconference of computer science and information systems 2009). Algarve, Portugal, pp 3–10

# Chapter 4
# Efficient Implementation of Computationally Complex Algorithms: Custom Instruction Approach

**Waqar Ahmed, Hasan Mahmood, and Umair Siddique**

## 1 Introduction

Cryptography plays a vital role in establishing secure links in modern telecommunication networks. Information is transformed and transmitted in such a way that a third party cannot extract valuable and pertinent data from a secure communication link. Many cryptographic algorithms have been proposed such as AES [1], DES [2], Twofish [3], Serpent [4], etc.

These cryptographic algorithms use permutation operations to make the information more secure. For example, there are six different permutation operations used in DES, two permutation operations in Twofish, and two permutations in Serpent. The efficient computer implementation of permutation algorithms has always been a challenging, an interesting, and an attractive problem for researchers. During the past 20 years, more than twenty algorithms have been published for generating permutations of $N$ elements [5]. The practical importance of permutation generation and its use in solving problems was described by Tompkins [6].

If we assume that the time taken to execute one permutation is 1 $\mu$s, then Table 4.1 shows the time required to complete the permutation from $N = 1$ to $N = 17$. For $N > 25$, the required time is far greater than the age of the earth. Therefore, it is very important to implement the permutation operation in the most efficient manner.

In order to improve the channel capacity, forward error correction techniques are used. These techniques add redundant information to the data bits and make a

W. Ahmed (✉) • H. Mahmood
Department of Electronics, Quaid-i-Azam University, Islamabad, Pakistan
e-mail: waqasat@ele.qau.edu.pk; hasan@qau.edu.pk

U. Siddique
Research Center for Modeling and Simulation, National University of Sciences
and Technology (NUST), Islamabad, Pakistan
e-mail: umair.siddique@rcms.nust.edu.pk

S.-I. Ao and L. Gelman (eds.), *Electrical Engineering and Intelligent Systems*,
Lecture Notes in Electrical Engineering 130, DOI 10.1007/978-1-4614-2317-1_4,
© Springer Science+Business Media, LLC 2013

**Table 4.1** Approximate time needed to compute the permutation of $N$ (1 µs per permutation) [5]

| $N$ | $N!$ | Time |
|---|---|---|
| 1 | 1 | – |
| 2 | 2 | – |
| 3 | 6 | – |
| 4 | 24 | – |
| 5 | 120 | – |
| 6 | 720 | – |
| 7 | 5,040 | – |
| 8 | 40,320 | – |
| 9 | 362,880 | – |
| 10 | 3,628,800 | 3 s |
| 11 | 39,916,800 | 40 s |
| 12 | 479,001,600 | 8 min |
| 13 | 6,227,020,800 | 2 h |
| 14 | 87,178,291,200 | 1 day |
| 15 | 1,307,674,368,000 | 2 weeks |
| 16 | 2,092,278,988,800 | 8 months |
| 17 | 3,556,894,280,960 | 10 years |

communication link reliable over noisy channels. There are many decoding algorithms, which effectively remove errors from the received information and some of these include Turbo decoder [7], Viterbi decoder [8], etc. One of the famous techniques is the Viterbi algorithm, in which Add, Compare, and Select (ACS) processes are called multiple times which makes it computationally complex. It is desirable to efficiently implement the ACS process in order to achieve high performance in decoding methods which use Viterbi algorithm.

We modify the DLX [9] and PicoJava II [10] Instruction Set Architecture (ISA) by adding new custom permutation instruction WUHPERM [11] in their instruction set. The performance of the new instruction is analyzed for execution time. We design and implement the new permutation instruction in CPUSIM 3.6.8 [12] and Mic-1 simulator [10], respectively. Similarly, by using the same approach of designing custom instructions, the ISA of DLX processor is enhanced by adding Texpand instruction, used in Viterbi decoding algorithm. We test the implementation and performance of this new instruction on the CPUSIM 3.6.8 simulator.

The chapter is organized as follows: Section 2 presents the modified AES algorithm, which is an enhanced version of the original AES algorithm and utilizes the permutation operation more intensively as compared to other algorithms. Section 3 provides information about the Viterbi algorithm. Section 4 presents details of the DLX processor architecture, and its microinstructions. Section 5 presents the detailed architecture of PicoJava II processor with assembly language instructions and microcodes. Section 6 presents the details of simulators used in this chapter. This section also explains the design of new permutation instruction and Texpand instruction. Section 7 presents the comparison of different implementations of the new proposed instructions. We discuss the related work in Section 8, and finally, the conclusions are presented in Section 9.

## 2  Modified AES Algorithm

The modified AES algorithm is an improvement in the original AES cryptographic method presented in [13]. AES is the first algorithm proposed by National Institute of Standards and Technology (NIST) in October 2000 and is published as FIPS 197 [1]. Currently, it is known as one of the most secure and popular symmetric key algorithms [14].

The Substitution box (S-box) plays a vital role in the AES algorithm, as it is widely used in the process of encryption and provides the confusion ability. Many cryptanalysts have studied the structural properties of AES. A simple algebraic structure within AES and its S-box was presented by Ferguson et al. [25]. The most important and essential algebraic structure within AES was further analyzed in [15] and a polynomial description of AES was introduced in [16].

A new $S_8$ S-box is obtained by using the action of symmetric group $S_8$ on AES S-box [17], and these new S-boxes are used to construct $40320^{40320}$ secret keys [13]. The creation of the encryption keys with the permutations of the existing S-boxes results in 40320 new S-boxes, which, in turn, enhances the security and allows the system to be more safe and reliable. As a result, the information can be transmitted more securely over an unsecure and open access channel.

The introduction of additional complexity to the existing AES algorithm increases the computation time in implementing the encryption algorithm; therefore, it is desirable to execute this algorithm in an efficient manner. We present a new instruction, which facilitates the efficient execution of the modified and more complex AES method.

## 3  Viterbi Algorithm

The Viterbi decoder is used extensively in digital communications applications such as cell phones, digital audio broadcasting, and digital video broadcasting. According to an estimate, $10^{15}$ bits/s are decoded every day by Viterbi algorithm in digital TV devices [18]. A typical communication system incorporates channel coding in order to correct transmission errors. The process of channel coding adds redundancy to the information bits. Many encoding methods are developed which are distinguished by various characteristics and most important is the error correcting capability. The choice of using a particular coding scheme depends on the type of channel noise and its characteristics. There are two major classes of codes: convolutional codes and block codes. Some of the well-known codes include Reed–Solomon codes, BCH codes, Reed–Muller codes, convolutional codes, etc. [7].

We describe the Viterbi decoding technique used for convolutional codes. The application of convolutional encoding in combination with Viterbi decoding is suitable for channels with Additive White Gaussian Noise (AWGN). The Viterbi decoding method was first introduced by the founder of Qualcomm Corporation,

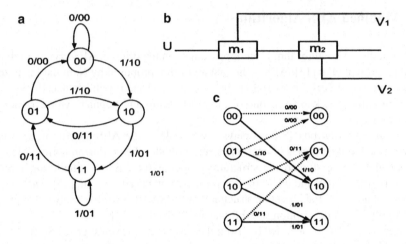

**Fig. 4.1** (**a**) State diagram of convolutional encoder. (**b**) A typical convolutional encoder. (**c**) Trellis for convolutional encoder

Andrew J. Viterbi [8]. After the introduction of Viterbi algorithm, many scientists and researchers extended this work by varying and improving the decoder design parameters in order to enhance the performance in software and hardware implementations.

The model of Viterbi algorithm is shown in Fig. 4.1. We use this model to implement the proposed custom instruction used in Viterbi algorithm. The Viterbi decoder utilizes the trellis structures to perform the decoding operation. The number of times the trellis expansion function is called, depends on the amount of decoding bits and the states in the trellis. The state diagram and the block diagram of a convolutional encoder are shown in Fig. 4.1a, b, respectively. This encoder has four states and there are two possible transitions from a state to two other states. In this design, the rate of the convolutional code is ½. The basic trellis structure is shown in Fig. 4.1c. This example is based on hard decision Viterbi algorithm. We take the information bits (110100), in which the first four are data bits and the last two are flush bits as an example to illustrate the behavior of convolutional encoder. The encoded information bits are (10 01 11 10 11 00). After passing through the channel the codeword is corrupted by noise and the third and seventh bits are in error. The received codeword is (101111001100). Based on the above design, the complete trellis diagram is shown in Fig. 4.2. This diagram describes a way to select the minimum weight path in trellis. At the top of trellis diagram in Fig. 4.2, the received bits are shown and in the left side corner, all possible states of the encoder are listed. The trellis expands from state (00) and only those paths survive which end at state (00). The dashed lines show the paths, which result by a "0" input bit, and solid lines result from "1" input bit. The path with the minimum weight is shown by a dark solid line.

In Viterbi algorithm, after a transition from a state, the weights are calculated for each possible path. Whenever there is a difference in a particular received bit and the state output bit, one is added to the previous accumulated weight in the

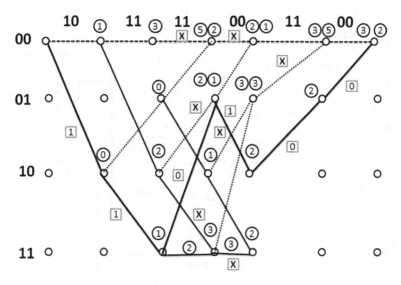

**Fig. 4.2** Trellis diagram for typical application

transition path. For example, if the received bits are (00) and output bits in a
particular state transition are (01), then there is a difference of one bit. One is
added to the cumulative weight at that particular path transition. Similarly, in the
case of difference of two bits we add two at that particular path transition. If more
than one path arrives at a particular state, the path with the lowest weight survives
and the remaining paths are deleted. In this case, if the weights of arriving paths are
equal, the path arriving from lower state survives. For example, if we have state
(00) and state (01) both transit to state (00) having same accumulated weight values,
we select only the state (00) path.

It can be seen from Fig. 4.2 that the trellis expand operation is called several
times as we progress in the decoding process. Therefore, it is desirable to create a
custom trellis expand instruction which executes this operation efficiently. We
enhance the ISA of DLX processor and create the Texpand instruction in CPUSIM
simulator in order to efficiently execute the trellis expand operation. The detailed
architecture of DLX processor is described in the following sections.

## 4  Architecture of DLX

The DLX architecture provides 32 general-purpose registers of 32 bits and each is
named as $R_0$–$R_{31}$. These registers have special roles. The value of register $R_0$ is always
zero. Branch instructions to subroutines implicitly use register $R_{31}$ to store the return
address. Memory is divided into words of 32 bits and is byte addressable. The DLX
microarchitecture is shown in Fig. 4.3 and its detailed data path can be seen in [9].

Some assembly language instructions with their microinstructions are shown in
Table 4.2.

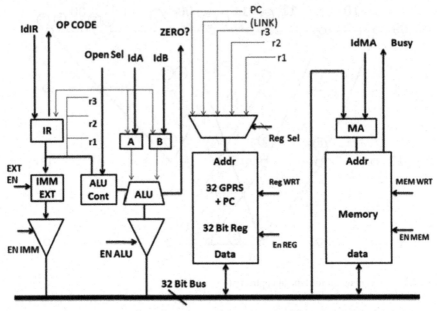

**MicroInstruction: Register to Register  Transfer (17 Control Signals)**

**Fig. 4.3** DLX microarchitecture [19]

**Table 4.2** DLX instructions and their corresponding microinstructions

| DLX instructions | Microinstructions |
|---|---|
| LD R4, 100(R1) | Ir(8-15)→mar |
| | Main[mar]→mdr |
| | Mdr→Ir(5-7) |
| | End |
| SW R4, 100(R1) | Ir(8-15)→mar |
| | Ir(5-7)→mdr |
| | Mdr→Main[mar] |
| | End |
| AND R1, R2, R3 | Ir(8-10)→B |
| | Ir(11-13)→A |
| | Acc←A & B |
| | Acc→Ir(5-7) |
| | End |
| SRL R1, R2, R3 | Ir(8-10)→B |
| | Ir(11-13)→A |
| | Acc←A≪B |
| | Acc→Ir(5-7) |
| | End |

## 5   Picojava II Architecture

PicoJava II is a 32 bit pipelined stack-based microprocessor which can execute the JVM instructions. There are about 30 JVM instructions that are microprogrammed and most of the JVM instructions execute in a single clock cycle. The block diagram of PicoJava II microarchitecture is shown in Fig. 4.4.

The pipeline of PicoJava II processor has six stages. The first stage is the instruction fetch which takes instruction from I-cache into 16 bytes instruction buffer. The second stage is decode and fold stage. The opcode and three register fields are decoded in the decode stage. In the fold stage, the instruction folding operation is performed, in which a particular sequence of instructions is detected and combined into one instruction [10]. In the third stage, the operands are fetched from the stack, that is, from the register file. The operands are then ready for the fourth stage, which is the execution stage. In the fifth stage, the results are stored in the cache. Some assembly language instructions and their microcode are shown in Table 4.3.

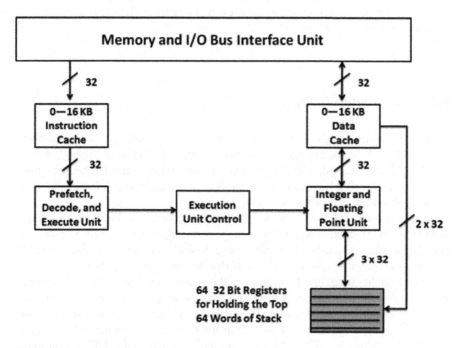

**Fig. 4.4**  PicoJava II architecture

**Table 4.3** Some assembly language instructions and their microinstructions [10]

| Mnemonic | Microcode | Description |
|---|---|---|
| iadd1 | MAR = SP = SP − 1; rd | Read in next-to-top word on stack |
| iadd2 | H = TOS | H = top of stack |
| iadd3 | MDR = TOS = MDR + H; wr; goto (MBR1) | Add top two words; write to new top of stack |
| iload1 | H = LV | MBR contains index; copy LV to H |
| iload2 | MAR = MBRU + H; rd | MAR = address of local variable to push |
| iload3 | MAR = SP = SP + 1 | SP points to new top of stack; prepare write |
| iload4 | PC = PC + 1; fetch; wr | Inc PC; get next opcode; write top of stack |
| iload5 | TOS = MDR; goto Main1 | Update TOS |

# 6 Simulators

In this chapter, we use the microprogramming technique to create new instructions. We use CPUSIM 3.6.8 and Mic-1 simulators to create and test the new instructions. These simulators have the ability to create custom instructions. Instructions are created by implementing the microprogramming code for each individual instruction.

The CPUSIM 3.6.8 simulator is created by Dale Skrien and is presented in [12]. The Mic-1 simulator is proposed by Andrew S. Tanenbaum in his book *Structured Computer Organization* [10].

## 6.1 Permutation Instruction

Many permutation algorithms have been proposed such as Heap Method, Johnson–Trotter Method, Loopless Johnson–Trotter Method, Ives Method, Alternate Ives Method, Langdon Method, Fischer–Krause Method, etc. The heap method runs faster and is simpler than other methods as presented in [5]. A ladder diagram for heap algorithm is depicted in Fig. 4.5.

In this chapter, we create custom permutation instructions based on this heap algorithm. The efficient implementation of the permutation operation for $S_8$ S-box to construct secure keys can be achieved by using these instructions as presented in [13].

The permutation operation in [13] is performed on 32-bit data. We divide these 32 bits into eight groups of 4-bit nibbles. In this chapter, we demonstrate the method used to create the permutation instructions for 4-bit nibbles and this technique can be further enhanced to 32-bit data. The heap algorithm for 4-bit data is shown in Fig. 4.5. It is apparent that we need four instructions to implement this algorithm. These instructions swap these 4 bits in the order as this ladder descends. At the end, we have 24 unique set of different permutations.

**Fig. 4.5** Heap ladder
diagram

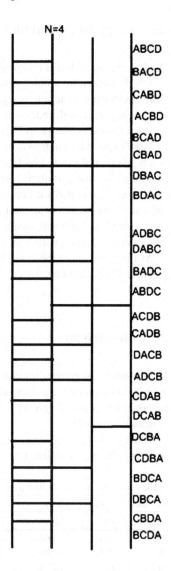

Therefore, we create four WUHPERM1, WUHPERM2, WUHPERM3, and WUHPERM4 instructions, which initially swap the first and second location bit, second and third location bit, third and fourth location bit, and fourth and first location bit, respectively.

The number of instructions used to calculate the permutation of a given data is given as:

$$N\ (I) = \mathbf{Log_2}\ (N),$$

where $N\ (I)$ is the number of instruction required and $N$ is the number of bits in data.

## 6.2  Texpand Instruction

The Viterbi algorithm is based on trellis expansion operation. During the execution of Viterbi algorithm implemented in assembly language programming, code for trellis expansion function is executed multiple times. For example, if there are 12 bits in a received codeword then this trellis function is called approximately nineteen times, as shown in Fig. 4.2.

The Texpand instruction is used to implement the trellis function which performs two fundamental tasks. Firstly, it performs add, compare, and select operation. In this operation, the cumulative weights of the arriving paths at a particular state are calculated. A comparison between the weights of the arriving paths is performed, and a select operation is applied to find out the surviving path. The second task is to keep track of the path with minimum weight that ultimately ends up at state (00), and to memorize state transition inputs used to trace back the output.

At the final stage, the trace back function is performed to determine the most probable transmitted sequence.

## 7  Comparison

The assembly language program is written for CPUSIM 3.6.8 and Mic-1 simulators in order to compare the performance. It is seen that the number of microinstructions for assembly code is greater than the microinstructions required to create the custom instruction.

Tables 4.4–4.7 show the comparison for an assembly language program which performs the permutation function, and the entire program when replaced by a

**Table 4.4** Permutation algorithm in CPUSIM simulator

| Assembly language program | | | | | |
|---|---|---|---|---|---|
| Bits | AI | MI | FI (I × 4) | Total MI | Time ($T$): Total MI × 4 |
| 4 | 90 | 367 | 90 | 457 | 1,828 |
| 8 | 174 | 706 | 174 | 880 | 3,520 |
| 16 | 342 | 1,384 | 342 | 1,726 | 6,904 |
| 32 | 678 | 2,740 | 678 | 3,418 | 1,372 |

**Table 4.5** Permutation instruction on CPUSIM simulator

| Assembly language program | | | | | |
|---|---|---|---|---|---|
| Bits | AI | MI | FI (I × 4) | Total MI | Time ($T$): Total MI × 4 |
| 4 | 33 | 158 | 132 | 290 | 1,160 |
| 8 | 58 | 298 | 232 | 530 | 2,120 |
| 16 | 112 | 575 | 448 | 1,023 | 4,096 |
| 32 | 220 | 1,131 | 880 | 2,011 | 8,044 |

**Table 4.6** Permutation algorithm on Mic-1 simulator

| Assembly language program | | | | | |
|---|---|---|---|---|---|
| Bits | AI | MI | FI (I × 4) | Total MI | Time (T): Total MI × 4 |
| 4 | 90 | 367 | 90 | 457 | 1,828 |
| 8 | 174 | 706 | 174 | 880 | 3,520 |
| 16 | 342 | 1,384 | 342 | 1,726 | 6,904 |
| 32 | 678 | 2,740 | 678 | 3,418 | 1,372 |

**Table 4.7** Permutation instruction on Mic-1 simulator

| WUHPERM instruction | | | | |
|---|---|---|---|---|
| MI | FI (I × 4) | Total MI | Time (T): (Total MI × 4) | %Performance (in %) |
| 62 | 1 | 63 | 252 | 752 |
| 112 | 1 | 113 | 452 | 779 |
| 212 | 1 | 213 | 852 | 810 |
| 412 | 1 | 413 | 1,652 | 827 |

custom permutation instruction, which swaps any two bits. In Tables 4.4 and 4.5, statistics are shown for the CPUSIM simulator program. It is seen that the time consumed by assembly language program is greater when compared to the system that implements the custom instruction in its algorithm. In Tables 4.6 and 4.7 analysis for Mic-1 simulator is presented. Here, we can see that the programs, which use custom instruction, have less execution time as compared to simple assembly language program. The performance is further enhanced if we compare these results with CPUSIM simulator output. This comparison can also be extended to all $N!$ unique permutations. In the case when $N$ is increased, it is seen that the performance substantially improves by using the custom instructions. It can be observed from Fig. 4.6a, b, that performance improvement rate on stack-based architecture (Mic-1 simulator) is greater as we increase the number of data bits. This increase is due to the redundancy in the assembly language program for this architecture, that is, we must transfer the data into the stack in order to execute arithmetic operations. The number of microinstruction used in Mic-1 simulator is greater than the instructions required for CPUSIM 3.6.8 simulator. This is because of the inherent property of RISC architecture, which takes less execution time in performing the same amount of work than a stack implementation for any processor.

Table 4.8 shows the comparison of the performance of Viterbi algorithm, in particular, trellis expansion function, when implemented in DLX assembly language and with custom Texpand instruction. This instruction is called nineteen times in the Viterbi algorithm for 6-bit decoding. The results show substantial improvement of three times when trellis expands function is implemented as a custom instruction.

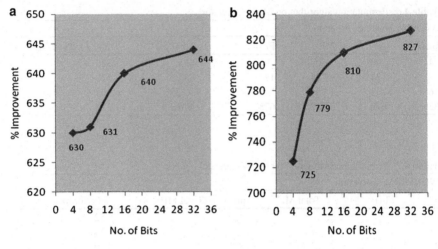

**Fig. 4.6** (a) Percentage performance for CPUSIM; (b) percentage performance for Mic-1

**Table 4.8** Comparison between trellis assembly language function and Texpand instruction

| Trellis assembly function | | Texpand instruction | |
|---|---|---|---|
| Assembly instruction (AI) | 63 | Assembly instruction (AI) | 1 |
| Microinstruction (MI) | 277 | Microinstruction (MI) | 100 |
| Fetched instruction (I × 4) | 63 | Fetched instruction (I × 4) | 1 |
| Function calls | 19 | Texpand instruction calls | 19 |
| Total MI ((MI + FI) × 19) | 6,460 | Total MI ((MI + FI) × 19) | 1,919 |
| Time (T) Total MI × 4 | 25,840 | Time (T) Total MI × 4 | 7,676 |
| | | Performance improvement | 337% |

# 8 Related Work

Various methodologies have been proposed to implement permutation operation using software and hardware. In software implementations, the permutation operation is achieved by EXTRACT and DEPOSIT instructions [20–21]. These instructions extract the bits individually by using AND mask and place the bits in a different order to produce permutation operation. In this work, we use only one instruction to perform the tasks of EXTRACT and DEPOSIT operations. A new instruction (WUHPERM) is added to achieve the permutation operation. This reduces the complexity in the code and allows easy implementation. Also, the number of additional fetch instructions is reduced by replacing multiple instructions with a single instruction. Hardware designs are also proposed to implement the permutation operation in an efficient way. A popular approach to achieve permutation is presented by Shi [22]. In the hardware approach, the cost of hardware increases and the data path becomes more complex. In a software approach, the hardware changes are negligible and there is no substantial increase in the cost of the microprocessor hardware.

Many different platforms are used to implement the Viterbi algorithm efficiently. In [23], the trellis custom instruction is created in Xtensa processor. Similarly, in [24], FPGA-based soft processor is used to accelerate the trellis function. Our approach is relatively simple and easy to implement because in [23, 24], additional logic elements are used to implement custom instruction. Care is needed to attach additional hardware, in order to avoid errors and increase the complexity of the existing data paths. In the presented approach, the new instructions are implemented by using the existing microinstructions which do not require any change in the existing architecture.

# 9   Conclusion

In this chapter, new instructions are presented which efficiently performs permutation operation and trellis expansion operation. The new instructions implement the mathematically intensive operations used by AES algorithms and the Viterbi algorithm, and as a result, enhancement in speed and performance is achieved. We report an enhancement in DLX processor instruction set for efficient implementation of Viterbi decoding algorithm and enhanced AES encryption algorithm. We also present results for enhanced AES encryption algorithm for PicoJava II processor. We create a custom permutation instruction (WUHPERM) and a custom trellis expansion instruction (Texpand) in CPUSIM simulator on RISC-based architecture. In addition, we implement the same WUHPERM instruction on Mic-1 simulator, which is based on JVM microarchitecture. The results show substantial improvements in the execution speed of approximately six times when the WUHPERM instruction is implemented in RISC architecture and eight times for stack-based architecture. The execution time is stupendously improved to approximately three times when Texpand instruction is designed for RISC architecture.

# References

1. Announcing the advanced encryption standard (AES). National Institute of Standard Technology, FIPS, 197
2. Schneier B (1996) Applied cryptography, 2nd edn. Wiley, New York
3. Schneier B, Kelsey J (1998). Twofish: A 128-bit block cipher. Available online http://www.schneier.com/twofish.html
4. Smith B, Anderson R, Biham E, Knudsen L (1998). Serpent: a proposal for the advanced encryption standard. Available online http://www.cl.cam.ac.uk/ftp/users/rja14/serpent.pdf
5. Sedwick R (June 1977) Permutation generation methods. Comput Surv 9(2):137–164

6. Tompkin C (1956) Machine attack on problems whose variable are permutations. In: Proceedings of symposium in applied math numerical analysis. McGraw Hill Inc., New York, US, pp 195–211
7. Bossert M (1999) Channel coding for telecommunications. Wiley, Chichester, UK
8. Viterbi AJ (1967) Error bounds for convolutional codes and an asymptotically optimum decoding algorithm. IEEE Trans Inform Theory IT-13(2):260–269
9. Hennessy J, Patterson D (1996) A Computer Architecture, A Quantitative Approach. Morgan Kaufmann publisher Inc., San Francisco
10. Tanenubaum AS (2005) Structured computer organization, 5th edn. Prentice Hall, Upper Saddle River, NJ
11. Ahmed W, Mahmood H, Siddique U (2011) The Efficient Implementation of S8 AES Algorithm. In: Proceedings of world congress on engineering 2011 (WCE 2011). International Conference of Computer Science and Engineering, London, UK, 6–8 July 2011, pp 1215–1219
12. Skrien D (2001) CPU Sim 3.1: A Tool for Simulating Computer Architectures for CS3 classes. ACM J Educ Resour Comput 1(4):46–591
13. Hussain I, Shah T, Mahmood H (2010) A new algorithm to construct secure keys for AES. Int J Contemp Math Sci 5(26):1263–1270
14. Tran MT, Bui DK, Duong AD (2008) Gray S-box for advanced encryption standard. In: Proceedings of international conference of computational intelligence and security (CIS'08), Suzhou, 13–17 Dec 2008, pp 253–258
15. Murphy S, Robshaw MJ (2002) Essential algebraic structure within the AES. In: Proceedings of the 22nd annual international cryptology conference 2002 (Crypto'02). LNCS 2442, Santa Barbara, CA, USA, 18–22 Aug 2002, pp 1–16
16. Rosenthal J (2003) A polynomial description of the Rijndael Advanced Encryption Standard. J Algebra Appl 2(2):223–236
17. Daemen J, Rijmen V (1999) AES proposal: Rijindael, AES submission, version 2. Available online http://csrc.nist.gov/archive/aes/rijndael/Rijndael-ammended.pdf
18. Forney GD (2005) The Viterbi algorithm: a personal history. In: Proceedings of Viterbi conference, LA, USA, 8–9 Mar 2005
19. Asanovic K (2002) DLX microprogramming slides. MIT Laboratory of computer science. Available online http://dspace.mit.edu/bitstream/handle/1721.1/35849/6-823Spring-2002/NR/rdonlyres/Electrical-Engineering-and-Computer-Science/6-823Computer-System-Architecture Spring2002/B1A470D6-9272-44BE-8E8D-B77FA84A7745/0/lecture04.pdf
20. Lee R (1989) Precision architecture. IEEE Comput 22(1):78–91
21. Lee R, Mahon M, Morris D (1992) Path length reduction features in the PA-RISC architecture. In: Proceedings of IEEE Compcon, San Francisco, CA, 24–28 Feb 1992, pp 129–135
22. Shi Z, Lee RB (2000) Bit permutation instructions for accelerating software cryptography. In: Proceedings of the IEEE international conference on application-specific systems. Architectures and Processors 2000 (ASAP 2000), Boston, MA, USA, 10–12 July 2000, pp 138–148
23. Convolutional coding on Xtensa processor application note. Tensilica, Inc., January 2009, Doc Number: AN01-123-04
24. Liang J, Tessier R, Geockel D (2004) A dynamically-reconfigurable, power-efficient turbo decoder. In: Proceedings of 12th annual IEEE symposium on field programmable custom computing machines (FCCM 2004), Napa, CA, USA, 20–23 April 2004, pp 91–100
25. Ferguson N, Schroeppel R, Whiting D (2001) A simple algebraic representation of Rijndael. In: Proceedings of Selected Areas in Cryptography 2011 (SAC01). LNCS 2259, London, UK, 16–17 Aug 2001, pp 103–111

# Chapter 5
# Learning of Type-2 Fuzzy Logic Systems by Simulated Annealing with Adaptive Step Size

Majid Almaraashi, Robert John, and Samad Ahmadi

## 1  Introduction

One of the features of fuzzy logic systems is that they can be hybridised with other methods such as neural networks, genetic algorithms and other search and optimisation approaches. These approaches have been proposed to add a learning capability to fuzzy systems to learn from data [6]. Fuzzy systems are good at explaining how they reach a decision but cannot automatically acquire the rules or membership functions to make a decision [10, p.2]. On the other hand, learning methods such as neural networks cannot explain how a decision was reached but have a good learning capability [10, p.2]. Hybridisation overcomes the limitations of each method in an approach such as neuro-fuzzy systems or genetic fuzzy systems. Soft Computing is a branch of computer science described as "a collection of methodologies aims to exploit the tolerance for imprecision and uncertainty to achieve tractability, robustness and low solution cost" [34]. In this research we are interested in the combination of type-2 fuzzy logic with simulated annealing to design a high performance and low-cost system. When designing a simple fuzzy system with few inputs, the experts may be able to use their knowledge to provide efficient rules but as the complexity of the system grows, the optimal rule-base and membership functions become difficult to acquire. So, researchers often use some automated tuning and learning methods and evaluate their solutions by some criteria [22].

Simulated annealing has been used in some fuzzy systems to learn or tune fuzzy systems (for example, see [8, 9, 22]). In addition, the combination of simulated annealing and type-1 Mamdani and TSK fuzzy systems exhibited good performance in forecasting Mackey–Glass time series as shown in [5] and [3]. In this

M. Almaraashi (✉) • R. John • S. Ahmadi
The Centre for Computational Intelligence, De Montfort University, Leicester LE1 9BH, UK
e-mail: almaraashi@dmu.ac.uk; rij@dmu.ac.uk; sahmadi@dmu.ac.uk

S.-I. Ao and L. Gelman (eds.), *Electrical Engineering and Intelligent Systems*,
Lecture Notes in Electrical Engineering 130, DOI 10.1007/978-1-4614-2317-1_5,
© Springer Science+Business Media, LLC 2013

paper, a forecasting method is proposed using an interval type-2 Mamdani model designed using simulated annealing which extends our previous work in [4]. The Mackey–Glass time series is a well-known bench mark data which will be used here as an application of forecasting. The rest of this paper starts by describing the data sets in Sect. 2 followed by a review of fuzzy systems (Sect. 3) and simulated annealing (Sect. 4). The methodology and the results of this work are detailed in Sect. 5 where the conclusion is drawn in Sect. 6.


## 2  Mackey–Glass Time Series

The Mackey–Glass Time Series is a chaotic time series proposed by Mackey and Glass [25]. It is obtained from this non-linear equation:

$$\frac{dx(t)}{dt} = \frac{\alpha * x(t - \tau)}{1 + x^n(t - \tau)} - \beta * x(t), \tag{5.1}$$

where $\alpha$, $\beta$ and $n$ are constant real numbers and $t$ is the current time where $\tau$ is the difference between the current time $t$ and the previous time $t - \tau$. To obtain the simulated data, the equation can be discretised using the Fourth-Order Runge–Kutta method. In the case where $\tau > 17$, it is known to exhibit chaos and has become one of the benchmark problems in soft computing [26, p.116].


## 3  Type-2 Fuzzy Systems

Type-1 fuzzy logic has been used successfully in a wide range of problems such as control system design, decision making, classification, system modelling and information retrieval [12, 31]. However, type-1 approach is not fully able to model uncertainties directly and minimise its effects [28]. These uncertainties exist in a large number of real-world applications. Uncertainties can be a result of [28]:

- Uncertainty in inputs
- Uncertainty in outputs
- Uncertainty that is related to the linguistic differences
- Uncertainty caused by the conditions change in the operation
- Uncertainty associated with the noisy data when training fuzzy logic controllers

All these uncertainties translate into uncertainties about fuzzy sets membership functions [28]. Type-1 fuzzy Logic cannot fully handle these uncertainties because type-1 fuzzy logic membership functions are totally precise which means that all kinds of uncertainties will disappear as soon as type-1 fuzzy set membership functions are used [11]. The existence of uncertainties in the majority of

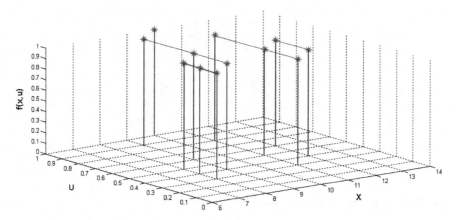

**Fig. 5.1** Interval type-2 fuzzy set "About 10"

real-world applications makes the use of type-1 fuzzy logic inappropriate in many cases especially with problems related to inefficiency of performance in fuzzy logic control [11]. Although, type-2 fuzzy sets need more computations than type-1 fuzzy sets, interval type-2 fuzzy sets can be used to reduce computational expenses. Type-2 fuzzy systems have, potentially, many advantages over type-1 fuzzy systems including the ability to handle numerical and linguistic uncertainties, allowing for a smooth control surface and response and giving more freedom than type-1 fuzzy sets [11]. Since last decade, type-2 fuzzy logic is a growing research topic with much evidence of successful applications [14] and [27].

A type-2 fuzzy set [28], denoted $\tilde{A}$, is characterised by a type-2 membership function $\mu_{\tilde{A}}(x, u)$ where $x \in X$ and $u \in J_x \subseteq [0, 1]$. For example:

$$\tilde{A} = ((x, u), \mu_{\tilde{A}}(x, u)) \mid \forall x \in X, \forall u \in J_x \subseteq [0, 1]$$

where $0 \le \mu_{\tilde{A}}(x, u) \le 1$. Set $\tilde{A}$ also can be expressed as:

$$\tilde{A} = \int_{x \in X} \int_{u \in J_x} \mu_{\tilde{A}}(x, u)/(x, u), J_x \in [0, 1]$$

where $\int$ denotes union. When universe of discourse is discrete, Set $\tilde{A}$ is described as

$$\tilde{A} = \sum_{x \in X} \sum_{u \in J_x} \mu_{\tilde{A}}(x, u)/(x, u), J_x \in [0, 1].$$

When all the secondary grades $\mu_{\tilde{A}}(x, u)$ equal 1, then, $\tilde{A}$ is an interval type-2 fuzzy set. Interval type-2 fuzzy sets are easier to compute than general type-2 fuzzy sets. Figure 5.1 shows an example of an interval type-2 fuzzy set. Figure 5.2 shows

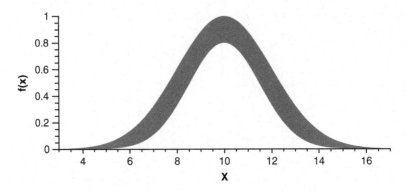

**Fig. 5.2** FOU for type-2 fuzzy set "About 10"

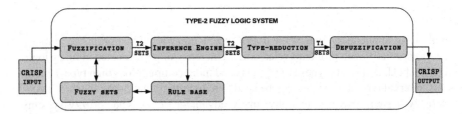

**Fig. 5.3** The components of a type-2 fuzzy logic system

a 2D representation of interval type-2 set called the footprint of uncertainty (FOU) represents the union of all primary membership grades and can be described easily by a lower and upper membership functions. The ease of computation and representation of an interval type-2 fuzzy sets is the main driver for their wide usage in real-world applications.

Type-2 fuzzy logic systems are rule-based systems that are similar to type-1 fuzzy logic systems in terms of the structure and components but a type-2 fuzzy logic system has an extra output process component which is called the type-reducer before defuzzification as shown in Fig. 5.3. The type-reducer reduces outputs type-2 fuzzy sets to type-1 fuzzy sets and then the defuzzifier reduces them to crisp outputs. The components of a type-2 Mamdani fuzzy logic system are [26]:

- *Fuzzifier*: Fuzzifier maps crisp inputs into type-2 fuzzy sets by evaluating the crisp inputs $x = (x_1, x_2, \ldots, x_n)$ based on the antecedent part of the rules and assigns each crisp input to its type-2 fuzzy set $\tilde{A}(x)$ with its membership grade in each type-2 fuzzy set.
- *Rules*: A fuzzy rule is a conditional statement in the form of IF-THEN where it contains two parts, the IF part called the antecedent part and the Then part called the consequent part.

- *Inference Engine*: Inference Engine maps input type-2 fuzzy sets into output type-2 fuzzy sets by applying the consequent part where this process of mapping from the antecedent part into the consequent part is interpreted as a type-2 fuzzy implication which needs computations of union and intersection of type-2 fuzzy sets. The inference engine in Mamdani system maps the input fuzzy sets into the output fuzzy sets then the defuzzifier converts them to crisp outputs. The rules in Mamdani model have fuzzy sets in both the antecedent part and the consequent part. For example, the $i$th rule in a Mamdani rule base can be described as follows:

$$R^i : \text{IF } x_1 \text{ is } \tilde{A}_1^i \text{ and } x_2 \text{ is } \tilde{A}_2^i \text{... and } x_p \text{ is } \tilde{A}_p^i \text{ THEN } y \text{ is } \tilde{B}^i$$

- *Output Processor*: There are two stages in the output process:

  - *Type-Reducer*: Type-reducer reduces type-2 fuzzy sets that have been produced by the inference engine to type-1 fuzzy sets by performing a centroid calculation [26]. For example, the centre of sets type-reducer replaces each rule type-2 consequent set by its centroid and then calculates a weighted average of these centroids to get a type-1 fuzzy set [17].
  - *Defuzzifier*: Defuzzifier maps the reduced output type-1 fuzzy sets that have been reduced by type-reducer into crisp values exactly as the case of defuzzification in type-1 fuzzy logic systems.

## 4  Simulated Annealing Algorithm

The concept of annealing in the optimisation field was introduced in [19]. Simulated annealing uses the Metropolis algorithm to imitate metal annealing in metallurgy where heating and controlled cooling of materials are used to reshape metals by increasing the temperature to higher values and then decreasing the temperature carefully to get the particles arranged until the system energy becomes minimal. Simulated annealing is a powerful randomised local search algorithm that has shown great success in finding optimal or near optimal solutions of combinatorial problems [2]. Simulated annealing is particularly useful in dealing with high-dimensionality problems as it scales well with the increase in variable numbers which allows simulated annealing to be a good candidate for fuzzy systems optimisation [8]. In general, simulated annealing can find good solutions for a wide range of problems but normally with the cost of high running times [2].

We now define the simulated annealing algorithm. Let $s$ be the current state and $N(s)$ be a neighbourhood of $s$ that includes alternative states. By selecting one state $s' \in N(s)$ and computing the difference between the current state cost and the selected state energy as $D = f(s') - f(s)$, $s'$ is chosen as the current state based on *Metropolis criterion* in two cases:

- If $D <= 0$ means the new state has a smaller or equal cost, then $s'$ is chosen as the current state as down-hills always accepted.
- If $D > 0$ and the probability of accepting $s'$ is larger than a random value $Rnd$ such that $e^{-D/T} > Rnd$ then $s'$ is chosen as the current state where $T$ is a control parameter known as *Temperature* which is gradually decreased during the search process making the algorithm more greedy as the probability of accepting uphill moves decreases over time. $Rnd$ is a randomly generated number, where $0 < Rnd < 1$. Accepting uphill moves is important for the algorithm to avoid being stuck in a local minima.

In the last case where $D > 0$ and the probability is lower than the random value $e^{-d/T} <= Rnd$, no moves are accepted and the current state $s$ continues to be the current solution. When starting with a large cooling parameter, large deteriorations can be accepted. Then, as the temperature decreases, only small deteriorations are accepted until the temperature approaches zero when no deteriorations are accepted. Therefore, adequate temperature scheduling is important to optimise the search. Simulated annealing can be implemented to find a closest possible optimal value within a finite time where the cooling schedule can be specified by four components [2]:

- Initial value of temperature.
- A function to decrease temperature value gradually.
- A final temperature value.
- The length of each homogeneous Markov chains. A Markov chain is a sequence of trials where the probability of the trial outcome depends on the previous trial outcome only and it is classified as a homogeneous when the transition probabilities do not depend on the trial number [1, p. 98].

The choice of good simulated annealing parameters is important for its success. For example, small initial temperatures could cause the algorithm to get stuck in local minimas as the first stages of the search are supposed to aim for exploration of regions while large ones could cause unnecessary excessive running times. In addition, appropriate cooling schedule is important for the same reason as fast cooling causing getting stuck in local minima and slow cooling causing the algorithm convergence to be very slow. In the fuzzy system literature, few researchers used adaptive step sizes such as [13] while most of the approaches reported were using small fixed step sizes [7]. In continuous optimisation problems, the adjustment of the neighbourhood range for simulated annealing might be important [29]. The step sizes should not be all equal for all inputs but rather it should be chosen based on its effects to the objective function [24]. One of the methods used to determine the step size during the annealing process was proposed in [30] which starts by using large step sizes and decreases them gradually. One of the methods to determine the initial temperature value proposed in [33] is to choose the initial temperature value within the standard deviation of the mean cost of a number of moves. When using finite Markov chains to model the simulated annealing mathematically, the temperature is reduced once for each Markov chain while the length of each chain should be related to the size of the neighbourhood in the problem [2].

# 5   Methodology and Results

The experiment can be divided into three steps: generating time series, constructing
the initial fuzzy system and optimising the fuzzy system parameters. Firstly, the time
series is generated with the following parameters in (5.1): $\alpha = 0.2, \beta = 0.1, \tau = 17,$
$\eta = 10$ as with other authors [13]. The input–output samples are extracted in the
form $x(t - 18), x(t - 12), x(t - 6) and x(t)$, where $t = 118$–1117 using a step size of
6. Samples of the generated data are depicted in Fig. 5.4. Then, the generated data
are divided into 500 data points for training and the remaining 500 data points for
testing. Using a step size of 6, the input values to the fuzzy system are the previous
data points $x(t - 18), x(t - 12), x(t - 12)$ and $x(t)$ while the output from the fuzzy
system is the predicted value $x(t + 6)$. Four initial input values $x(114)$ and $x(115)$ and
$x(116)$ and $x(117)$ are used to predict the first four training outputs.

Two fuzzy systems have been chosen: type-1 and type-2 fuzzy logic systems.
The fuzzy model consists of four input fuzzy sets $A_1, A_2, A_3$ and $A_4$ and one
independent output fuzzy set $B_i$ for each rule. Gaussian membership functions
were chosen to define the fuzzy sets. Any other types of membership functions
can be chosen. The parameters of the Gaussian membership function are the mean
$m$ and the standard deviation $\sigma$ in type-1 fuzzy logic system. For type-2 fuzzy logic
system, each fuzzy set is represented by two means and one standard deviation as
shown in Fig. 5.5. All the means and standard deviations are initialised for all the
input fuzzy sets by dividing each input space into four fuzzy sets and enabling
enough overlapping between them. The fuzzification process is based on the
product t-norm while the centre-of-sets has been chosen for type-reduction and

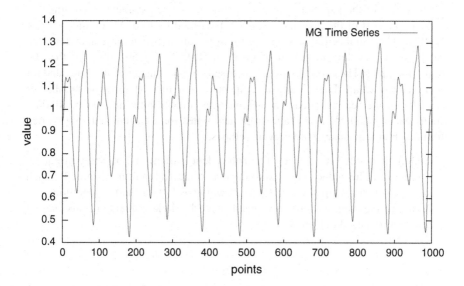

**Fig. 5.4**   Mackey–Glass time series when Tau $= 17$

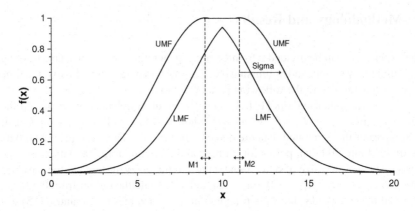

**Fig. 5.5** The parameters in interval type-2 Gaussian membership function

defuzzification. Type-reduction uses KM algorithm proposed in [16] to calculate the centre of sets. The training procedure aims to learn the parameters of the antecedent parts and the consequent parts of the fuzzy system rules. Then, the found parameters are used to predict the next 500 testing data points. By using four inputs and two fuzzy sets for each input, we end up with 16 rules and 8 input fuzzy sets representing all possible combinations of input values with input fuzzy sets. While each rule in Mamdani is linked with one independent output set. The total number of parameters in type-1 fuzzy logic system is $8 + 8 + (16*2) = 48$ while it is 72 parameters in type-2 fuzzy logic system. In type-2 fuzzy logic system, $(8*2) + 8 = 24$ parameters in the antecedent part and $(16*2) + 16 = 48$ in the consequent part.

The learning process is done using simulated annealing that searches for the best configuration of the parameters by trying to modify one parameter each time and evaluate the cost of the new state which is measured by Root Mean Square Error (RMSE). The simulated annealing algorithm is initialised with a temperature that equals to the standard deviation of mean of RMSE's for 500 runs for the 500 training samples. The cooling schedule is based on a cooling rate of 0. 9 updated for each Markov chain. Each Markov chain has a length related to the number of variables in the search space. The search ends after a finite number of Markov chains namely 400 Markov chains. The new states for a current state are chosen from neighbouring states randomly as follows:

- Adding a number to one of the antecedent parameters or the consequent parameters. This value is related to the maximum and minimum value for each input space and is equal to $= maximumminimum/50$ for the first iteration.
- Adapting the step size for each input at each Markov chain by this scaling function proposed in [30]:

$$s_n = \frac{2s_0}{1 + \exp^{\frac{\beta n}{n_{max}}}}$$

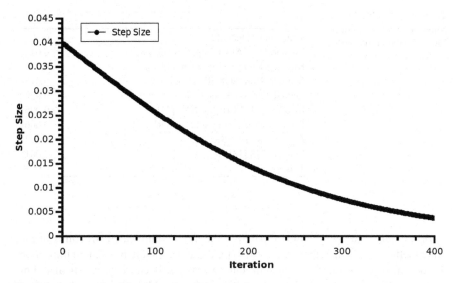

**Fig. 5.6** An example of the adaptation of the step size

**Table 5.1** The forecasting results for Mackey–Glass time series

| Experiment | System | $RMSE_{average}$ | $RMSE_{minimum}$ | $Time_{seconds}$ |
|---|---|---|---|---|
| Training Results | T1 | 0.02536 | 0.01334 | 4,329 |
| Training Results | T2 | 0.0139 | 0.009 | 21,628 |
| Testing Results | T1 | 0.02501 | 0.01335 | – |
| Testing Results | T2 | 0.01383 | 0.00898 | – |

where $s_0$ is the initial step size, $n$ the current iteration, $s_n$ the step size at current iteration, $n_{max}$ the maximum number of iteration (Markov chains), and $\beta$ is an adaptation constant, a value of 7 has been chosen.

The adaptation of the step size is proposed to reduce the computation as the fixed small step sizes need long time. An example of the adaptation of the step size is shown in Fig. 5.6. The new state is evaluated by examining the 500 data samples outputs. The experiment has been carried out 15 times and the average and the minimum RMSE of the training and testing samples have been calculated. It is shown from the results in Table 5.1 that type-2 system outperforms type-1 system with an average RMSE of 0. 01383 and a minimum of 0. 00898 compared to an average of 0. 02501 and a minimum of 0. 01335 in T1FLS. These results agree with some previous findings about the ability of the type-2 fuzzy sets to handle uncertain data better than type-1 fuzzy sets [11, 26]. To show how good the combination is, the comparison between our SA-T2FLS results with others in Table 5.2 shows that our result of (RMSE) = 0.0089 is one of the closest result to the best result obtained by ANFIS, GEFREX and Kukolj work despite the difference in structures. Our method is using a simple structure which is a combination of a general search

**Table 5.2** Results comparison for predicting Mackey–Glass time series

| Method | RMSE |
| --- | --- |
| Wang and Mendel [21] | 0.08 |
| Lin and Lin/FALCON-ART [21] | 0.04 |
| Kim and Kim/GA Ensemble [18] | 0.026 |
| Juang and Lin/SONFIN [15] | 0.018 |
| Lo and Yang/TSK model [23] | 0.0161 |
| Russo/GEFREX (GA + NN) [32] | 0.0061 |
| Kukolj/Fuzzy cluster + LS + WRLS [20] | 0.0061 |
| Almaraashi/SA-TSK | 0.0037 |
| Jang/ANFIS [13] | 0.0015 |
| This Model | 0.0089 |

algorithm and a fuzzy system compared to the more complicated structures of the other methods. Note that the fuzzy system model and structure has a great impact on the performance of the fuzzy system and normally it is chosen heuristically. For example, TSK might give better results than some Mamdani models for this problem [5]. In this work, we chose Mamdani model with dependent fuzzy sets for each input as our aim is to unveil the potential for simulated annealing to be a good candidate for tuning fuzzy systems rather than finding the best fuzzy system components. On the other hand, the time taken to learn T2FLS was very long compared to T1FLS. This is expected due to the extra computations associated with T2FLS especially in the defuzzification phase. We believe that by investigating more about the best formalisation of simulated annealing structure that suit type-2 fuzzy logic systems, we could have improved results.

## 6    Conclusion

Simulated annealing is used in this work to design Mamdani fuzzy logic systems by searching for the best parameters of the antecedent and the consequent parts of the fuzzy system to predict a well-known time series. Both type-1 and type-2 fuzzy logic systems have been compared in their ability to handle uncertainty. The result shows the ability for simulated annealing to be a good candidate for type-2 fuzzy logic systems to model uncertain data. In addition, type-2 fuzzy logic system shows more ability to handle uncertain data than type-1 fuzzy logic system. However, the time taken to learn type-2 fuzzy logic systems was very long compared to type-1 fuzzy logic systems due to the extra computations associated with type-2 fuzzy logic systems. In addition, the paper describes a method to reduce the computations associated with simulated annealing using adaptive step sizes.

# References

1. Aarts E, Lenstra JK (2003) Local search in combinatorial optimization. Princeton University Press, Princeton, NJ
2. Aarts EHL, Eikelder HMM Ten (2002) Handbook of applied optimization, chapter Simulated Annealing. Oxford University Press, NY, pp 209–220
3. Majid A, Robert J (2010) Tuning fuzzy systems by simulated annealing to predict time series with added noise. In: Proceedings of UKCI, Essex
4. Majid A, Robert J (2011) Tuning of type-2 fuzzy systems by simulated annealing to predict time series. Lecture notes in engineering and computer science: proceedings of the world congress on engineering 2011, WCE 2011, vol 2. U.K. Newswood Limited, London, pp 976–980
5. Majid A, Robert J, Simon C, Adrian H (2010) Time series forecasting using a tsk fuzzy system tuned with simulated annealing. In: Proceedings of FUZZ-IEEE2010 world congress on computational intelligence, Barcelona
6. Cordon O, Gomide F, Herrera F, Hoffmann F, Magdalena L (2004) Ten years of genetic fuzzy systems: current framework and new trends. Fuzzy Sets Syst 141(1):5–32
7. Dadone P (2001) Design optimization of fuzzy logic systems. PhD thesis, Virginia Polytechnic Institute and State University
8. Drack L, Zadeh HS (2006) Soft computing in engineering design optimisation. J Int Fuzzy Syst 17(4):353–365
9. Garibaldi JM, Ifeachor EC (1999) Application of simulated annealing fuzzy model tuning to umbilical cord acid-base interpretation. IEEE Trans Fuzzy Syst 7(1):72–84
10. Goonatilake S, Khebbal S (1995) Intelligent hybrid systems. Wiley, New York, NY
11. Hagras H (2007) Type-2 flcs: a new generation of fuzzy controllers. Comput Intel Mag IEEE 2 (1):30–43
12. Hoffmann F (2001) Evolutionary algorithms for fuzzy control system design. Proc IEEE 89 (9):1318–1333
13. Jang JSR (1993) Anfis: adaptive-network-based fuzzy inference system. IEEE Trans Syst Man Cybernet 23(3):665–685
14. John R, Coupland S (2007) Type-2 fuzzy logic: a historical view. Comput Intel Mag IEEE 2:57–62
15. Chia-Feng J, Chin-Teng L (1998) An online self-constructing neural fuzzy inference network and its applications. IEEE Trans Fuzzy Syst 6(1):12–32
16. Karnik NN, Mendel JM (2001) Centroid of a type-2 fuzzy set. Inform Sci 132(1):195–220
17. Karnik NN, Mendel JM, Liang Q (1999) Type-2 fuzzy logic systems. IEEE Trans Fuzzy Syst 7 (6):643–658
18. Kim D, Kim C (2002) Forecasting time series with genetic fuzzy predictor ensemble. IEEE Trans Fuzzy Syst 5(4):523–535
19. Kirkpatrick S, Gelatt C, Vecchi M (1983) Optimization by simulated annealing, 1983. Science 220:671–680
20. Dragan K (2002) Design of adaptive takagi-sugeno-kang fuzzy models. Appl Soft Comput 2(2):89–103
21. Cheng-Jian L, Chin-Teng L (1997) An art-based fuzzy adaptive learning control network. IEEE Trans Fuzzy Syst 5(4):477–496
22. Liu G, Yang W (2000) Learning and tuning of fuzzy membership functions by simulated annealing algorithm. In: The 2000 IEEE Asia-Pacific conference on circuits and systems, 2000. IEEE APCCAS 2000, pp 367–370
23. Ji-Chang L, Chien-Hsing Y (1999) A heuristic error-feedback learning algorithm for fuzzy modeling. IEEE Trans Syst Man Cybernet Part A: Syst Humans 29(6):686–691
24. Locatelli M (2002) Simulated annealing algorithms for continuous global optimization. Handbook Global Opt 2:179–229

25. Mackey MC, Glass L (1977) Oscillation and chaos in physiological control systems. Science 197(4300):287–289
26. Mendel JM (2001) Uncertain rule-based fuzzy logic systems: introduction and new directions. Prentice Hall, Englewood Cliffs, NJ
27. Mendel JM (2007) Advances in type-2 fuzzy sets and systems. Inform Sci 177(1):84–110
28. Mendel JM, John RIB (2002) Type-2 fuzzy sets made simple. IEEE Trans Fuzzy Syst 10(2):117–127
29. Miki M, Hiroyasu T, Ono K (2002) Simulated annealing with advanced adaptive neighborhood. In: Second international workshop on intelligent systems design and application, Dynamic Publishers, pp 113–118
30. Nolle L, Goodyear A, Hopgood A, Picton P, Braithwaite N (2001) On step width adaptation in simulated annealing for continuous parameter optimisation. Computational intelligence. Theory and applications. Springer, Heidelberg, pp 589–598
31. Timothy JR (2004) Fuzzy logic with engineering applications. Wiley, New York, NY
32. Russo M (2000) Genetic fuzzy learning. IEEE Trans Evol Comput 4(3):259–273
33. White SR (1984) Concepts of scale in simulated annealing. In: American institute of physics conference series, vol 122, pp 261–270
34. Zadeh LA (1994) Soft computing and fuzzy logic. IEEE Software 11(6):48–56

# Chapter 6
# Genetic Algorithms for Integrated Optimisation of Precedence-Constrained Production Sequencing and Scheduling

Son Duy Dao and Romeo Marin Marian

## 1 Introduction

Precedence-constrained problems belong to a class of problems commonly found in project management, logistics, routing, assembly flow, scheduling and networking [1]. Precedence constraints are generally divided into two categories: hard precedence constraints and soft precedence constraints. A hard precedence constraint is a constraint that makes the solution illegal or infeasible if violated, while a soft precedence constraint is a constraint that imposes a penalty if violated rather than rendering the sequence and schedule infeasible. In this study, production sequencing and scheduling problems associated with soft precedence constraints are considered and described as follows:

There is a manufacturing company having a number of different production lines able to produce a number of different products, each taking a certain amount of time. Production cost and labour requirement in each production line, selling price, deadline of completion and material requirement of each product are known. However, the labour, material and working capital of the company available for production are limited. Any product changeover/setup takes a given amount of time and product changeover in each production line is different from each other. Beside the production cost, there are overheads. Each production line can produce any product of the company. Moreover, any product made after its deadline incurs a penalty as a percentage of the selling price per day of delay and also affects the customer satisfaction index of the company. If the delay is greater than a preset number of days, 10 for example, the product will not be accepted by customer and it will be returned to the company. The company also faces other constraints such as minimum number of different products to be made in a certain time horizon.

S.D. Dao (✉) • R.M. Marian
School of Advanced Manufacturing and Mechanical Engineering, University of South Australia,
Mawson Lakes Campus, Mawson Lakes, SA 5095, Australia
e-mail: daosd001@mymail.unisa.edu.au; romeo.marian@unisa.edu.au

S.-I. Ao and L. Gelman (eds.), *Electrical Engineering and Intelligent Systems*,
Lecture Notes in Electrical Engineering 130, DOI 10.1007/978-1-4614-2317-1_6,
© Springer Science+Business Media, LLC 2013

A solution for the above class of problems is a selection of products to produce, which production line to be used and in what order, so as to maximise the total profit of the company as well as customer satisfaction index while satisfying simultaneously all the given constraints.

Clearly, this class of problems requires a triple optimisation, so-called integrated optimisation, at the same time: allocating available resources to each production line, production line sequencing and production line scheduling. It is a multi-dimensional, precedence-constrained, knapsack problem. The knapsack problem is a classical NP-hard problem, and it has been thoroughly studied in the last few decades [2].

## 2   Integration of Precedence-Constrained Task Sequencing and Scheduling

Assembly and disassembly scheduling is one of popular precedence-constrained problems, in which the precedence relations result from geometric and physical constraints of the assembled items. Chen [3] applied AND/OR precedence-constrained sequencing approach to solve assembly scheduling problem while Marian [4, 5] used GA to optimise precedence-constrained assembly sequences. For similar problems, Duman and Or [6] developed an approach which initially ignores precedence relations and solves the problem as a pure travelling salesman problem (TSP), and then, it is applied to eliminate component damage in the resulting TSP tour. Recently, a precedence-constrained sequencing problem (PCSP) in disassembly scheduling was formulated by modifying the two-commodity network flow model, carried out by Lambert [7].

For PCSP in supply chain, Moon et al. [8] proposed GA with a priority-based encoding method to solve the problem. For problems with sequence-dependent changeover cost and precedence constraints, He and Kusiak [9] developed a heuristic algorithm. A unique reasoning approach to solve PCSPs based on case-based reasoning with evolutionary algorithm was developed by Su [10]. In addition, GA approach based on a topological sort-based representation procedure was proposed to solve precedence-constrained sequencing problems, which aims at locating the optimal sequence with the shortest travelling time among all feasible sequences [11]. Clearly, this problem can be formulated as a travelling salesman problem with precedence constraints.

For single-machine sequencing with precedence constraints, Morton and Dharan [12] developed a heuristic—algoristics to find optimal solutions while Chekuri and Motwani [13] developed an efficient combinational two-approximation algorithm for precedence-constrained scheduling to minimise the sum of weighted completion times on a single machine. In addition, Azar et al. [14] considered this problem as a special case of the vertex cover problem and then solved it by means of vertex cover theory.

For scheduling precedence-constrained tasks on a multiprocessor system, Jonsson and Shin [15] proposed a parametrised branch-and-bound algorithm to minimise the maximum task lateness in the system. In addition, Yen et al. [1] developed priority-list scheduling method to minimise the makespan in problems of scheduling with a set of precedence constraint tasks onto a finite number of identical processors with and without communication overhead. Moreover, heuristic algorithms to obtain near optimal schedules in a reasonable amount of computation time for scheduling precedence-constrained task graphs with non-negligible intertask communication onto multiprocessors were developed by Selvakumar and Siva [16]. For scheduling the precedence-constrained task allocation for pipelined execution, Hary and Ozguner [17] used point-to-point networks. Mouhamed and Maasarani [18] proposed a class of global priority-based scheduling heuristics, called generalised list scheduling, to schedule precedence-constrained computations on message-passing systems.

For integration of precedence-constrained production sequencing and scheduling, the authors [19–21] proposed a GA with variability of chromosome size to optimise the class of problems for one as well as two objective functions in one as well as multiple production line environments.

It is evident that literature on optimising precedence-constrained production sequencing and scheduling simultaneously, especially for the large-size or real-life problems, is very limited.

This chapter presents a new methodology, which is based on the previous works of the authors [19–21] on developing the GA capable to optimise a class of soft precedence constraint production sequencing and scheduling problems. This methodology is general and capable to take into account all relevant production data, constraints and objective functions. Other important feature of the methodology includes the ability to research a global optimum.

The rest of this chapter is organised as follows: Sect. 3 introduces the research methodology; Sect. 4 is a case study; Sect. 5 gives conclusions and presents future work.

# 3   Research Methodology

## 3.1   Modelling the Problem

Considering:

- A manufacturing company has $N$ different production lines and produces $n$ different products in a certain time horizon
- Deadline of product completion and requirements of the customer such as penalties for late delivery and returned product are known
- Capacity of the company such as the labour, material and working capital available for production is limited
- Producing cost and labour requirement of a product made in different production line are different

- Any product changeover takes a certain amount of time and it is different for different production lines

Determine:

- Which products to be produced (selection)
- What is the order to produce the selected products (sequencing)
- Allocation of the selected products to which production line (scheduling)

So that:

Profit of the company as well as its customer satisfaction index is maximised while all of the given constraints are simultaneously satisfied.

Conditions:

- Beside the producing cost, each hour of running company also costs $H$ dollar and this expense has to be paid from the working capital
- The proceedings from selling products will only be available for next period of time, so they should be ignored for the current planning horizon—current period of time
- The company can produce any mix of products
- The company must produce at least $D$ different products in the next period of time

## 3.2   Optimisation Methodology

In this chapter, a GA with new features in chromosome encoding, mutation and crossover operations is proposed to optimise the precedence-constrained production sequencing and scheduling problems for multiple production lines. In this problem, one of the parameters to be optimised is resource allocation to each production line. Once resources such as labour, working capital and material are allocated to each production line, two other optimised parameters which are product scheduling and sequencing in each production line can be determined. As a result and to still do a proper optimisation for the resource allocation as a variable, optimisation by using this GA is initially done for a certain resource allocation case. After a local optimum/near optimum solution has been found for that particular resource allocation case (in the first stage), the algorithm is run for a number of different resource allocation cases (second stage). This approach can quickly point out towards the most promising zone of the solution space and permits to determine a global optimum [22].

The major components of the GA are presented as follows:

### 3.2.1   Chromosome Encoding

Each production line can be encoded as a string corresponding to products allocated to each production line and their producing sequences. Accordingly, each chromosome encoding a solution for the problem consists of $N$ strings corresponding to the $N$ production lines. It should be noted that the length of each string may be different from each other because of constraints. In order to generate a feasible chromosome, the following steps are proposed.

The first step is to generate at random $N$ strings of numbers ranging from 1 to $n$, which represent the corresponding products to be made and their producing sequences. It is noted that the length of each string must be large enough to account, later, for the different constraints and to be trimmed to length. Such set of $N$ strings generated in this step is called *initial chromosome*.

The second step is to cut or remove some products allocated at the end of each sequence in the *initial chromosome* based on the production-line constraints in allocated labour, material and working capital. After the cutting operation, the $N$ strings of number will become an *intermediary chromosome*.

The third step is to check the *intermediary chromosome* against all constraints. The chromosome is adjusted in length so that it satisfies all constraints and becomes a *feasible chromosome*, which, subsequently, can undergo genetic operators.

### 3.2.2  Fitness Function

Fitness function is the sum of total profit of the company and customer satisfaction index. It should be noted that a weight coefficient should be introduced to determine the contribution of each function. The fitness function is calculated for each *feasible chromosome*.

### 3.2.3  Genetic Operators

Due to variability in the size of *feasible chromosomes*, new crossover and mutation operations, which are not applied to *feasible chromosome* but to *initial chromosome* are developed. After each crossover or mutation operation applied to *initial chromosome*, constraint-based cutting operation must be applied to make sure that the off-spring chromosomes are feasible.

### 3.2.4  Optimisation Implementation

With classical structure, the proposed GA is implemented in Matlab to search the optimal/good solutions for the problem. This study considers the problem associated with soft precedence constraints, which will incur a penalty if violated rather rendering the sequence and schedule infeasible. A penalty implies that the respective chromosome is less likely to pass in the next generation, but still may have very valuable characteristics to pass on through the evolution process.

The proposed approach is illustrated with the aid of a comprehensive case study.

# 4  Case Study

## 4.1  Modelling of the Planning Problem

To illustrate the proposed approach and demonstrate its capability, a complex case study was developed.

There is a manufacturing company having three different production lines, which are able to produce 50 different products, say P1, P2, ..., P50, in the next month. Information about the products is detailed in Table 6.1. For the next month, there are available: 800 h of labour for production, 1,000 kg of material and $1,700 K as working capital available.

Additionally:

- Apart from the producing cost, overheads for running the company also cost $300/h and this expense has to be paid from the working capital.
- Any product changeover takes 0.5, 1.0 and 1.5 h in production line 1, 2 and 3, respectively.
- Any product made after the deadline incurs a penalty of 5% of the initial price per day of delay.
- Any product made after the deadline for more than 10 days will not be accepted by the customer and it will be returned to the company.
- Each day after the deadline of a product incurs a penalty of 1 point of customer satisfaction index.
- Proceedings from selling products will be available for next month, so they should be ignored for the current planning horizon—current month.
- The company can select any mix of products to produce in the next month, as long as its selection contains at least 20 different ones.
- The company can work 24 h/day, 7 days/week.

**Table 6.1** Product information

| Product | Cost (K$) Line 1 | Line 2 | Line 3 | Labour (h) Line 1 | Line 2 | Line 3 | Price (K$) | Material (Kg) | Deadline (day of month) |
|---|---|---|---|---|---|---|---|---|---|
| P1 | 37 | 34 | 36 | 8 | 7 | 8 | 176 | 15 | 17 |
| P2 | 39 | 45 | 44 | 5 | 9 | 12 | 110 | 16 | 16 |
| P3 | 14 | 21 | 17 | 3 | 8 | 7 | 107 | 20 | 13 |
| P4 | 19 | 21 | 18 | 3 | 6 | 6 | 125 | 4 | 10 |
| P5 | 11 | 11 | 10 | 10 | 15 | 18 | 211 | 13 | 29 |
| P6 | 25 | 32 | 39 | 8 | 12 | 17 | 176 | 1 | 21 |
| P7 | 26 | 23 | 28 | 8 | 12 | 17 | 163 | 7 | 12 |
| P8 | 30 | 36 | 36 | 5 | 6 | 11 | 132 | 8 | 25 |
| P9 | 11 | 7 | 12 | 6 | 10 | 12 | 101 | 15 | 10 |
| P10 | 10 | 6 | 9 | 9 | 14 | 15 | 197 | 11 | 12 |

(continued)

**Table 6.1** (continued)

| Product | Cost (K$) | | | Labour (h) | | | Price (K$) | Material (Kg) | Deadline (day of month) |
|---|---|---|---|---|---|---|---|---|---|
| | Line 1 | Line 2 | Line 3 | Line 1 | Line 2 | Line 3 | | | |
| P11 | 26 | 22 | 29 | 9 | 12 | 13 | 113 | 17 | 26 |
| P12 | 28 | 32 | 26 | 4 | 7 | 9 | 175 | 9 | 5 |
| P13 | 14 | 13 | 15 | 5 | 10 | 9 | 175 | 7 | 19 |
| P14 | 12 | 19 | 12 | 2 | 5 | 10 | 178 | 15 | 29 |
| P15 | 33 | 27 | 27 | 8 | 11 | 10 | 130 | 3 | 23 |
| P16 | 25 | 29 | 35 | 9 | 12 | 14 | 133 | 14 | 26 |
| P17 | 14 | 18 | 14 | 9 | 14 | 18 | 117 | 19 | 21 |
| P18 | 15 | 9 | 11 | 9 | 13 | 15 | 172 | 4 | 25 |
| P19 | 13 | 14 | 12 | 9 | 8 | 13 | 199 | 19 | 5 |
| P20 | 16 | 12 | 7 | 4 | 7 | 9 | 159 | 4 | 14 |
| P21 | 38 | 41 | 36 | 6 | 8 | 13 | 213 | 5 | 21 |
| P22 | 26 | 21 | 20 | 2 | 5 | 4 | 147 | 3 | 20 |
| P23 | 30 | 36 | 30 | 2 | 5 | 10 | 128 | 10 | 30 |
| P24 | 12 | 11 | 12 | 6 | 8 | 11 | 102 | 13 | 5 |
| P25 | 37 | 31 | 32 | 7 | 12 | 13 | 111 | 1 | 7 |
| P26 | 20 | 22 | 25 | 5 | 9 | 12 | 129 | 19 | 12 |
| P27 | 15 | 17 | 11 | 5 | 4 | 7 | 205 | 3 | 16 |
| P28 | 40 | 46 | 46 | 5 | 4 | 4 | 148 | 15 | 19 |
| P29 | 34 | 28 | 31 | 9 | 13 | 13 | 130 | 11 | 5 |
| P30 | 25 | 25 | 27 | 8 | 8 | 13 | 208 | 12 | 25 |
| P31 | 37 | 31 | 32 | 10 | 15 | 16 | 118 | 17 | 23 |
| P32 | 13 | 20 | 15 | 4 | 3 | 7 | 215 | 3 | 9 |
| P33 | 37 | 30 | 23 | 4 | 6 | 5 | 123 | 1 | 17 |
| P34 | 36 | 43 | 39 | 10 | 12 | 14 | 154 | 12 | 8 |
| P35 | 24 | 27 | 31 | 2 | 3 | 2 | 156 | 3 | 23 |
| P36 | 14 | 8 | 5 | 4 | 4 | 3 | 148 | 3 | 25 |
| P37 | 26 | 30 | 30 | 8 | 10 | 9 | 105 | 4 | 18 |
| P38 | 35 | 40 | 45 | 2 | 5 | 9 | 153 | 7 | 29 |
| P39 | 19 | 16 | 22 | 2 | 2 | 1 | 152 | 14 | 18 |
| P40 | 27 | 28 | 32 | 4 | 8 | 12 | 158 | 14 | 15 |
| P41 | 20 | 19 | 17 | 7 | 11 | 14 | 163 | 6 | 29 |
| P42 | 20 | 19 | 16 | 4 | 4 | 8 | 149 | 11 | 27 |
| P43 | 32 | 38 | 43 | 6 | 11 | 15 | 153 | 10 | 19 |
| P44 | 22 | 20 | 14 | 4 | 3 | 3 | 215 | 7 | 22 |
| P45 | 29 | 24 | 29 | 9 | 13 | 15 | 200 | 1 | 10 |
| P46 | 26 | 26 | 30 | 8 | 7 | 10 | 153 | 13 | 16 |
| P47 | 13 | 8 | 8 | 3 | 3 | 5 | 148 | 15 | 6 |
| P48 | 26 | 22 | 15 | 8 | 8 | 7 | 164 | 3 | 29 |
| P49 | 13 | 10 | 4 | 2 | 7 | 11 | 157 | 18 | 12 |
| P50 | 32 | 31 | 32 | 6 | 7 | 9 | 217 | 20 | 20 |

**Table 6.2** Normalised Taguchi orthogonal array design

| | Taguchi orthogonal array design | | | | Normalised Taguchi orthogonal array design | | | |
|------|--------|--------|--------|-----|--------|--------|--------|-----|
| Case | Line 1 | Line 2 | Line 3 | Sum | Line 1 | Line 2 | Line 3 | Sum |
| 1 | 1 | 1 | 1 | 3 | 0.333 | 0.333 | 0.333 | 1 |
| 2 | 1 | 2 | 2 | 5 | 0.200 | 0.400 | 0.400 | 1 |
| 3 | 1 | 3 | 3 | 7 | 0.143 | 0.429 | 0.429 | 1 |
| 4 | 2 | 1 | 2 | 5 | 0.400 | 0.200 | 0.400 | 1 |
| 5 | 2 | 2 | 3 | 7 | 0.286 | 0.286 | 0.429 | 1 |
| 6 | 2 | 3 | 1 | 6 | 0.333 | 0.500 | 0.167 | 1 |
| 7 | 3 | 1 | 3 | 7 | 0.429 | 0.143 | 0.429 | 1 |
| 8 | 3 | 2 | 1 | 6 | 0.500 | 0.333 | 0.167 | 1 |
| 9 | 3 | 3 | 2 | 8 | 0.375 | 0.375 | 0.250 | 1 |

The problem is to do the planning for next month by (1) selecting what products to produce, (2) allocating the selected products to which production line and (3) selecting the producing sequence in three production lines to maximise the profit of the company as well as its customer satisfaction index while satisfying simultaneously all constraints above.

## 4.2    Optimisation with Genetic Algorithm

### 4.2.1    Resource Allocation to Each Production Line

As mentioned earlier, resource allocation to each production line is one of the three parameters to be optimised. It is very difficult to use resource allocation coefficients as random variables because it takes a great deal of time to obtain a global optimal solution or even a good solution for the problem. Therefore, at this stage, normalised Taguchi orthogonal array design (NTOAD) is proposed to systematically cover a number of the resource allocation cases. Accordingly, the optimisation algorithm is designed for the fixed resources allocated to each production line, and then, the algorithm is run for different resource allocation cases as NTOAD. Clearly, the best solution among the solution obtained should be selected and can be considered as global optimum. As a result, the algorithm still does a proper optimisation for the resource allocation as a variable. In this study, the proposed resource-allocated coefficient based on NTOAD is as shown in Table 6.2.

### 4.2.2    Optimisation Criteria

Criteria optimisation for this problem are overall profit of the company and its customer satisfaction index, converted into the fitness function.

**Table 6.3**  Procedure of chromosome generation

| | A | B | C | D | E | F | G | H | I | J | K | L | M | N | O | P | Q | R | S | T | U | V | W | X | Y | Z | AA | AB | AC | AD | AE | AF | AG | AH | AI | AJ | AK | AL | AM | AN | AO |
|---|---|---|---|---|---|---|---|---|---|---|---|---|---|---|---|---|---|---|---|---|---|---|---|---|---|---|---|---|---|---|---|---|---|---|---|---|---|---|---|---|---|
| 1 | | Chromosome | | | | | | | | | | | | | | | | | | | | | | | | | | | | | | | | | | | | | | | |
| 2 | Sequence | 1 | 2 | 3 | 4 | 5 | 6 | 7 | 8 | 9 | 10 | 11 | 12 | 13 | 14 | 15 | 16 | 17 | 18 | 19 | 20 | 21 | 22 | 23 | 24 | 25 | 26 | 27 | 28 | 29 | 30 | 31 | 32 | 33 | 34 | 35 | 36 | 37 | 38 | 39 | 40 |
| 3 | Line1 | 4 | 5 | 41 | 8 | 49 | 23 | 5 | 20 | 4 | 21 | 15 | 49 | 19 | 18 | 4 | 22 | 21 | 35 | 35 | 7 | 3 | 34 | 24 | 43 | 10 | 7 | 20 | 15 | 41 | 18 | 45 | 14 | 22 | 10 | 30 | 35 | 4 | 33 | 36 | 17 |
| 4 | Line 2 | 34 | 27 | 41 | 33 | 33 | 22 | 8 | 42 | 21 | 33 | 22 | 9 | 11 | 48 | 37 | 28 | 49 | 34 | 34 | 50 | 28 | 10 | 49 | 33 | 22 | 30 | 30 | 31 | 49 | 30 | 44 | 30 | 16 | 22 | 24 | 32 | 17 | 21 | 48 | 6 |
| 5 | Line 3 | 3 | 6 | 36 | 26 | 40 | 41 | 9 | 40 | 27 | 32 | 2 | 6 | 25 | 46 | 14 | 47 | 16 | 27 | 10 | 9 | 44 | 19 | 9 | 19 | 25 | 12 | 13 | 14 | 37 | 6 | 41 | 2 | 9 | 6 | 35 | 3 | 27 | 41 | 27 | 31 |

### 4.2.3  Chromosome Definition

Let numbers from 1 to 50 denote the corresponding products P1 to P50. Each chromosome is generated by three steps.

The first step is to randomly generate the *initial chromosome*, which looks like as shown in region from cells B3 to AO5 in Table 6.3.

The second step is to cut the *initial chromosome* based on three constraints of labour, working capital and material in each production line. Output of this step is *intermediary chromosome,* which looks like as shown in the cells shown in the highlighted cells in Table 6.3.

The third step is to check the total number of different products in the *intermediary chromosome*. As shown in Table 6.3, for instance, there are 45 different products in the *intermediary chromosome*. Obviously, this number of different products is greater than 20 so that the *intermediary chromosome* satisfies the constraint of the required minimum variety of products. Therefore, the *feasible chromosome* that satisfies all of the given constraints is the same as the *intermediary chromosome*—the highlighted cells in Table 6.3.

It should be noted that, in the other case, if the number of different products in *intermediary chromosome* is less than 20 as required minimum number of different products, the three-step proposed must be repeated until all of the given constraints are satisfied, which means that *feasible chromosome* is achieved. In addition, the length of each string corresponding to each production line in the *feasible chromosome* is different from the one in *initial chromosome* and different from time to time.

### 4.2.4  Crossover Operation

Crossover, in principle, is a simple cut and swap operation. Due to the nature of constraints, a modified crossover operation is required. In this study, to make sure the offspring chromosomes are feasible, crossover operation is not applied to *feasible chromosome*, but to the corresponding *initial chromosome*. After crossover operation, the corresponding off-springs will be cut to satisfy constraints and to ensure all offsprings are feasible. This study uses one-point crossover, illustrated as follows:

• Randomly select two *feasible chromosomes* and their corresponding *initial chromosomes* as shown in Table 6.4, for example. It is noted that the *feasible chromosomes* are shown in the highlighted regions in Table 6.4.

**Table 6.4** Two feasible chromosomes and their corresponding initial chromosomes

| | | | | | | | | | | | | | | | Parent chromosome 1 | | | | | | | | | | | | | | | | | | | | | | | | | |
|---|---|---|---|---|---|---|---|---|---|---|---|---|---|---|---|---|---|---|---|---|---|---|---|---|---|---|---|---|---|---|---|---|---|---|---|---|---|---|---|---|
| Sequence | 1 | 2 | 3 | 4 | 5 | 6 | 7 | 8 | 9 | 10 | 11 | 12 | 13 | 14 | 15 | 16 | 17 | 18 | 19 | 20 | 21 | 22 | 23 | 24 | 25 | 26 | 27 | 28 | 29 | 30 | 31 | 32 | 33 | 34 | 35 | 36 | 37 | 38 | 39 | 40 |
| Line1 | 4 | 5 | 41 | 8 | 49 | 23 | 5 | 20 | 4 | 21 | 15 | 49 | 19 | 18 | 4 | 22 | 21 | 35 | 35 | 7 | 3 | 34 | 24 | 43 | 10 | 7 | 20 | 15 | 41 | 18 | 45 | 14 | 22 | 10 | 30 | 35 | 4 | 33 | 36 | 17 |
| Line 2 | 34 | 27 | 41 | 33 | 33 | 22 | 8 | 42 | 21 | 33 | 22 | 9 | 11 | 48 | 37 | 28 | 49 | 34 | 34 | 50 | 28 | 10 | 49 | 33 | 22 | 30 | 30 | 31 | 49 | 30 | 44 | 30 | 16 | 22 | 24 | 32 | 17 | 21 | 48 | 6 |
| Line 3 | 3 | 6 | 36 | 26 | 40 | 41 | 9 | 40 | 27 | 32 | 2 | 6 | 25 | 46 | 14 | 47 | 16 | 27 | 10 | 9 | 44 | 19 | 9 | 19 | 25 | 12 | 13 | 14 | 37 | 6 | 41 | 2 | 9 | 6 | 35 | 3 | 27 | 41 | 27 | 31 |
| | | | | | | | | | | | | | | | Parent chromosome 2 | | | | | | | | | | | | | | | | | | | | | | | | | |
| Sequence | 1 | 2 | 3 | 4 | 5 | 6 | 7 | 8 | 9 | 10 | 11 | 12 | 13 | 14 | 15 | 16 | 17 | 18 | 19 | 20 | 21 | 22 | 23 | 24 | 25 | 26 | 27 | 28 | 29 | 30 | 31 | 32 | 33 | 34 | 35 | 36 | 37 | 38 | 39 | 40 |
| Line1 | 3 | 7 | 43 | 33 | 44 | 3 | 2 | 3 | 43 | 33 | 23 | 23 | 1 | 2 | 4 | 5 | 6 | 9 | 11 | 12 | 12 | 34 | 45 | 47 | 47 | 23 | 31 | 45 | 41 | 25 | 37 | 39 | 28 | 38 | 39 | 35 | 2 | 1 | 34 | 50 |
| Line 2 | 45 | 50 | 23 | 24 | 43 | 42 | 41 | 50 | 23 | 25 | 27 | 28 | 21 | 20 | 19 | 15 | 15 | 14 | 19 | 33 | 22 | 34 | 35 | 39 | 41 | 43 | 30 | 21 | 32 | 43 | 50 | 34 | 35 | 43 | 23 | 2 | 3 | 4 | 9 | 12 |
| Line 3 | 21 | 32 | 44 | 50 | 3 | 2 | 9 | 11 | 14 | 12 | 4 | 6 | 19 | 14 | 13 | 42 | 32 | 43 | 23 | 5 | 6 | 45 | 46 | 49 | 43 | 32 | 9 | 22 | 23 | 45 | 43 | 32 | 21 | 19 | 17 | 4 | 3 | 3 | 24 | 43 |

**Table 6.5** Constraint-based cutting operations for feasible offspring

| | | | | | | | | | | | | | | | Offspring chromosome 1 | | | | | | | | | | | | | | | | | | | | | | | | | |
|---|---|---|---|---|---|---|---|---|---|---|---|---|---|---|---|---|---|---|---|---|---|---|---|---|---|---|---|---|---|---|---|---|---|---|---|---|---|---|---|---|
| Sequence | 1 | 2 | 3 | 4 | 5 | 6 | 7 | 8 | 9 | 10 | 11 | 12 | 13 | 14 | 15 | 16 | 17 | 18 | 19 | 20 | 21 | 22 | 23 | 24 | 25 | 26 | 27 | 28 | 29 | 30 | 31 | 32 | 33 | 34 | 35 | 36 | 37 | 38 | 39 | 40 |
| Line1 | 4 | 5 | 41 | 8 | 49 | 23 | 5 | 20 | 4 | 21 | 15 | 49 | 19 | 18 | 4 | 5 | 6 | 9 | 11 | 12 | 12 | 34 | 45 | 47 | 47 | 23 | 31 | 45 | 0 | 0 | 0 | 0 | 0 | 0 | 0 | 0 | 0 | 0 | 0 | 0 |
| Line 2 | 34 | 27 | 41 | 33 | 33 | 22 | 8 | 42 | 21 | 33 | 22 | 9 | 11 | 48 | 19 | 15 | 15 | 14 | 19 | 33 | 22 | 34 | 35 | 39 | 0 | 0 | 0 | 0 | 0 | 0 | 0 | 0 | 0 | 0 | 0 | 0 | 0 | 0 | 0 | 0 |
| Line 3 | 3 | 6 | 36 | 26 | 40 | 41 | 9 | 40 | 27 | 32 | 2 | 6 | 25 | 46 | 13 | 42 | 32 | 43 | 23 | 5 | 6 | 45 | 46 | 49 | 43 | 0 | 0 | 0 | 0 | 0 | 0 | 0 | 0 | 0 | 0 | 0 | 0 | 0 | 0 | 0 |
| | | | | | | | | | | | | | | | Offspring chromosome 2 | | | | | | | | | | | | | | | | | | | | | | | | | |
| Sequence | 1 | 2 | 3 | 4 | 5 | 6 | 7 | 8 | 9 | 10 | 11 | 12 | 13 | 14 | 15 | 16 | 17 | 18 | 19 | 20 | 21 | 22 | 23 | 24 | 25 | 26 | 27 | 28 | 29 | 30 | 31 | 32 | 33 | 34 | 35 | 36 | 37 | 38 | 39 | 40 |
| Line1 | 3 | 7 | 43 | 33 | 44 | 3 | 2 | 3 | 43 | 33 | 23 | 23 | 1 | 2 | 4 | 22 | 21 | 35 | 35 | 7 | 3 | 34 | 24 | 43 | 10 | 7 | 20 | 15 | 41 | 18 | 0 | 0 | 0 | 0 | 0 | 0 | 0 | 0 | 0 | 0 |
| Line 2 | 45 | 50 | 23 | 24 | 43 | 42 | 41 | 50 | 23 | 25 | 27 | 28 | 21 | 20 | 37 | 28 | 49 | 34 | 34 | 50 | 28 | 10 | 49 | 33 | 22 | 30 | 30 | 0 | 0 | 0 | 0 | 0 | 0 | 0 | 0 | 0 | 0 | 0 | 0 | 0 |
| Line 3 | 21 | 32 | 44 | 50 | 3 | 2 | 9 | 11 | 14 | 12 | 4 | 6 | 19 | 14 | 14 | 47 | 16 | 27 | 10 | 9 | 44 | 19 | 9 | 19 | 25 | 0 | 0 | 0 | 0 | 0 | 0 | 0 | 0 | 0 | 0 | 0 | 0 | 0 | 0 | 0 |

- Randomly select the cut point within the "feasible region"—the highlighted region, the 15th gene for example, as highlighted in Table 6.4.
- Swap the two parts.
- Two offsprings obtained up to now will be cut to satisfy constraints and guarantee feasibility. As a result, the output of crossover operation is finalised as shown in Table 6.5.

### 4.2.5 Mutation Operation

Once again, due to the constraints, the modified mutation operation is not applied to *feasible chromosome*, but to the corresponding *initial chromosome*. Clearly, after any mutation operation, all off-spring chromosomes must be checked and repaired to guarantee feasibility. The mutation operation used in this study is as follows:

- Randomly select one *feasible chromosome* and its corresponding *initial chromosome* from the population and then randomly select two genes in the "feasible region" in the corresponding *initial chromosome* to swap as shown in the first sub-table in Table 6.6.
- After swapping the two selected genes, constraint-based cutting operations are then applied to every production line in the obtained chromosome to ensure feasibility. As a result, a feasible chromosome output of mutation operation is achieved as shown in the last sub-table in Table 6.6.

**Table 6.6** Modified mutation procedure

| Parent chromosome | | | | | | | | | | | | | | | | | | | | | | | | | | | | | | | | | | | | | | | |
|---|---|---|---|---|---|---|---|---|---|---|---|---|---|---|---|---|---|---|---|---|---|---|---|---|---|---|---|---|---|---|---|---|---|---|---|---|---|---|---|
| Sequence | 1 | 2 | 3 | 4 | 5 | 6 | 7 | 8 | 9 | 10 | 11 | 12 | 13 | 14 | 15 | 16 | 17 | 18 | 19 | 20 | 21 | 22 | 23 | 24 | 25 | 26 | 27 | 28 | 29 | 30 | 31 | 32 | 33 | 34 | 35 | 36 | 37 | 38 | 39 | 40 |
| Line1 | 4 | 5 | 41 | 8 | 49 | 23 | 5 | 20 | 4 | 21 | 15 | 49 | 19 | 18 | 4 | 22 | 21 | 35 | 35 | 7 | 3 | 34 | 24 | 43 | 10 | 7 | 20 | 15 | 41 | 18 | 45 | 14 | 22 | 10 | 30 | 35 | 4 | 33 | 36 | 17 |
| Line 2 | 34 | 27 | 41 | 33 | 33 | 22 | 8 | 42 | 21 | 33 | 22 | 9 | 11 | 48 | 37 | 28 | 49 | 34 | 34 | 50 | 28 | 10 | 49 | 33 | 22 | 30 | 30 | 31 | 49 | 30 | 44 | 30 | 16 | 22 | 24 | 32 | 17 | 21 | 48 | 6 |
| Line 3 | 3 | 6 | 36 | 26 | 40 | 41 | 9 | 40 | 27 | 32 | 2 | 6 | 25 | 46 | 14 | 47 | 16 | 27 | 10 | 9 | 44 | 19 | 9 | 19 | 25 | 12 | 13 | 14 | 37 | 6 | 41 | 2 | 9 | 6 | 35 | 3 | 27 | 41 | 27 | 31 |

| Offspring chromosome | | | | | | | | | | | | | | | | | | | | | | | | | | | | | | | | | | | | | | | |
|---|---|---|---|---|---|---|---|---|---|---|---|---|---|---|---|---|---|---|---|---|---|---|---|---|---|---|---|---|---|---|---|---|---|---|---|---|---|---|---|
| Sequence | 1 | 2 | 3 | 4 | 5 | 6 | 7 | 8 | 9 | 10 | 11 | 12 | 13 | 14 | 15 | 16 | 17 | 18 | 19 | 20 | 21 | 22 | 23 | 24 | 25 | 26 | 27 | 28 | 29 | 30 | 31 | 32 | 33 | 34 | 35 | 36 | 37 | 38 | 39 | 40 |
| Line1 | 4 | 5 | 41 | 8 | 49 | 23 | 5 | 20 | 4 | 21 | 15 | 49 | 19 | 18 | 4 | 22 | 21 | 33 | 35 | 7 | 3 | 34 | 24 | 43 | 10 | 7 | 20 | 15 | 0 | 0 | 0 | 0 | 0 | 0 | 0 | 0 | 0 | 0 | 0 | 0 |
| Line 2 | 34 | 27 | 41 | 33 | 35 | 22 | 8 | 42 | 21 | 33 | 22 | 9 | 11 | 48 | 37 | 28 | 49 | 34 | 34 | 50 | 28 | 10 | 49 | 33 | 22 | 30 | 0 | 0 | 0 | 0 | 0 | 0 | 0 | 0 | 0 | 0 | 0 | 0 | 0 | 0 |
| Line 3 | 3 | 6 | 36 | 26 | 40 | 41 | 9 | 40 | 27 | 32 | 2 | 6 | 25 | 46 | 14 | 47 | 16 | 27 | 10 | 9 | 44 | 19 | 9 | 19 | 25 | 12 | 13 | 0 | 0 | 0 | 0 | 0 | 0 | 0 | 0 | 0 | 0 | 0 | 0 | 0 |

*Note*: The lengths of the offspring after mutation as well as crossover might be different from those of the parent chromosomes due to feasibility constraints.

### 4.2.6  Evaluation Operation

The fitness function of the *feasible chromosome* that is used to evaluate the quality of the solution is sum of total profit of the company and its customer satisfaction index with given weight coefficient, which is calculated as follows:

$$F = w^*[I - (CP + CR + CD + CRT)] + (1 - w)^*S$$

where $F$ is the fitness value, $I$ is the total income, CP is the total cost of producing products, CR is the total cost of running company, CD is the total cost associated with penalty due to products made after deadline, CRT is the total cost due to returned products or not accepted by customer because they are too late, $S$ is the total points of customer satisfaction index, and $w$ is the weight coefficient, assumed $w = 0.7$ in this case study.

### 4.2.7  Selection Operation

In this study, roulette wheel approach is used to select the population for the next generations. This approach belongs to the fitness-proportional selection and selects a new population based on the probability distribution associated with fitness value [23].

**Table 6.7** The fitness values achieved by different cases and runs

| Case Run | 1 | 2 | 3 | 4 | 5 | 6 | 7 | 8 | 9 |
|---|---|---|---|---|---|---|---|---|---|
| 1 | 14,022 | 13,400 | 11,736 | 12,793 | 13,797 | 13,504 | 12,460 | 12,564 | 13,739 |
| 2 | 14,244 | 12,938 | 11,573 | 13,231 | 13,727 | 12,433 | 12,471 | 12,794 | 12,857 |
| 3 | 14,211 | 13,551 | 12,654 | 12,967 | 13,454 | 13,031 | 11,633 | 12,614 | 13,967 |
| 4 | 14,287 | 11,893 | 12,276 | 13,407 | 13,816 | 12,380 | 12,129 | 13,033 | 13,739 |
| 5 | 14,235 | 13,007 | 11,888 | 13,377 | 13,142 | 12,621 | 11,968 | 12,831 | 12,857 |
| 6 | 14,202 | 13,434 | 12,028 | 13,494 | 13,177 | 13,504 | 12,322 | 12,794 | 13,967 |
| 7 | 14,259 | 13,345 | 12,223 | 13,184 | 12,601 | 12,433 | 12,609 | 12,614 | 13,132 |
| 8 | 14,172 | 13,283 | 12,450 | 13,282 | 14,053 | 13,031 | 11,367 | 13,033 | 14,042 |
| 9 | 14,259 | 13,253 | 12,380 | 13,333 | 13,643 | 12,380 | 11,727 | 12,831 | 13,885 |
| 10 | 14,172 | 12,939 | 12,654 | 13,257 | 13,746 | 12,310 | 12,340 | 13,190 | 13,615 |
| Average | 14,206.3 | 13,104.3 | 12,186.2 | 13,232.5 | 13,515.6 | 12,762.7 | 12,102.6 | 12,829.8 | 13,580.0 |
| Max | 14,287 | 13,551 | 12,654 | 13,494 | 14,053 | 13,504 | 12,609 | 13,190 | 14,042 |

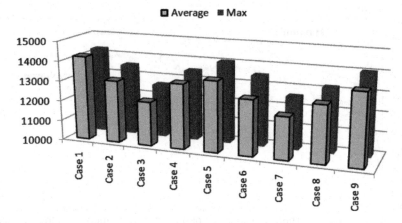

**Fig. 6.1**  Maximum and average fitness values in different cases

### 4.2.8   Genetic Algorithm Implementation

The proposed GA has been successfully implemented in Matlab and has been comprehensively tested for convergence and consistence of the solutions. Experiments show that the proposed GA with the following parameters gives the good solutions: population size of 200, crossover rate of 50% and mutation rate of 50%, and it typically converges after less than 2,000 generations taking less than 40 minutes by an Intel Dual Core laptop, CPU T3400, 2.17 GHz, 3 GB RAM. It should be noted that the optimisation process takes quite a long time because a large number of different constraint-based cutting operations must be applied to all the production lines in every single chromosome in initial population as well as in crossover and mutation operations in every generation.

It is evident that GA cannot guarantee to find best solution after only one run. However, it is very good at finding good/best solution(s) reasonably quickly [4]. Therefore, it is easy to evaluate the quality of solution by comparing different runs. Accordingly, the best solution among the ones obtained from different runs should be selected. If the number of runs is large enough, this validates the optimality of the solution. In this case study, for each resource allocation case, the GA is run for 10 times and results are shown in Table 6.7 and visualised in Fig. 6.1.

The result shown in Table 6.7 indicates that the resource allocation case 1 gives the highest fitness value—on average of 14,206.3. The best solution among those solutions obtained is as highlighted in Table 6.7. With this solution, the achieved fitness value is 14,287. The convergence of the GA for that solution is shown in Fig. 6.2 and detail of the solution is shown in Table 6.8.

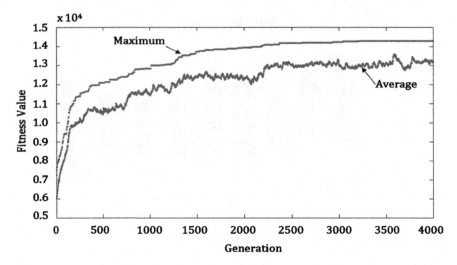

**Fig. 6.2** Evolution of fitness value—output of the GA

**Table 6.8** Optimal solution

| | | | | | | | | | | | | | | | | | | | | | | | | | Best chromosome | | | | | | | | | | | | | | | | | | | | | | | |
|---|---|---|---|---|---|---|---|---|---|---|---|---|---|---|---|---|---|---|---|---|---|---|---|---|---|---|---|---|---|---|---|---|---|---|---|---|---|---|---|---|---|---|---|---|---|---|
| Sequence | 1 | 2 | 3 | 4 | 5 | 6 | 7 | 8 | 9 | 10 | 11 | 12 | 13 | 14 | 15 | 16 | 17 | 18 | 19 | 20 | 21 | 22 | 23 | 24 | 25 | 26 | 27 | 28 | 29 | 30 | 31 | 32 | 33 | 34 | 35 | 36 | 37 | 38 | 39 | 40 | 41 | 42 | 43 | 44 | 45 | 46 | 47 |
| Line1 | 32 | 32 | 32 | 32 | 32 | 32 | 3 | 32 | 5 | 32 | 32 | 32 | 32 | 17 | 26 | 32 | 32 | 32 | 32 | 32 | 32 | 32 | 32 | 6 | 10 | 32 | 32 | 32 | 10 | 5 | 32 | 32 | 5 | 32 | 13 | 18 | 10 | 0 | 0 | 0 | 0 | 0 | 0 | 0 | 0 | 0 | 0 |
| Line 2 | 10 | 47 | 44 | 9 | 39 | 44 | 47 | 47 | 49 | 12 | 50 | 31 | 22 | 16 | 5 | 47 | 24 | 44 | 18 | 44 | 36 | 47 | 18 | 30 | 20 | 46 | 18 | 44 | 36 | 19 | 18 | 36 | 36 | 0 | 0 | 0 | 0 | 0 | 0 | 0 | 0 | 0 | 0 | 0 | 0 | 0 | 0 |
| Line 3 | 10 | 44 | 44 | 44 | 36 | 36 | 44 | 44 | 27 | 47 | 5 | 44 | 27 | 36 | 49 | 36 | 14 | 36 | 36 | 44 | 44 | 44 | 36 | 44 | 44 | 44 | 44 | 44 | 44 | 44 | 48 | 44 | 41 | 4 | 44 | 49 | 27 | 36 | 36 | 36 | 36 | 24 | 44 | 0 | 0 | | |

# 5   Conclusion and Future Work

In this chapter, a class of soft precedence-constrained production sequencing and scheduling problems for multiple production lines has been modelled. Due to the nature of constraints, the multi-objectives GA with new strategies for chromosome encoding, feasibility of chromosome, crossover as well as mutation operations have been developed to optimise the model.

The proposed approach has been illustrated, and its robustness has been verified in the complex and realistic case study which contains the most important constraints currently encountered in a typical manufacturing company. The proposed GA has been extensively tested, for various input parameters. The evolution of the output is consistently convergent towards the optimum. It is evident that the proposed method can easily accommodate much larger and more complex problems in this class.

Further work should be conducted in the following areas:

- Developing the GA for extended problem, e.g. adding more constraints such as variety of material used, different product changeover for different pairs of products, etc.
- Incorporation of stochastic events into the model and investigating their influence on the optimality.
- Comprehensively investigating the effect of resource allocation strategy on the optimality.

# References

1. Yen C, Tseng SS, Yang CT (1995) Scheduling of precedence constrained tasks on multiprocessor systems. In: Proceedings of IEEE first international conference on algorithms and architectures for parallel processing, Brisbane, QLD, Australia, 19–21 April 1995, pp 379–382
2. Zou D, Gao L, Li S, Wu J (2011) Solving 0–1 knapsack problem by a novel global harmony search algorithm. Appl Soft Comput 11:1556–1564
3. Chen CLP (1990) AND/OR precedence constraint traveling salesman problem and its application to assembly schedule generation. In: Proceedings of IEEE international conference on systems, man and cybernetics, Los Angeles, CA, USA, 4–7 November 1990, pp 560–562
4. Marian RM (2003) Optimisation of assembly sequences using genetic algorithms. PhD Thesis, University of South Australia, Adelaide
5. Marian RM, Luong LHS, Abhary K (2006) A genetic algorithm for the optimisation of assembly sequences. Comput Ind Eng 50(4):503–527
6. Duman E, Or I (2004) Precedence constrained TSP arising in printed circuit board assembly. Int J Prod Res 42:67–78
7. Lambert AJD (2006) Exact methods in optimum disassembly sequence search for problems subject to sequence dependent costs. Omega 34:538–549
8. Moon C, Kim J, Choi G, Seo Y (2002) An efficient genetic algorithm for the traveling salesman problem with precedence constraints. Eur J Oper Res 140:606–617
9. He W, Kusiak A (1992) Scheduling manufacturing systems. Comput Ind 20:163–175
10. Su Q (2007) Applying case-based reasoning in assembly sequence planning. Int J Prod Res 45:29–47
11. Yun Y, Moon C (2009) Genetic algorithm approach for precedence-constrained sequencing problems. J Intell Manuf 20:1–10
12. Morton TE, Dharan BG (1978) Algoristics for single-machine sequencing with precedence constraints. Manag Sci 24:1011–1020
13. Chekuri C, Motwani R (1999) Precedence constrained scheduling to minimize sum of weighted completion times on a single machine. Discrete Appl Math 98:29–38
14. Ambuhl C, Mastrolilli M (2006) Single machine precedence constrained scheduling is a vertex cover problem. In: Proceedings of the 14th conference on annual European symposium, Zurich, Switzerland, 11–13 September 2006, pp 28–39
15. Jonsson J, Shin KG (1997) A parametrized branch-and-bound strategy for scheduling precedence-constrained tasks on a multiprocessor system. In: Proceedings of the international conference on parallel processing, Bloomington, IL, USA, 11–15 August 1997, pp 158–165
16. Selvakumar S, Siva RMC (1994) Scheduling precedence constrained task graphs with non-negligible intertask communication onto multiprocessors. IEEE Trans Parallel Distrib Syst 5:328–336

17. Hary SL, Ozguner F (1999) Precedence-constrained task allocation onto point-to-point networks for pipelined execution. IEEE Trans Parallel Distrib Syst 10:838–851
18. Mmouhamed MA, Maasarani AA (1994) Performance evaluation of scheduling precedence-constrained computations on message-passing systems. IEEE Trans Parallel Distrib Syst 5:1317–1321
19. Dao SD, Marian R (2011) Optimisation of precedence-constrained production sequencing and scheduling using genetic algorithm. In: Proceedings of the international multiconference of engineers and computer scientists, Hong Kong, 16–18 March 2011, pp 59–64
20. Dao SD, Marian R (2011) Modeling and optimisation of precedence-constrained production sequencing and scheduling using multi-objective genetic algorithm. In: Lecture notes in engineering and computer science: Proceedings of the world congress on engineering 2011. WCE 2011, London, 6–8 July 2011, pp 1027–1032
21. Dao SD, Marian R (2011) Modeling and optimisation of precedence-constrained production sequencing and scheduling for multiple production lines using genetic algorithm. J Comput Technol Appl, 2(6):487–499
22. Marian RM, Luong LHS, Akararungruangkul R (2008) Optimisation of distribution networks using genetic algorithms. Part 2—The genetic algorithm and genetic operators. Int J Manuf Technol Manag 15(1):84–101
23. Gen M, Cheng R (1997) Genetic algorithms and engineering design. Wiley, New York

# Chapter 7
# A Closed-Loop Bid Adjustment Method of Dynamic Task Allocation of Robots

**S.H. Choi and W.K. Zhu**

## 1 Introduction

Task allocation for multiple robots addresses the problem of assigning a set of tasks to the corresponding robots while achieving some specific objectives of team performance. Research in this field dates back to the late 1980s. Task allocation of multiple robots in dynamic environments is core to multi-robot control for a number of real-world applications, such as military [1], transport services [2], search and rescue [3], etc.

The working environments of task allocation can be static or dynamic [4]. Static task allocation assumes completely known information about the environment, such as the number of tasks and robots, the arrival time of tasks, and the process of task execution. Traditionally, applications in multi-robot domains have largely remained in static scenarios, with an aim to minimise a cost function, such as total path length, or execution time of the team. Obviously, static approaches cannot adapt to changes in a dynamic environment.

Dynamic task allocation, on the other hand, makes decisions based on real-time information and is therefore more adaptive to changes. This chapter assumes a set of dynamically released tasks to be completed by a team of robots, and the conditions of the work process keep changing during task execution. This kind of dynamic working environment is ubiquitous in real-life applications, such as exploring and mapping by robots in unknown environments, unexpected adversarial targets in combats, stochastic pickups and delivery transport services, etc. [5].

With reference to the taxonomy of Gerkey and Mataric [6], multi-robot task allocation problems can be classified along three dimensions. In the dimension of robot, it can be a single-task robot or a multi-task one. A single-task robot is capable

S. Choi (✉) • W. Zhu
Department of Industrial and Manufacturing Systems Engineering,
The University of Hong Kong, Hong Kong, China
e-mail: shchoi@hku.hk; wenkaizhu@hku.hk

S.-I. Ao and L. Gelman (eds.), *Electrical Engineering and Intelligent Systems*,
Lecture Notes in Electrical Engineering 130, DOI 10.1007/978-1-4614-2317-1_7,
© Springer Science+Business Media, LLC 2013

of executing exactly one task at a time, while a multi-task robot can handle more than one task simultaneously. In the dimension of task, it can be a single-robot task or a multi-robot task. A single-robot task requires only one robot to execute it, while a multi-robot task requires more than one robot to work on it at the same time. In terms of planning horizon, task allocation can be instantaneous assignment and time-extended assignment. Instantaneous assignment only considers the tasks currently available. Time-extended assignment elaborates the effect of current assignment on future assignment, involving task dependency and schedule. Instantaneous assignment is commonly used since it needs less computation on task sequencing algorithms and is particularly practicable in dynamic situations where tasks are randomly released [7]. The task allocation problem in this chapter is restricted to a single-task robot, single-robot task, and instantaneous assignment.

Multi-robot task allocation is a typically NP-hard problem. Its challenges become even more complicated when operations in uncertain environments, such as unexpected interference between robots, stochastic task requests, inconsistent information, and various component failures, are considered [4]. In such cases, it is not worth spending time and resources to secure an optimal solution, if the solution keeps changing as operations go on. Moreover, if time-window constraints are imposed, there may not be enough time to compute an exact and global solution.

The basic objective of a multi-robot task allocation problem is to have tractable planning that produces efficient and practicable solutions. Auction-based, or market-based, approaches manage this by assembling team information at a single location to make decisions about assigning tasks over the team to produce practicable solutions quickly and concisely [8].

In an auction, a set of tasks are offered by an auctioneer in the announcement phase. Broadcast with the auction announcement, each of the robots estimates the cost of a task separately and submits a bid to the auctioneer. Once all bids are received or a pre-specified deadline has passed, the auction is cleared. In the winner determination phase, the auctioneer decides with some selection criteria which robot wins which task [9]. This chapter adopts cost minimisation, in that an auctioned task is awarded to a robot offering the lowest bid price.

Single-item auctioning is simple yet commonly used to offer one task at a time [10]. Combinatorial auctioning, on the other hand, is more complex in that multiple tasks are offered and participants can bid on any combination of these tasks. Since the number of combinations to be considered increases exponentially, auction administration, such as bid valuation, communication, and auction clearing, would soon become intractable [5]. Therefore, sequential single-item auction is a practicable approach when tasks are dynamically released, and is adopted in this chapter.

The earliest example of auction-based multi-robot coordination, called contract net protocol [11], appeared about 30 years ago. Auction-based multi-robot coordination approaches have been growing in popularity in recent years. They have been successfully implemented in a variety of domains, such as robotic transport [12, 13], mapping and exploration [14], house cleaning [15], and reconnaissance [16]. Auction-based approaches are preferable in online applications in that they can quickly and concisely assemble team information at a single location to make decisions, significantly reducing the combinatorial nature of task assignment problems [17]. The solution quality, although not optimal, is guaranteed in most cases. Auction-based approaches are suitable for dynamic and uncertain applications since they can accommodate new information through frequent auctioning of tasks [5].

However, some issues of auction-based task allocation have yet to be further investigated [5]. Firstly, a clear conceptual understanding of auction-based coordination approaches is needed. Further works should be devoted to studying how components, such as performance assessment mechanism, bidding strategy, and auction clearing mechanism, can be implemented effectively in different multi-robot applications. Secondly, the fundamental premise of success in an auction relies on the ability of individual robots to make reasonable cost estimation and submit acceptably accurate bid prices. However, robots generally do not have sufficient information for reliable cost calculation, which requires an accurate model of the environment and computation-expensive operations. Thus, heuristics and approximation algorithms are commonly used, such as the first-come-first-served and the shortest-distance-first. Some progress has been made to improve the accuracy of bid pricing. Duvallet and Stentz [18] applied an imitation learning technique to bias the bid prices in auctions to make better solutions. Two simulated scenarios were presented to demonstrate the applicability of this technique, including three fire-fighting agents putting out fires in buildings, and eight players in an adversarial game trying to outscore their opponents. This approach needed a considerable amount of training samples and time to reach a reasonable solution, and the learning rate should be skillfully tuned. Khan et al. [19] developed a platform to simulate a team of five robots cooperating to move some objects to the specified goals, based on an auction-based task assignment method. The number of robots involved in an operation could be adaptively changed. They tried to improve the accuracy of cost estimation by an elaborated bidding function, which considered a set of environment conditions, such as the distance between a task and a robot, the velocity and orientation of a robot, obstacles in the path between a robot and a task, and the possible success rate of a task. While the simulation results were encouraging, it seemed that the selection criteria of environmental factors needed further justification.

Nevertheless, the above-mentioned approaches to improving cost estimation, like most of the current auction-based methods, are open-looped [20]. They cannot assess whether or not a bidder has kept its commitment to a task, because they do not have a mechanism to evaluate the bidder's performance after winning the task. Human bidders are self-interested in auctions, and would sometimes deliberately offer over-optimistic bid prices. Robots, on the other hand, are assumed to be honest in estimating the costs before offering the bid prices. However, there are often discrepancies between the bid prices and the actual costs in real-life applications, particularly in dynamic working environments. Discrepancies between the bid prices and the actual costs are usually caused by the uncertainties of a dynamic environment, such as unexpected task requests, changing traffic conditions, communication delay, inconsistent information, and stochastic component failures [5]. Unfortunately, these uncertainties are difficult to model explicitly in advance. By submitting either over-estimated or under-estimated bids, robots may not be able to realise their task promises. As a result, the overall team performance would be significantly hampered.

This chapter therefore presents a closed-loop bid adjustment mechanism for auction-based multi-robot task allocation in light of operational uncertainties, with which a team of robots can evaluate and improve their bids, respectively, and hence enhance the overall team performance. Each of the robots in a team maintains an array of track records for different corresponding types of tasks it has ever executed. After a robot has completed a specific type of task, it assesses and logs its performance to the corresponding track record, which reflects the discrepancy between the submitted bid price and the related actual cost of the task. These track records serve as closed-loop feedback information to adjust and improve the bid prices in future auctions. Moreover, when adjusting the bid price of a task, a series of performance records, with time-discounting factors, are taken into account to damp out fluctuations. As such, bid prices can be regulated and fine-tuned to alleviate some deviations of cost estimation due to operational uncertainties. Tasks are more likely allocated to competent robots that offer more accurate and reliable bids, resulting in significant improvement in the overall team performance.

Section 2 elaborates on the proposed bid adjustment mechanism, while Section 3 describes its implementation for simulation of free-range automated guided vehicles serving at a container terminal to demonstrate the effectiveness of the adjustment mechanism. Section 4 draws conclusion and discusses the future work.

# 2   The Closed-Loop Bid Adjustment Method

## 2.1   The Task Auction Architecture

Figure 7.1 shows the task auction architecture. Task allocation in this chapter is restricted to a single-task robot, single-robot task, and instantaneous assignment problem. Hence, we adopt sequential single-item auctioning for situations where different types of tasks are stochastically released for auction during operation. A central processor is the auctioneer who auctions these tasks one by one. All the idle robots bid for a task being auctioned, and the one that submitted the lowest bid price wins the task.

## 2.2   The Closed-Loop Bid Adjustment Mechanism

Each of the robots maintains an array of track records for the corresponding types of tasks that it has ever executed. Figure 7.2 shows a block diagram of this bid

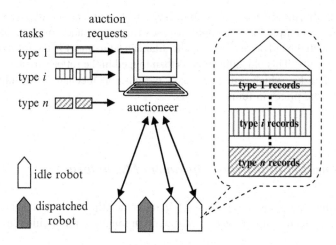

**Fig. 7.1**  Task auction architecture

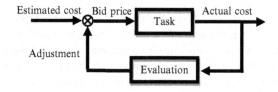

**Fig. 7.2**  The closed-loop bid adjustment mechanism

adjustment mechanism. After a robot has completed a specific type of task, it evaluates its own performance and records a reward or a penalty accordingly. This track record facilitates adjustment of the bid price that the robot in question will subsequently submit for another task of the same type.

## 2.3    Task Auction Procedure

When a task is stochastically released for auction, the central auctioneer is requested to initiate an auction. The auctioneer broadcasts this specific type of task to the robot team. Some robots in the team may be occupied with their ongoing tasks, and would simply ignore this auction announcement. Only the idle robots will take part in this auction. Each of the idle robots estimates the cost of the task by some heuristics or approximation methods, such as assuming a straight route with no obstacles in the way. Then, it adjusts the estimated cost, based on its track records for this type of tasks, to obtain a bid price. Subsequently, the resulting bid price is submitted to the auctioneer before the clearing time. After receiving all the bids from the robot team, the auctioneer assesses the bid prices and awards the task to the robot offering the lowest bid price. The robot awarded with the bid executes the task. After completion, the robot assesses its performance by comparing the actual cost with the bid price of this task to attain an adjustment in the form of either reward or penalty. This bid adjustment is then logged in the robot's track record for this type of tasks, which will serve as a closed-loop mechanism to adjust and improve the bid prices in future auctions.

## 2.4   Algorithm of the Bid Adjustment Mechanism

For a specific type of tasks, we denote $Actual_k$ as the $k$th record of actual cost, and $Bid_k$ as the $k$th record of bid price. Adjustments are in the form of either rewards or penalties:

$$Adjust_k = Actual_k - Bid_k. \tag{7.1}$$

If the actual cost of the task is smaller than the proposed bid price, it means that the robot has kept its commitment to the task or has even outperformed its promise. The robot will get a reward with a negative adjustment value. Similarly, a penalty with a positive adjustment value will be imposed to this robot if the actual cost is larger than the proposed bid price, which means the robot has under-estimated the cost of task and has not completed it as promised.

When a robot bids for a next task of the same type, it first estimates the cost, and then tunes the bid price based on the previous adjustment:

$$Bid_{k+1} = Cost_{k+1} + Adjust_k, \tag{7.2}$$

where $Cost_{k+1}$ is the $(k + 1)$th estimated cost, which can be acquired by other heuristics or approximation methods.

To damp out huge fluctuations and to reflect more reliable estimations, a series of previous adjustments should be taken into account. Moreover, since the working environment is changing dynamically, older track records are deemed relatively obsolete as time elapses. Hence, a time-discounting factor, $\alpha$, where $0<\alpha<1$, is introduced to weigh the track records. The averaged bid adjustment is

$$\frac{\sum_{j=0}^{k-1} \alpha^j Adjust_{k-j}}{\sum_{j=0}^{k-1} \alpha^j}.$$

In practise, the latest three terms are sufficient for adjustment of the bid price. The complete form of the proposed bid adjustment mechanism is given in (7.3).

$$Bid_{k+1} = Cost_{k+1} + \frac{\sum_{j=0}^{k-1} \alpha^j Adjust_{k-j}}{\sum_{j=0}^{k-1} \alpha^j}. \qquad (7.3)$$

The task being auctioned is therefore assigned to the robot that submitted the lowest adjusted bid price, based on (7.3). As such, this closed-loop bid adjustment mechanism can improve bidding accuracy, considerably enhancing the overall team performance.

## 2.5   Robustness Analysis of the Algorithm

With the closed-loop feedback mechanism, the bid price submitted by a robot for a task being auctioned can be regulated and fine-tuned to mitigate deviations of cost estimation due to operational uncertainties. Moreover, a series of adjustment values are averaged with related time-discounting factors to damp out possible fluctuations of adjustments, further safeguarding the robustness of the closed-loop regulation mechanism. Therefore, the stability of the proposed approach can be effectively secured.

## 2.6   Workflow of Serial Auctions for All Tasks

The complete workflow of task auctions during operation is listed as follows:

Step 1: A task is released and a request for auction is sent to the auctioneer to announce.

Step 2: For each idle robot to participate in the auction:

(2a): If this type of task has NOT been executed before, sets the bid adjustment to 0 and creates a track record for this type of tasks;
else
Reads the bid adjustment from the track record.
(2b): Estimates the cost of the task.
(2c): Adjusts bid price by adding bid adjustment to the estimated cost.
(2d): Submits adjusted bid price before clearing time.

Step 3: The auctioneer assesses all the bid prices received and awards the task to the robot offering the lowest bid price.

Step 4: The winning robot executes its awarded task until it is finished.

Step 5: The robot compares the actual cost with the proposed bid price, and updates the bid adjustment.

Step 6: The robot logs the bid adjustment into its related track record and calculates the averaged adjustment value.

Step 7: Repeats from Step 1 until no more task is released.


# 3   Implementation and Case Study

The closed-loop bid adjustment mechanism is incorporated with a multi-robot dynamic task allocation module in a simulator, which also includes a module for motion planning of a fleet of robots. The details of this motion planning approach can be found in [21].


## 3.1   Implementation of the Simulator

The simulator is developed with the Player/Stage [22] and C++ programming language. The Player/Stage is an open-source package widely used for multi-robot control and simulation. It runs in a Linux-based operating system called Fedora 13, and consists of two sub-packages, namely Player and Stage. Player provides a network interface to a variety of physical robots and sensors. Player's client/server model allows robot control programmes to be written in a number of programming languages and to run on any computer with a network connection to physical robots. Stage is a Player plug-in simulation package which simulates a population of mobile robots moving and sensing in a 2D bit-mapped environment. Various sensor models are provided. Virtual devices of Stage present a standard Player interface, and hence few or no changes are required to move between simulation and hardware. Controllers designed in Stage have been demonstrated to work on various physical robots.


## 3.2   Scenario Setup

An automated guided vehicle (AGV), with autonomous control and sensing devices, can be regarded as an autonomous robot. A team of AGVs at a container terminal transporting containers from the quay-side to the yard-side is used to verify the practicability of the proposed task allocation approach, as shown in Fig. 7.3. There are two vessels berthed at the quay-side. Each vessel is served by five quay cranes which unload the containers from the vessels. Small rectangles in grey on quay-side and yard-side represent containers. The containers beside the vessels are ready to be picked up, while those being handled by the quay cranes are not shown in the figure. The racks at the quay-side are labelled as 1, 2, . . . , 10, while those at the yard-side are labelled as $A, B, . . . , J$.

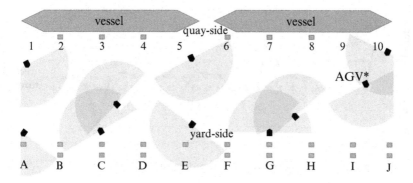

**Fig. 7.3** Simulated working environment of free-range AGVs at a container terminal

Traditionally, most AGVs use fixed guide-paths, such as loops and networks. The fixed routing approaches allow for reliable automation of vehicles. Such AGVs are however less manoeuvrable. With the advent of more powerful onboard processors and advanced sensors, it is now possible for AGVs to navigate without physical guide-paths. Some experimental systems have indeed been developed [23], showing improved transport capacity.

The AGVs work in an area of 600 m × 150 m. Each AGV measures 12 m × 4.5 m × 1.5 m and weighs 25 tonnes. The maximum velocity and the maximum acceleration of an AGV are $V_{max} = 7ms^{-1}$ and $a_{max} = 1ms^{-2}$, respectively. Inertial measurement units (IMU) are used to estimate the relative positions, velocities, and accelerations of vehicles. Sonars are used for detection of obstacles. The sensing field of view is 180°, with a scanning range $R = 50$ m.

There are two major operational uncertainties for AGVs at container terminals, namely dynamic task requirements and uncertain traffic conditions. Dynamic task requirements are mainly due to the variation of vessel arrival time, the handling time of quay cranes, and the characteristic of containers to be transported. Uncertain traffic conditions are mainly due to stochastic interferences between AGVs [24]. It is assumed that each AGV can only carry one container at a time, and obviously a container should only be transported by one AGV. Whenever a container is put onto a rack from a quay crane, it is ready for auction. This is a single-task robot, single-robot task, and instantaneous assignment problem. Hence, this scenario of a team of decentralised free-range AGVs working at a dynamic container terminal is deemed a good test-bed to validate the proposed bid adjustment mechanism for dynamic multi-robot task allocation.

A specific type of tasks is described by the pickup location and the destination of delivery, as $T(n, x)$, where $n$ specifies the label of the pickup location at quay-side ($n = 1, 2, \ldots, 10$), and $x$ specifies the label of the destination at yard-side ($x = A, B, \ldots, J$). For example in Fig. 7.3, $T(7, J)$ is a type of tasks requiring to transport containers from rack 7 at quay-side to rack $J$ at yard-side. The cost of a task in the simulation is the time consumed in units of minutes to handle and transport a container.

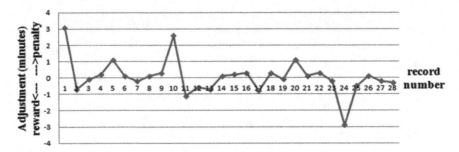

**Fig. 7.4**  Adjustment records of task type $T(7, J)$ performed by AGV*

## 3.3  Case Study

A case study involves simulation of ten AGVs to transport 600 containers, with two changes of task requirement during operation.

Containers with different types of task requests are dynamically released from the quay cranes onto the racks. When a container is up for auction, the central auctioneer announces the specific type of task to the robot team, and all the idle AGVs participate in this auction. Each idle robot calculates and adjusts its bid price based on its track record for the same type of tasks, and submits the bid to the auctioneer. After receiving all the bids from the AGV team, the auctioneer assesses the bid prices and awards the task to the AGV that proposed the lowest bid price. The winning AGV proceeds to pickup the container, transports it to the specified location at the yard-side, and subsequently drops it onto the rack. After completing this task, the AGV compares the actual cost with the corresponding bid price to attain an adjustment in term of a reward or a penalty. This bid adjustment is then logged in the track record for this type of tasks, which serves as a closed-loop mechanism to adjust the bid prices in future auctions.

Figure 7.4 shows the track records of twenty-eight tasks of type $T(7, J)$ ever performed by AGV* labelled in Fig. 7.3. The time-discounting factor, $\alpha$, was set to be 0.5. For the first time after AGV* had finished a $T(7, J)$ type of task, the adjustment record was a penalty of about three minutes. It indicates that AGV* under-estimated the cost of the task and offered a bid price which turned out to be about three minutes less than the actual cost incurred afterwards. In other words, AGV* did not keep its commitment to this task. Being imposed with this penalty, AGV* modified the bid price for the task type $T(7, J)$ hereafter. The subsequent second to ninth adjustment values of this type of tasks were within the accuracy of $\pm 1.1$ min band.

Since these adjustment values indicate the discrepancies between the bid prices and the related actual costs, it verifies that, with the closed-loop bid adjustment mechanism in auctions, the discrepancies between the actual costs and the bid prices were effectively reduced. With the improved bid prices, tasks were assigned to the competent robots that proposed more reliable bid prices, accordingly enhancing the overall team performance.

The robustness of the proposed bid adjustment mechanism was tested by intentionally modifying the characteristic of task type $T(7, J)$ after some time in the operation. In this case, heavier containers were released to transport. An AGV carrying a heavier container should generally be more sluggish to evade other AGVs in the way and move at a lower speed. Hence, the actual cost of task completion should be higher than that before the modification. However, the bidding AGVs were only broadcast with the task information with pickup and drop-off locations, having no idea that the containers were heavier. The idle AGVs involved in auctions still offered the previously adjusted bid prices. A winning and dispatched AGV, like AGV*, could not complete the task as promised, and got a penalty of about 2.6 min. Being imposed with this penalty, AGV* adjusted the bidding price for task type $T(7, J)$ in future auctions. The subsequent 11th to 23rd adjustment values were within the accuracy of $\pm 1.1$ min band.

Sometime later during operation, another deliberate change of the characteristic of task type $T(7, J)$ was introduced, and this type of containers became lighter to transport. An AGV carrying a lighter container should move at a higher speed and be more flexible to evade the obstacles in the way. Hence, the actual cost of task fulfillment should be lower than that before the modification. A winning and dispatched AGV, like AGV*, was able to complete the task earlier than expected, and got a reward of about 2.9 min. With this reward, the AGV* adjusted the bidding price for task type $T(7, J)$ in the future auctions. The subsequent 25th to 28th adjustments were effectively reduced again.

This simulation verifies that the closed-loop bid adjustment mechanism can stably reduce the discrepancies between the bidding prices and the actual costs, even with some dynamic situations during operation.

With the closed-loop feedback mechanism, the bid price can be adjusted and fine-tuned to suppress some disturbances due to operational uncertainties. More-over, a series of adjustment values are averaged with related time-discounting factors to damp out the fluctuations of adjustment values, further enhancing the robustness of the adjustment mechanism. Therefore, stability of the proposed approach can be effectively retained.

Figure 7.5 presents a comparison of the overall team performances, in terms of operational time. The one without bid adjustment took 436.6 min, while the other one with bid adjustment consumed 362.4 min. With the proposed bid adjustment mechanism, the bid prices for different types of tasks were adjusted and improved according to the dynamic conditions during operation. As a result, the discrepancies between the bid prices and the related actual costs could be effectively reduced. Competent AGVs that offered more reliable bid prices were awarded the auctioned containers. A considerable improvement of 17% in overall team performance was achieved.

**Fig.7.5** Comparison of
performances, with and
without bid adjustment

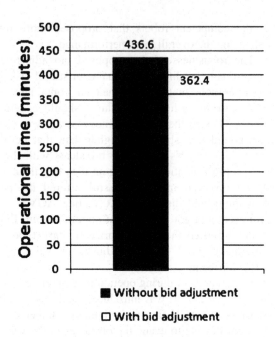

■ Without bid adjustment

□ With bid adjustment

## 4 Conclusion and Future Work

This chapter presents an auction-based approach with the closed-loop bid adjustment mechanism to dynamic task allocation for robots. The bid adjustment mechanism fine-tunes bid prices based on the performance track records of each robot. A simulator is developed, with a case study of AGVs transporting containers, to validate this task allocation approach. Simulation results show that the bid adjustment mechanism can effectively reduce the discrepancies between the submitted bid prices and the corresponding actual costs of tasks. The stability of the approach is also verified in light of some operational uncertainties. This bid adjustment mechanism enhances the likelihood of allocating tasks to competent robots that submit more accurate bids, and as a result, improves the overall team performance substantially.

Despite the advantageous features above, the auction-based approach can be enhanced to further reduce overshot values and transient fluctuations, which are inevitable phenomena caused by stochastic operational conditions. Our approach can already damp out some fluctuations shown in Fig. 7.4, but it would be beneficial to further alleviate the discrepancies between the bid prices and the actual costs. For this purpose, some techniques of learning theory and adaptive regulation are now under active consideration, although their responsiveness to dynamic situations should be investigated in detail.

# References

1. Richards A, Bellingham J, Tillerson M, How J (2002) Co-ordination and control of multiple UAVs. In: Proceedings of AIAA conference on guidance, navigation, and control, CA, USA, pp 251–256
2. Murray RM (2007) Recent research in cooperative control of multivehicle systems. J Dyn Syst Meas Contr 129(5):571–583
3. Parker LE (2009) Path planning and motion coordination in multiple mobile robot teams. Encyclopedia Complex Syst Sci 2009:5783–5800
4. Parker LE (2008) Multiple mobile robot systems. In: Bruno S, Oussama K (eds) Springer Handbook of Robotics. Springer, The Netherlands, pp 921–941
5. Dias MB, Zlot R, Kalra N, Stentz A (2006) Market-based multi-robot coordination: a survey and analysis. Proc IEEE 94(7):1257–1270
6. Gerkey BP, Mataric MJ (2004) A formal analysis and taxonomy of task allocation in multi-robot systems. Int J Robot Res 23(9):939–954
7. Lerman K, Jones C, Galstyan A, Mataric MJ (2006) Analysis of dynamic task allocation in multi-robot systems. Int J Robot Res 25(3):225–241
8. Mosteo AR, Montano L (2007) Comparative experiments on optimization criteria and algorithms for auction based multi-robot task allocation. In: Proceedings of IEEE international conference on robotics and automation, Roma, Italy, pp 3345–3350
9. Jones EG, Dias MB, Stentz A (2007) Learning-enhanced market-based task allocation for oversubscribed domains. In: Proceedings of the 2007 IEEE/RSJ international conference on intelligent robots and systems, San Diego, USA, pp 2308–2313
10. Schoenig A, Pagnucco M (2010) Evaluating sequential single-item auctions for dynamic task allocation. Lect Notes in Comput Sci 646(11):506–515
11. Smith RG (1980) The contract net protocol: high-level communication and control in a distributed problem solver. IEEE Trans Comput 29(12):1104–1113
12. Jones EG, Dias MB, Stentz A (2011) Time-extended multi-robot coordination for domains with intra-path constraints. Auton Robot 30(4):41–56
13. Herrero-P'erez D, Mat'ınez-Barber'a H (2008) Decentralized coordination of autonomous AGVs in flexible manufacturing systems. In: Proceedings of the IEEE/RSJ international conference on intelligent robots and systems, Nice, France, pp 3674–3679
14. Schneider J, Apfelbaum D, Bagnell D, Simmons R (2005) Learning opportunity costs in multi-robot market based planners. In: Proceedings of the 2005 IEEE international conference on robotics and automation, Spain, pp 1151–1156
15. Kalra N, Ferguson D, Stentz A (2005) Hoplites: a market-based framework for planned tight coordination in multirobot teams. In: Proceedings of IEEE international conference on robotics and automation, Spain, pp 1170–1177
16. Bertuccelli LF, Choi H-L, Cho P, How JP (2009) Real-time multi-UAV task assignment in dynamic and uncertain environments. In: Proceedings of the american institute of aeronautics and astronautics conference on guidance, navigation, and control, Chicago, USA, pp 389–395
17. Gerkey BP, Mataric MJ (2002) Sold!: Auction Methods for Multirobot Coordination. IEEE Trans Robot Autom 18(5):758–768
18. Duvallet F, Stentz A (2010) Imitation learning for task allocation. In: Proceedings of international conference on intelligent robots and systems, Taiwan, pp 3568–3573
19. Khan MT, Imanuel T, de Silva CW (2010) Autonomous market-based multi-robot cooperation. In: Proceedings of international conference on intelligent and advanced systems, Malaysia, pp1–6
20. Zhu WK, Choi SH (2011) An auction-based approach with closed-loop bid adjustment to dynamic task allocation in robot teams. In: Proceedings of the world congress on engineering 2011 (WCE 2011). Lecture Notes in Engineering and Computer Science, London, UK, 6–8 July 2011, pp 1016–1066

21. Choi SH, Zhu WK (2011) A bio-inspired intelligent approach to motion planning of mobile robots. Comput-Aided Des Appl 8(5):773–783
22. Player/Stage Project. Available online http://playerstage.sourceforge.net/
23. Duinkerken MB, van der Zee M, Lodewijks G (2006) Dynamic free range routing for automated guided vehicles. In: Proceedings of IEEE international conference on networking, sensing, and control, FL, USA, pp 312–317
24. Vis IFA (2006) Survey of research in the design and control of automated guided vehicle systems. Eur J Oper Res 170(5):677–709

# Chapter 8
# Reasoning on OWL2 Ontologies with Rules Using Metalogic

Visit Hirankitti and Trang Xuan Mai

## 1 Introduction

Ontologies have played an important role in the semantic web (or shortly "SW"). An ontology forms vocabularies and sentences, it expresses knowledge to be shared on the web. OWL was accepted by W3C as a language for representing a web ontology. Its core, OWL-DL, is essentially an XML encoding of an expressive Description Logic (DL) built upon Resource Description Framework (RDF) with a substantial fragment of RDF Schema (RDFS). The vocabularies defined in such an ontology consist of classes and properties; in logic classes can be treated as unary predicates, while properties as binary predicates, and all these predicates represent relations. OWL was successfully applied to SW in the past. However, some SW knowledge should be formulated more naturally in rules.

Knowledge representation using rules is common in logic programming. It is also proposed to use in SW to complement to other means of knowledge representation in OWL. The realization of rules allows a means to deduce and combine information. This leads to a way for enhancing content, and supporting reasoning capabilities, on OWL ontologies.

The extension of SW ontologies with rules has recently attracted much attention in the SW research, and many approaches have been proposed for it. One of them is to combine DL with first-order Horn-clause rules. This is the basis of the Semantic Web Rule Language (SWRL) [1], a language for rule formulation and rule extension

V. Hirankitti (✉)
School of Computer Engineering, King Mongkut's Institute of Technology Ladkrabang, Ladkrabang, Bangkok 10520, Thailand
e-mail: khvisit@kmitl.ac.th

T.X. Mai
International College, King Mongkut's Institute of Technology Ladkrabang, Ladkrabang, Bangkok 10520, Thailand
e-mail: trangmx@gmail.com

S.-I. Ao and L. Gelman (eds.), *Electrical Engineering and Intelligent Systems*, Lecture Notes in Electrical Engineering 130, DOI 10.1007/978-1-4614-2317-1_8, © Springer Science+Business Media, LLC 2013

to OWL. However, inferences on SWRL rules can lead to undecidability even though the rules are assumed to be function-free [1]. To make the inferences decidable, some restrictions were put upon the rule language in the form of DL-safe rules [2] or of Description Logic Programs (DLP) [3]. An improvement on OWL was recently developed by W3C, namely "OWL 2". OWL 2 is more expressive as it allows rule formulation based on DL $\mathcal{SROIQ}$ [4], in which DL rules can be completely ensured as decidable fragment of SWRL.

In our previous work [5], we have developed a meta-logical approach for reasoning with SW ontologies expressed in OWL 2, and in [6] we went further by extending the framework so that it can reason with SW ontologies and rules in OWL 2. This paper is an extended and revised version of it.

The remainder of this paper is organized as follows. Section 2 reviews some concepts of ontologies with a rule extension, and shows how rules can be expressed in OWL 2. Accordingly, we extend our previous work in order to reason with ontologies and rules expressed in OWL 2 in Sect. 3. Section 4 demonstrates how our framework can reason with ontologies and rules. Next we discuss related works and finally conclude this work.

## 2    Extending Ontologies with Rules

### 2.1    SW Ontologies with Rules

Adding rules to ontologies expressed in OWL could be regarded as a step forward in SW research, as inferences can now be made upon SW ontologies; and many research proposals have been proposed ranging from hybrid to homogeneous approaches. The idea behind the former was that the predicates in rules and predicates in ontologies are made distinguished, and suitable interface between them is provided. The works in this direction are, such as, $\mathcal{AL}$-log language, an integration of Datalog and Description Logic $\mathcal{ALC}$ [7], CARIN, an integration of Datalog with different DLs [8], and an integration of OWL DL (or more precisely the DL $\mathcal{SHOIN}$) with normal rules under answer set semantics [9].

For the latter approaches both rules and ontologies are combined into a single logical language without making distinction between the rule predicates and the ontology predicates. Two examples are DLP [3] and SWRL [1]. In [3] DLP was proposed by combining DLs, the basis of ontology languages, with Logic Programs (LP), the basis of rule languages. It supports a bidirectional translation of premises and inferences between the fragment of DL and LP, and vice versa. This translation enables one to construct rules on top of SW ontologies. Later SWRL was proposed in [1] as a new language for an integration of rules and ontologies, in which OWL was extended with Horn-clause rules expressed in RuleML.

Recently SWRL became widely used for describing ontologies with rules. However, simple addition of the rules to ontologies leads to undecidability when

**Table 8.1** DL-FOL equivalence

| Expression | DL | FOL |
|---|---|---|
| subclassOf | $C \sqsubseteq D$ | $\forall x.C(x) \rightarrow D(x)$ |
| subpropertyOf | $P_1 \sqsubseteq P_2$ | $\forall x,y.P_1(x,y) \rightarrow P_2(x,y)$ |
| transitiveProperty | $P^+ \sqsubseteq P$ | $\forall x,y,z.(P(x,y) \wedge P(y,z)) \rightarrow P(x,z)$ |
| functionalPropery | $\top \leq 1\ P$ | $\forall x,y,z.(P(x,y) \wedge P(x,z)) \rightarrow y = z$ |
| inverseProperty | $P \equiv Q^-$ | $\forall x,y.P(x,y) \rightarrow Q(y,x)$ |
| intersectionOf | $C_1 \sqcap \ldots \sqcap C_n$ | $C_1(x) \wedge \ldots \wedge C_n(x)$ |
| unionOf | $C_1 \sqcup \ldots \sqcup C_n$ | $C_1(x) \vee \ldots \vee C_n(x)$ |
| complementOf | $\neg C$ | $\neg C(x)$ |

reasoning with SWRL ontologies. To retain decidability, some restrictions are put upon SWRL rules, and these restricted rules can be rewritten as a set of DL axioms using features introduced in $\mathcal{SROIQ}$. This technique was presented in [10]. As OWL 2 was developed based on DL $\mathcal{SROIQ}$, OWL 2 can therefore express rules in the form of DL axioms.

## 2.2 Expressing Rules in OWL 2

SW ontology languages such as DAML + OIL and OWL were developed based on DL $\mathcal{SHOIQ}$ [11]. $\mathcal{SHOIQ}$ provides a variety of constructors for building class expressions. The DL class expressions correspond to first order logic (FOL) which is used to formulate rules. Table 8.1 shows the correspondence between FOL formulae and the DL class expressions [3].

According to this correspondence, an FOL sentence can be expressed in DL as well as in DAML + OIL or OWL. For example, a rule of the form: `C(x)` $\wedge \neg D(x) \rightarrow E(x) \vee F(x)$ can be rewritten as a DL axiom: $C \sqcap \neg D \sqsubseteq E \sqcup F$, and a rule of the form: `C(x)` $\wedge$ `R(x,y)` $\rightarrow$ `E(x)` can be rewritten in DL as the axiom: $C \sqcap R.\top \sqsubseteq E$. However, with some rules such as:

(A) `hasParent(x,y)` $\wedge$ `hasBrother(y,z)` $\rightarrow$ `hasUncle(x,z)`
(B) `Man(x)` $\wedge$ `hasChild(x,y)` $\rightarrow$ `fatherOf(x,y)`.

There is no correspondence, so they cannot be rewritten as DL axioms, however to be able to do so we need some extra axioms in DL $\mathcal{SROIQ}$, these are some new features introduced in OWL 2.

OWL 2 which is based on DL $\mathcal{SROIQ}$ [4] supports the Role Inclusion Axiom (RIA), or namely the "property chain" axiom, and the Self Concept; these can be used to express more forms of rules, and the previous example.

The RIA is a construct of the form `R o S` $\sqsubseteq$ `T` where o is a binary composition operator. This is equivalent to an FOL formula: $\forall x,y,z.(R(x,y) \wedge S(y,z)) \rightarrow T(x,z)$. By adopting this, rule A can easily be rewritten as a DL $\mathcal{SROIQ}$ axiom:

`hasParent o hasBrother` $\sqsubseteq$ `hasUncle.`

The Self Concept allows one to express a "local reflexive" property, e.g. $R(x,x)$, in which a role $R$ relates an individual $x$ to itself. The Self Concept can be used to transform a property $R(x,x)$ into a class $C_R$ and vice versa, and this is due to a DL $\mathcal{SROIQ}$ axiom $C_R \equiv \exists R.Self$.

Therefore to derive the DL $\mathcal{SROIQ}$ axioms corresponding to rule B, we first transform a class Man into a property $P_{Man}$ by introducing an axiom Man $\equiv \exists P_{Man}.Self$, we then apply the previous RIA. As a result, rule B will be equivalent to the DL $\mathcal{SROIQ}$ axioms:

$$Man \equiv \exists P_{Man}.Self,$$
$$P_{Man} \text{ o has Child} \sqsubseteq fatherOf.$$

With the DL-FOL and DL $\mathcal{SROIQ}$-FL mappings we can express a rather wide range form of rules in DL axioms of OWL 2 and vice versa, if they satisfy certain restrictions [10]. In the next section we extend our previous framework [5] so that it can reason with ontologies and rules in OWL 2.

## 3  Our Meta-Logical Approach

### 3.1  Our Framework

Our framework forms a logical system consisting of meta-programs and an inference engine. The former is in the form of logical sentences representing a meta-level description of an SW ontology, i.e. the ontology described by OWL 2 is transformed into a meta-logical representation. The latter is a meta-interpreter, in the form of a demo (meta-)program, which is used to draw conclusions from the former. The meta-interpreter can also communicate to the Internet to obtain SW ontologies, communicate with the user to get SW information, and draw inferential consequences for the user.

In this paper our previous framework [5], which was designed to support reasoning with OWL 2 ontologies, is now enhanced with ability to reason with ontologies and rules expressed in OWL 2. The system can simply be illustrated in Fig. 8.1. In order to support rules, which are expressed by using the new features in OWL 2, the meta-program in our previous framework has to be extended with some new forms of meta-statements.

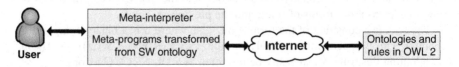

**Fig. 8.1** Our meta-logical system

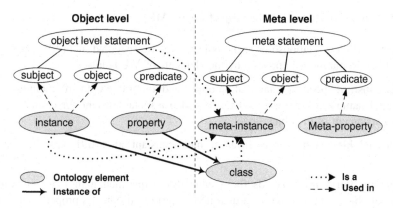

**Fig. 8.2**  Object level and meta level of ontology elements

To explain our framework, in the next three sub-sections we first introduce our meta-language for formulating the meta-programs of ontologies and rules, then explain the meta-programs and finally our meta-interpreter.

## 3.2   Meta-Languages for an OWL 2 Ontology

The language elements of an SW ontology are classes, properties, instances, and relationships between/among them described in the object level and the meta-level as depicted in Fig. 8.2. At the object level, an instance can be an individual or a literal of a domain, e.g. john, and property is a relationship between individuals, or is an individual's attribute, e.g. hasSon, type. At the meta-level, a meta-instance can be an individual name, a property name, a class, or an object-level statement. A meta-property is a property to describe a meta-instance's attribute or a relationship between/among meta-instances, e.g. reflexive.

According to SW convention, to make a unique name in an ontology, one qualifies it with a namespace as <namespace>:<name>, e.g. 'f':'son', 'f':'hasSon'. Henceforth this qualified name will be used throughout. Since names mentioned in an ontology are meta-language elements, this SW naming convention is supported at the meta-level. We therefore adopt :(namespace, name) as the function for name composition, the function is used for composing a qualified name, i.e. <namespace>:<name>.

In our framework, for an SW ontology we distinguish between its object and meta levels, and also its object and meta languages. The object language specifies objects and their relationships in the real world. The meta-language describes the syntactic form of the object language. As in our framework we develop a meta-interpreter to manipulate both the object and meta languages at the meta-level, we will formulate two meta-languages: one discussing objects and their relationships we call it "meta-language for the object level (**ML**)", and the other "meta-language for the meta-level (**MML**)," which discusses classes, instances, properties, and their relationships.

### 3.2.1   Meta-Language for the Object Level (ML)

Objects and their relationships at the object level are specified in an SW ontology and this information is expressed by the elements of **ML** below.

**Meta-constant** specifies a name of an object and a literal, e.g. 'son', including its qualified name, e.g. 'f':'son'. 'true' is also a meta-constant for expressing a true statement of the object-level.

**Meta-variable** stands for a different meta-constant at a different time, e.g. Person.

**Meta-function symbol** stands for a name (or a qualified name) of a relation between objects, or a name (or a qualified name) of an object's property—i.e. an object-level predicate name, such as 'f':'hasSon'. It also stands for other meta-level function symbols to represent logical connectives, e.g. ' ← ', ' ∧ ', '¬'.

**Meta-term** is either a meta-constant or a meta-variable or meta-function symbol applied to a tuple of meta-terms, e.g. 'f':'hasSon'('f':'fa','f':'son').

The meta-term of the form P(S,O), where P is an object-level predicate name, S and O are meta-constants or meta-variable, is used to express an object-level predicate (an object-level atomic statement) at the meta-level e.g. 'f':'hasSon'('f':'fa','f':'son'), and its negation is '¬P(S,O). For an object-level statement, it is represented by a meta-term of the form W ' ← ' 'true', where W is a term representing a positive or negative object-level predicate. Let all meta-variables appearing in this term be universally quantified.

**Meta-statement for the object level** reflects an object-level sentence to its existence at the meta-level. It has the form: *statement(object-level-sentence)*, e.g.

statement('f':'hasSon'('f':'fa','f':'son')' ← '"true').

### 3.2.2   Meta-Language for the Meta Level (MML)

Apart from the object language, in an SW ontology one also defines classes, properties, and their relationships, as well as class-instance relations, and we argue that this information is *meta-information of the object level language of the SW ontology*. Here we express this information by **MML** which includes:

**Meta-constant** specifying a name of an instance, a property, a class, a literal, including its qualified name. 'true' is also a meta-constant for expressing a true statement of the meta-level language.

**Meta-variable** standing for a different meta-constant at a different time.

**Meta-function symbol** standing for a logical connective, e.g. ' ← ', ' ∧ ', ' : ', '¬'; or a name of set operators applied on classes such as union; or a meta-predicate name being a name (or a qualified name) of a relation between entities; or a name (or a qualified name) of characteristic of a property, which may fall into one of the following categories:

*Class-class relations*: equivalent class of, etc.
*Class-instance relations*: instance of, class of, etc.
*Property-property relations*: property chain of, etc.
*Class-property relations*: keys, etc.
*Relations between literals and instances/classes/properties*: we can take these relations as attributes of instances, of classes, or of properties, e.g. comment.
*Characteristics of properties*: reflexive, asymmetric, etc.

**Meta-term** being either a meta-constant or a meta-variable or meta-function symbol applied to a tuple of meta-terms, e.g. 'f': 'fatherOf'('f':'f1','f':'s1'),'owl':'equiva-lentClass'(C,EC).

When a meta-term expresses a meta-level predicate stating a relation between entities, it has the form **Pred(Sub,Obj)**, and when it expresses a meta-level predicate stating a characteristic of a property, it has the form **Pred(Prop)**, where **Pred** is a meta-predicate name, **Sub**, **Obj**, and **Prop** (a property) are meta-constants or meta-variables. For the negation, it has the form '¬'**Pred(Sub,Obj)** or '¬'**Pred(Prop)**, where '¬' is a function symbol, e.g. '¬''rdf':'type'(I,C), where I is a meta-constant specifying an individual, and C is a meta-constant specifying a class.

The meta-term expressing a *meta-level sentence* is a term **Pred(Sub,Obj)** (or its negation) or **Pred(Prop)** (or its negation) or a logical-connective function symbol applied to the tuple of these terms. Let all meta-variables appearing in the meta-level sentence be universally quantified. One form of the sentence is a Horn-clause **meta-rule**, e.g.

> 'owl':'propertyDisjointWith'(P,DP) ' ← '
> 'owl':'propertyDisjointWith'(DP,P).

**Meta-statement** being a meta-predicate or meta-predicates connected by logical connective. It has two forms meta_statement(meta-level-sentence) and axiom(meta-level-sentence), the latter represents a rule for a mathematical axiom, e.g.:

> meta_statement('owl':'propertyDisjointWith'
>     ('f':'likes','f':'dislikes') ' ← ' 'true').
> axiom('owl':'propertyDisjointWith'(P,DP) ' ← '
>     'owl':'propertyDisjointWith'(DP,P)).

## 3.3   Meta-Programs for an Ontology with Rules

Each OWL2 ontology will be transformed into a meta-program containing a (sub-) meta-program expressed in **ML**, called **MP**, and a (sub-)meta-program expressed in **MML**, called **MMP**. Another meta-program expresses mathematical axioms for classes and properties in **MML** called **AMP** is also provided for the inference engine in order to reason with **MP** and **MMP**.

### 3.3.1 Meta-Program for the Object Level (MP)

**MP** contains meta-statements for the object level in the forms of `statement` `(P(S,O)'` ← "true'). This meta-program expresses information about instances and their relationships. In terms of the rule system, it can be understood as "facts". Here is an example:

> `statement('f':'hasFather'('f':'M02','f':'M01')'` ← "true').

In order to work with the *negative* property assertion in OWL 2, we use a meta-statement of the form `statement('¬'P(S,O)'` ← "true'), where P expresses (positive) property. For example, to declare person M01 is not the father of person M03, we will have a meta-statement in **MP**: `statement` `('¬"f':'fatherOf'('f':'M01', 'f':'M03')'` ← "true').

### 3.3.2 Meta-Program for the Meta Level (MMP)

**MMP** contains meta-statements for classes, properties, their relationships, and class-instance relations in terms of meta-rules. The **MMP** is represented in the following forms:

> `meta_statement(P(S,O)'` ← "true'),
> `meta_statement(P(S,Os)'` ← "true'),
> `meta_statement(C(Prop)'` ← "true'),

where P, S, O are predicate, subject, and object of a triple (S,P,O) defined in the ontology. C is a characteristic of a property Prop. Os is a tuple composing of several objects. Here are some typical examples:

*Some meta-statement about classes and their relationships:*

`meta_statement('rdfs':'subClassOf'(C,SC)'` ← ''true').
// *C is a sub-class of class SC.*
`meta_statement('rdfs':'equivalentClass'(C,EC)'` ← ''true').
// *Classes C and EC are equivalent.*
`meta_statement('owl':'disjoinwith'(C,DC)'` ← ''true').
// *Classes C and DC are disjoint.*
`meta_statement('owl':'intersectionOf'(C,Cs)'` ← ''true').
// *Class C is the result of intersection of all classes in Cs.*
`meta_statement('owl':'unionOf'(C,Cs)'` ← ''true').
// *Class C is the result of union of all classes in Cs.*
`meta_statement('owl':'complementOf'(C,CC)'` ← ''true').
// *Class C is the complement of class CC.*
`meta_statement('rdf':'type'(I,C)'` ← ''true').
// *I is an instance of class C.*

*Some meta-statements about properties and their relationships:*

```
meta_statement('owl':'inverseOf'(P,IP)' ← ' 'true').
```
*// Property P is an inversion of property IP.*
```
meta_statement('owl':'symmtric'(P)' ← ' 'true').
```
*// Property P is symmetric.*
```
meta_statement('rdfs':'domain'(P,D)' ← ' 'true').
```
*//The domain of property P is D.*

The new features in OWL 2 that referred to in Sect. 2 can be translated to the following meta-statements in **MMP**:

```
meta_statement('owl':'propertyChainOf'(P,[P1,P2])' ← "true').
```
        *// RIA: Property P is composition of properties P1, P2.*
```
   meta_statement('owl':'objectHasSelf'(C, Pc)' ← "true').
```

*// the Self Concept: C is a class of individuals which are related to themselves under role $P_C$.*

With such meta-statements we can transform DL rules into a meta-program. Here are examples of **MMP** that correspond to the rules we listed in Sect. 2.2:

Rule "$C \sqcap \neg D \sqsubseteq E \sqcup F$" is transformed into **MMP**:

```
meta_statement('rdfs':'subClassOf'(M,N)' ← ' 'true').
meta_statement('rdfs':'unionOf'(N, [E,F])' ← ' 'true').
meta_statement('rdfs':'intersectionOf'(M, [C,D'])' ← ' 'true).
meta_statement('rdfs':'complementOf'(D',D)  ' ← ' 'true').
```

**Rule** "*hasParent o hasBrother $\sqsubseteq$ hasUncle*" **is transformed into**
```
meta_statement('owl':'propertyChainOf'('f':'hasUncle',
            ['f':'hasParent','f':'hasBrother'])' ← ' 'true').
```

**Rule** "Man(x) ∧ hasChild(x,y) → fatherOf(x,y)" **is transformed into**
```
meta_statement('owl':'objectHasSelf'('f':'Man',P_Man)
                                        ' ← ' 'true').
meta_statement('owl':'propertyChainOf'
        ('f':'fatherOf', [P_Man,'f':'hasChild'])' ← ' 'true').
```

### 3.3.3   Meta-Program for the Axioms (AMP)

**AMP** contains axioms for classes and properties, they are expressed in the meta-rule form. In [5], we had several axioms in **AMP** to support for OWL. In order to work with ontologies and rules expressed in OWL 2, we *add more axioms* to manipulate with the new features in OWL 2, and an axiom for a set complement. Here we list all the axioms corresponding to the formulae in Table 8.1 and to the new features in OWL 2:

axiom(P(S,O)`← ' 'rdfs':'subPropertyOf'(SP,P) ∧ SP(S,O)).     (acsp)
// *Axiom to handle subproperty formulae.*

axiom(P(S,O)`← ' 'owl':'transitive'(P∧P(S,O1) ∧ P(O1,O)).     (actp)
// *Axiom to handle the transitive property*

axiom(P(S,O)`← '                                               (acfp)
'owl':'functional'(P) ∧ P(S,O1) ∧ 'owl':'sameAs'(O,O1)).
// *Axiom to handle the functional property*

axiom(P(S,O)`← ' 'owl':'inverseOf'(P,IP) ∧ IP(O,S)).          (acip)
// *Property IP is an inverse property of P.*

axiom(P(S,O)`← '                                               (acpc)
'owl':'propertyChainOf'(P,[P1,P2]) ∧ P1(S,O1) ∧ P2(O1,O)).
// *Axiom to handle the chain property.*

axiom(P(S,S)`← '                                               (acsc)
'owl':'objectHasSelf'(C,P) ∧ 'rdf':'type'(S,C)).
// *Axiom to handle the Self Concept.*

axiom('rdf':'type'(I,C)`← '                                    (acic)
'owl':'intersectionOf'(C,Cs) ∧ 'intertype'(I,Cs)).
'intertype'(I,[H|T])`← ' 'rdf':'type'(I,H) ∧ 'intertype'(I,T).
// *Axiom to handle a set intersection.*

axiom('rdf':'type'(I,C)`← '                                    (acuc)
'owl':'unionOf'(C,Cs) ∧ 'unionType'(I,Cs)).
'unionType'(I,[H|T])`← ' 'rdf':'type'(I,H).
'unionType'(I,[H|T])`← ' 'unionType'(I,T).
// *Axiom to handle a set union.*

axiom('¬''rdf':'type'(I,C)`← '                                 (accc)
'owl':'complementOf'(C, Cc) ∧ 'rdf':'type'(I,Cc)).
// *Axiom to handle a set complement.*

## 3.4   The Meta-Interpreter

The meta-interpreter in our framework is constructed for reasoning with the meta-programs **MPs**, **MMPs**, and **AMPs**. It is defined by a demo predicate of the form demo(A). With this predicate we can infer the answer A from the meta-programs. Our meta-interpreter adapts the Vanilla meta-interpreter in [12] in order for reasoning with the meta-programs transformed from ontologies and rules where we have defined three kinds of meta-level statements: (1) statement(A ← B) for the object-level of an ontology, (2) meta_statement(A ← B) for the meta-level of an ontology (including rules), and (3) $\texttt{axiom(A} \leftarrow \texttt{B)}$ for the mathematical axioms. The definition of $\texttt{demo/1}$ is:

```
demo('true').                                      (true)
demo(A' ∧ 'B) ← demo(A) ∧ demo(B).                 (conj)
demo(A) ← statement(A' ← 'B) ∧ demo(B).            (ost)
demo(A) ← meta_statement(A' ← 'B) ∧ demo(B).       (mst)
demo(A) ← axiom(A' ← 'B) ∧ demo(B).                (ast)
```

The first clause (true) is the basic case for proving that an atom is true. The second clause (conj) is used for proving a conjunction goal. Three last clauses (ost), (mst), and (ast) are used for interpreting three meta statements of the three meta-programs **MP**, **MMP**, and **AMP** respectively.

## 4  Query Answering in Our Framework

In our framework ontologies and rules expressed in OWL 2 are transformed into meta-programs, which are formed by three sub meta-programs **MP**, **MMP** and **AMP**. The meta-programs are used as inputs for the meta-interpreter which is implemented in Prolog. The meta-interpreter is an inference engine for reasoning with meta-programs to derive conclusions.

The family ontology [13] is used as an example to demonstrate our framework. Due to its lack of rules, three OWL2 rules are added to it. After the ontology is transformed into meta-programs, some parts of them are:

- The **MP** program

```
statement('f':'hasParent'('f':'M02','f':'M01')' ← ''true').    (1)
statement('f':'hasParent'('f':'F02','f':'M01')' ← ''true').    (2)
statement('f':'hasParent'('f':'M03','f':'F02')' ← ''true').    (3)
```

- The **MMP** program

```
meta_statement('rdf':'type'('f':'M01','f':'Man')' ← ''true').(1')
meta_statement('rdf':'type'('f':'M02','f':'Man')' ← ''true').(2')
meta_statement('rdf':'type'('f':'F02','f':'WoMan')        (3')
                                     ' ← ''true').
meta_statement('owl':'complementOf'('f':'Man','f':'WoMan')  (4')
                                     ' ← ''true').
meta_statement('owl':'unionOf'                              (5')
               ('f':'Human', ['f':'Man','f':'WoMan'])'←''true').
```

The first rule $\texttt{hasParent(x,y)} \wedge \texttt{hasParent(z,y)} \rightarrow \texttt{siblingOf}$ $\texttt{(x,z)}$ added is expressed by $\texttt{hasParent o hasParent}^- \sqsubseteq \texttt{siblingOf}$

DL axiom, where `hasParent⁻` is the inverse property of `hasParent`. This is transformed into the following meta-statements in **MMP**:

```
meta_statement('owl':'inverseOf'                                    (6')
        ('f':'parentOf','f':'hasParent') '←' 'true').
meta_statement('owl':'propertyChainOf'('f':'siblingOf',             (7')
   ['f':'hasParent','f':'parentOf']) '←' 'true').
```

The next one `Man(x) ∧ siblingOf(x,y) → brotherOf(x,y)` is transformed into the meta-statements in **MMP**:

```
meta_statement('owl':'objectHasSelf'                                (8')
                        ('f':'Man',P_Man) '←' 'true').
meta_statement('owl':'propertyChainOf'                              (9')
        ('f':'brotherOf',[P_Man,'f':'siblingOf']) '←' 'true').
```

The third rule `brotherOf(x,y) ∧ parentOf(y,z) → uncleOf(x,z)` is transformed into the following meta-statements in **MMP**:

```
meta_statement('owl':'propertyChainOf'('f':'uncleOf',              (10')
                ['f':'brotherOf','f':'parentOf']) '←' 'true').
```

Now we can pose some queries to the meta-interpreter to get answers:

```
?- demo('rdf':'type'('f':'M02',X)).                                (q1)
   X = 'f':'Man';
   X = 'f':'Human'.
```
*//1st answer is supported by (mst), (1'), and (true), and 2nd answer is supported by (ast), (acuc), (conj), (5'), (1'), and (true).*
```
?- demo('¬''rdf':'type'('f':'F02',X)).                             (q2)
   X = 'f':'Man'.
```
*//The adopted clauses are (ast), (accc), (conj), (mst), (4'), (3') and (true).*
```
?- demo('f':'siblingOf'('f':'M02',X)).                             (q3)
   X = 'f':'F02'.
```
*//The adopted clauses are (ast), (acpc), (conj), (7'), (true), (ost), (1), (mst), (acip), (6'), and (2).*
```
?- demo('f':'brotherOf'(X, 'f':'F02')).                            (q4)
   X = 'f':'M02'.
```
*//The adopted clauses are (ast), (acpc), (conj), (9'), (true), (acsc), (8'), (mst), (2'), as well as the clauses adopted for answering q3.*
```
?- demo('f':'uncleOf'('f':'M02',X)).                               (q5)
   X = 'f':'M03'.
```
*//The adopted clauses are (ast), (acpc), (conj), (10'), (true), as well as the clauses adopted for answering q4, together with (3).*

## 5 Related Works

We now look at other approach that enhances ontologies with rules. Grosof et al.[3] proposed the DLP; this approach supports bi-directional translation between logical sentences from DLP fragment of Description Logic and logic programs. Every concept referred to in an ontology is mapped into a unary relation with a concept name becoming a name of the relation and an individual name becoming an argument. Every instance-property-instance relationship is mapped into a binary relation, concepts and property constructor statements are converted into rules. The distinction between this approach and ours is the following.

Firstly this approach was designed to support a subset of DAML + OIL, and provides only a mapping from RDFS and DAML + OIL to logic programs, but does not support the new features in OWL 2, such as the property chain axiom and the Self Concept. Secondly, this approach has a weakness when representing an ontology in a logic program. For example, to represent the statement "a is union of $b_1$, $b_2$, ..., $b_n$", it requires n number of rules, i.e. $a(X):-b_1(X)$, ..., $a(X):-b_n(X)$. However, in our representation, this requires only one statement, i.e. our (acuc) axiom, which is more compact.

More importantly, their representation of logic program is at the object level only. Thus, the names of concepts, or the names of roles, of an ontology which are meta-terms cannot be accessed and reasoned from their logic programs and therefore cannot be queried by an inference engine. For example, in their representation, statement (1) and (1') in Sect. 4 would be presented as facts hasParent ('f':'M02','f':'M01') and Man('F':'M01'). With these we can ask only a question who is a man or who is a parent of 'M02', but it is impossible to get answers to a question like which class 'M01' is an instance of, or what a relationship between 'M02' and 'M01' is. This is because a predicate name is a meta-level information that cannot be reasoned and queried at the object-level by a Prolog interpreter. However, in our approach since we separate the meta-level from the object-level knowledge in an ontology, such queries can be asked and be answered via the demo predicate, that is:

```
?-demo(P('f':'M02','f':'M01')).
  P='f':'hasParent'.

?-demo('rdf':'type'('f':'M01','f':X)).
  X='Man'.
```

## 6 Conclusion

In this paper we have presented a meta-logical framework for reasoning with ontologies and rules expressed in OWL 2. The logical system of our framework consists of meta-programs transformed from ontologies and rules expressed in

OWL 2, and an inference engine defined by a demo predicate with some new extra auxiliary axioms proposed in the paper.

**Acknowledgment** We gracefully acknowledge the financial support for this research from the Japan International Corporation Agency (JICA) under the AUN/SEED-Net Program.

# References

1. Horrocks I, Patel-Schneider PF (2004) A proposal for an OWL rules language. In: Proceedings of the 13th international conference on the WWW, ACM
2. Motik B, Sattler U, Studer R (2005) Query answering for OWL-DL with rules. J Web Semant Sci Serv Agents World Wide Web 41—60
3. Grosof BN, Horrocks I, Volz R, Decker S (2003) Description Logic Programs: combining logic programs with Description Logic. In: Proceedings of the 12th international conference on the WWW, ACM, pp 48—57
4. Horrocks I, Kutz O, Sattler U (2006) The even more irresistible $\mathcal{SROIQ}$. In: Proceedings of the 10th international conference on principles of knowledge representation and reasoning. AAAI Press, pp 57—67
5. Hirankitti V, Mai TX (2010) A Meta-logical approach for reasoning with an OWL 2 ontology. In: Proceedings of the 2010 IEEE international conference on computer sciences: research, innovation & vision for the future, Vietnam, pp 35—40
6. Hirankitti V, Mai TX (2011) A meta-logical approach for reasoning with ontologies and rules in OWL 2. In: Proceedings of the world congress on engineering 2011, WCE 2011. Lecture notes in engineering and computer science. 6—8 July, 2011, London, UK, pp 960—965
7. Donini FM, Lenzerini M, Schaerf A (1998) $\mathcal{AL}$-log: integrating Datalog and Description Logics. J Intell Inform Syst 10:227—252
8. Levy AY, Rousset M (1998) Combining Horn rules and Description Logics in CARIN. Artif Intell 165—209
9. Eiter T, Lukasiewicz T, Schindlauer R, Tompits H (2004) Combining answer set programming with Description Logics for the semantic web. In: Proceedings of the ninth international conference of principles of knowledge representation and reasoning (KR2004), pp 141—151
10. Gasse F, Sattler U, Haarslev V (2008) Rewriting rules into $\mathcal{SROIQ}$ axioms. Poster at 21st international workshop on DLs
11. Patel-Schneider PF, Hayes P, Horrocks I (2004) OWL Web Ontology Language: semantics and abstract syntax, W3C recommendation, http://www.w3.org/TR/owl-semantics/
12. Kowalski RA, Kim JS (1991) A metalogic programming approach to multi-agent knowledge and belief. In: AI and mathematical theory of computation, pp 231—246
13. The family ontology http://www.owldl.com/ontologies/family.owl

# Chapter 9
# Bayesian Approach to Robot Group Control

Tomislav Stipancic, Bojan Jerbic, and Petar Curkovic

## 1 Introduction

Robots in their essence have the purpose to replace human labor, not only in industrial applications but also in other human activities. There are many applications connected to robotics, as for example in medicine, rescue operations, research, aiding the disabled people, etc. In order to fulfill the requirements of everyday life, robotic systems are inevitably becoming more and more sophisticated. The level of complexity demands novel or different research perspectives to be considered.

Ubiquitous computing (Ubicomp) [1] is a post-desktop model of human–computer or computer–computer interaction in which information processing has been thoroughly integrated into everyday objects and activities [2]. The environment then becomes a space constantly analyzed by devices in order to detect significant changes that can trigger the system to react, depending on its original functionality. Ubicomp applications are normally envisioned to be sensitive to context, where context can include an object's location, activity, goals, resources, state of mind, and nearby people and things. Ubicomp involves many different research areas, e.g., distributed computing, mobile computing [3], sensor networks, human–computer interaction, artificial intelligence [4], etc.

In an automatic assembly, the control of a system is usually connected to the control of the working environment. An uncontrolled situation is any situation where any object or subject is not completely defined from the aspects of position, orientation, action, and/or process. Every environment is naturally unstructured, which can be revealed if it is observed under a fine enough scale. In other words, it is not possible to completely determine any environment, no matter how tight the applied

T. Stipancic (✉) • B. Jerbic • P. Curkovic
Faculty of Mechanical Engineering and Naval Architecture, University of Zagreb,
Ivana Lucica 5, 10000 Zagreb, Croatia
e-mail: tomislav.stipancic@fsb.hr; bojan.jerbic@fsb.hr; petar.curkovic@fsb.hr

S.-I. Ao and L. Gelman (eds.), *Electrical Engineering and Intelligent Systems*, 109
Lecture Notes in Electrical Engineering 130, DOI 10.1007/978-1-4614-2317-1_9,
© Springer Science+Business Media, LLC 2013

tolerance ranges may be. Unconstrained environment is usually introduced to the system through the application of tolerances. This is connected with issues of sensitivity and instability and may result in malfunctioning, even if small environmental changes occur [5]. For such reasons, the system that cannot perceive the environment has to be programmed for a limited range of actions foreseen in advance by the system developer. By default, it cannot act in any unpredicted situation. In the best scenario, it could send a signal that an unpredicted situation had occurred. Dynamic information control, based on the contextual perception of the environment, needs less predetermined operational and structural knowledge. That requires system adaptation skills and some level of decision-making capabilities.

Each object, process or condition is unique by its very nature. Therefore, the context of space and time becomes an important task in autonomous system development. If an agent is able to make decisions about an action that is not completely restrained in the workspace, using certain perceptions, knowledge and intelligence, it can be said that the system is controlled [6]. In order to get an automatic system controlled, the corresponding knowledge about all relevant components and processes must exist. For this reason, the research of new methodologies and paradigms is directed toward the development of adaptive, anthropmatic, and cognitive agent capabilities [7].

A model that relies on ontology for defining an industrial assembly/disassembly domain, description logic reasoning (DL Reasoning) for planning an adaptive behavior and bayesian network reasoning (BN Reasoning) for probabilistic action planning can enable adaptive and autonomous behavior for all agents in the group. By utilizing the proposed model, a group of robots would become capable to convert their ordinary environment to a ubiquitous one. Behavioral patterns produced by the model can in their essence be compared to one that uses living beings for contextual understanding.

## 2 The Model Overview

The transition from free to controlled/known spatial state is one of the most demanding tasks in automatic processes. A programmer that tries to control the system has to cope with many certain and uncertain situations. Although it is possible, it is very hard to model a complex system to predict all possible outcomes that the environment is able to produce due to its nondeterministic nature. A system designed in such a manner can be called reactive because it reacts to environmental stimulus. The reactive system can be very fragile if something unexpected occurs, because it usually does not have self-recovery capabilities that would, by default, prevent errors arising from unexpected situations. On the contrary, the system that is able to realize a context of an environment can act depending on contextual information. Such a system can potentially do both: It can act reactively and it can comprehend the present and predict results of its future actions. The contextual

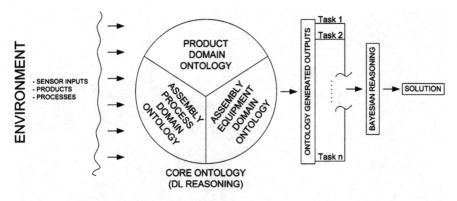

**Fig. 9.1** Cognitive—behavioral model for robot group control

perception implies understanding of a problem domain much broader than a single agent could feel and it can be carried out with the interaction between the agent and the environment together with related objects, other agents, processes, and events.

Figure 9.1 shows the proposed cognitive—behavioral model for robot group control. As it can be seen in the figure, the model tends to be used in industrial environments for robot assembly/disassembly operations.

The model includes the following components: information gathering by means of sensors integrated into the environment, domain ontology together with DL, and BN Reasoning mechanisms used for finding the solution, depending on the ontology generated outputs. The synergy of all model components can ensure the adaptive behavior of the robot working in the group. By predicting behavioral patterns of other robots in the group, a single robot becomes capable to plan its own behavior.

Another challenge is to develop a collaborative robot group work in real-life scenarios. By using probabilities, it is possible to give an agent the capability to behave in seemingly non-predictable scenarios. First of all, the agent has to learn behavioral patterns of other agents by means of BN with respect to DL Reasoning and previously or simultaneously collected stimulus coming from the environment. After the learning phase, the agent can become capable of predicting actions coming from other agents and depending on its DL Reasoning mechanisms.

## 2.1   Environment

How, when, and which data to collect are important questions for all further steps leading to contextual perception of the environment. Answers depend mainly on application goals and require a thorough analysis to be made. The analysis should reveal spatial and temporal dependencies and characteristics of processes, equipment, and objects. Information obtained by sensors is expected to change constantly. Figure 9.2 shows the environment for model development and testing.

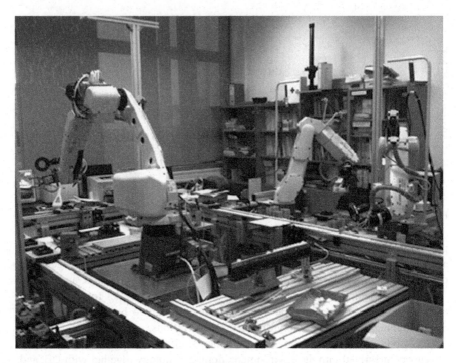

**Fig. 9.2** The environment used for model developing and testing

The used environment is characteristic for assembly/disassembly industrial applications, and it contains different types of sensors for providing an information flow for decision-making processes with respect to the context. In robotic/automatic assembly, information about the position and orientation of work pieces and all other relevant in-process objects and processes represents essential data for program control. The change in information is a consequence of different environmental conditions, for example—change of position of the part to be assembled or change in information about the available part quantity. This can be called expected informational change. The system has to decide only how to use information gathered autonomously with respect to used ontology.

## 2.2  Core Ontology

Ontologies denote a formal representation of entities (classes) along with associated attributes (objects) and their mutual relations [8]. Since ontologies allow representation of an arbitrary domain and can simplify work for end users, they have been proven to be extremely convenient. With time, ontologies have also been proven to have certain disadvantages. By designing a certain domain coming from his own field of expertise, an expert uses personal knowledge and impressions, which can

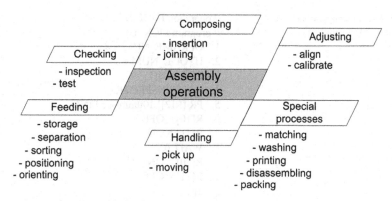

**Fig. 9.3**  Assembly operations taxonomy represents the part of Core Ontology

differ when compared with other experts in the field. Such an approach can disable other ontologies representing the same domain to exchange information mutually. This means that gaps, overlaps, and inconsistencies will continue to exist when independently developed ontologies are used together [9].

Projects MASON [10] and ONTOMAS [11] represent an endeavor of researchers to build ontology for describing the field of production activities. The design process of the Core Ontology was guided by a concern about integration with other similar ontologies because the ontology developed through ONTOMAS project and the Core Ontology uses the same integrated assembly model for domain knowledge description, originally proposed by Rampersad [12]. Figure 9.3 shows a part of the Core Ontology and represents the taxonomy of assembly operations.

The next step after the taxonomy definition is to transform it to ontology by using DL. The developed ontology can then be integrated into the Core Ontology along with all other domain definitions.

The model is designed in a way that every assembly operation represents an algorithm for controlling the robot. For example, the "pick up" operation is a part of "handling" the group and contains predefined actions for picking the part previously seen by means of the vision system. Every algorithm has its own simple program. In this particular application programs are written for robots produced by FANUC Robotics, as it is shown in Fig. 9.4.

An agent, or let us say a robot, would start the programs depending on the current contextual information derived from interactions between all domain components respecting the Core Ontology. One of the most important features is the model's ability to support decision-making processes. A well-defined Core Ontology contains taxonomies with corresponding decision-making mechanisms. DL Reasoning would assure a correct order of the program executions for reactive robot group behavior.

To design the Core Ontology along with all corresponding components (taxonomy, DL, etc.) it is used OWL-DL [13]. OWL-DL is one of the OWL [14] dialects that support knowledge sharing and reuse, which can be very important if we want to add new knowledge to the model.

**Fig. 9.4** An example of an algorithm for a simple pick-up operation

```
// operation PICK_UP
// program PICK_UP
1.  UFRAME_NUM = 8
2.  UTOOL_NUM = 2
3.  R[1] = DI[1]
4.  CALL INSPECTION
5.  PR [R[2]] 100mm/sec FINE
6.  R[10] = OFF
7.  R[11] = ON
8.  WAIT 0.1
9.  R[10] = ON
10. R[11] = OFF
[End]
// operation INSPECTION called from the
// PICK_UP operation program INSPECTION
1.  VISION RUN_FIND 'OBSERVER'
2.  VISION GET_OFFSET 'OBSERVER' VR[1]
3.  R[2] = VR[1].MES[1]
R[End]
```

**Fig. 9.5** A partial interpretation of the working environment

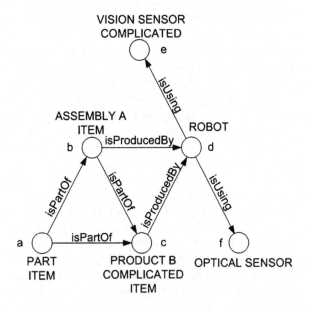

Figure 9.5 shows the interpretation designed to define a part of the working environment characteristic for industrial assembly applications. Such interpretations have to contain all relevant components along with their corresponding properties. The interpretation is presented with a set of mutually connected nods. Typically, nodes are used to characterize concepts, i.e., sets or classes of individual objects, and links are used to characterize relationships among

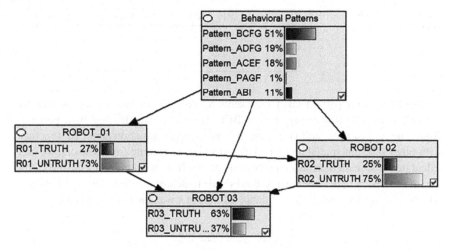

**Fig. 9.6** BN along with corresponding stages for an environment with three robots

them [15]. So-called atomic concepts are represented by literals a, b, . . ., f and include other classes and individuals. For example, object PRODUCT B is a member of three other classes (c, COMPLICATED, and ITEM). DL enables to produce other classes (concepts) derived from interpretations by using, so-called DL descriptions. Such complex descriptions can contain other concepts (classes), roles (properties), and constructors. The same DL methodologies are used within Protege-OWL editor [17, 18] to derive proper system responses according to the information gathered from the working environment.

## 2.3   Bayesian Network Reasoning

Probability techniques can help the system work in uncertain situations or scenarios. With respect to the proposed model, DL Reasoning can provide more than one possible solution, which can enable indecisive behavior of robots. In a nondeterministic world, the deterministic way of seeing the world is often not expressive enough to address real-world problems [16]. Mathematically, a Bayesian Network is a directed acyclic graph in which a set of random variables makes up the nodes in the network. A set of directed links connects pairs of nodes, and each node has a conditional probability table that quantifies the effects of parents on it.

The ontology can produce ambiguities in its solutions by suggesting more than one solution for solving the task. BN can be used in determining the best option with respect to current stimulus coming from sensors placed in the environment. Within the BN nodes represent each of the variables. A single nod in the network contains a set of probable values called states for each variable. Figure 9.6 shows one BN along with corresponding stages for Behavioral Patterns and three robots.

**Table 9.1** Table of influences

|              | BP  | ROBOT 01 | ROBOT 02 | ROBOT 03 |
|--------------|-----|----------|----------|----------|
| ROBOT 01     | +   |          |          |          |
| ROBOT 02     | +   | +        |          |          |
| ROBOT 03     | +   | +        | +        |          |

The behavioral pattern (BP) represents a sequence of elementary operations derived from DL Reasoning, e.g., ABCD sequence can denote: A-inspection, B-pick up, C-insertion, and D-testing. By gathering information about BP together with other relevant spatial and temporal information coming form sensors, we can calculate the so-called posterior probability $p(x|y)$. If we want to infer a quantity $x$ by means of sensory data $y$, we can use Bayes' rule to determine the inverse probability, which specifies the probability of data $y$ assuming $x$ was the case, as in (9.1).

$$p(x|y) = \frac{p(y|x) \times p(x)}{p(y)} \tag{9.1}$$

In robotics, this inverse probability is often coined "generative model," since it describes, at some level of abstraction, how state variables $X$ cause sensor measurements $Y$ [16].

Terms a prior and a posterior are usually used in philosophy to distinguish two types of knowledge, justification or arguments. A priori knowledge or justification is independent of experience as opposed to a posteriori knowledge which makes reference to experience and can be used for decision making processes [17].

By the given environmental conditions, the ontology can suggest a set of different behavioral patterns. The second problem is very interesting and it can be seen in mutual interactions of the robots belonging to the group. Collisions are more than certain because all robots use the same ontology for planning reactive behaviors. For example, if the ontology offers the group the behavioral pattern ACDG, which indicates a sequence of operations for solving the current task, the question is how to decide which robot should carry out the task. The Cognitive—behavioral model for robot group control uses BN to solve the riddle. By anticipating the next steps of other robots in the group, a single robot can plan its future actions. Before they are ready, robots in the group should take the learning stage. At the beginning, the group should contain only one member to react in the manner derived from the ontology. Before joining the group, the second robot should learn how to predict behavioral patterns of the first robot by observing it. If ontology proposes a couple of behavioral patterns for a certain action scenario, the second robot should determine its behavioral pattern respecting the ontology solutions and the pattern used by the first robot. The process of integration is the same for every new robot joining the group with a small difference. The new robot should observe both the behavioral pattern used by the first and by the second robot (Fig. 9.6). The table of influences (Table 9.1) shows the mutual influences between all system components.

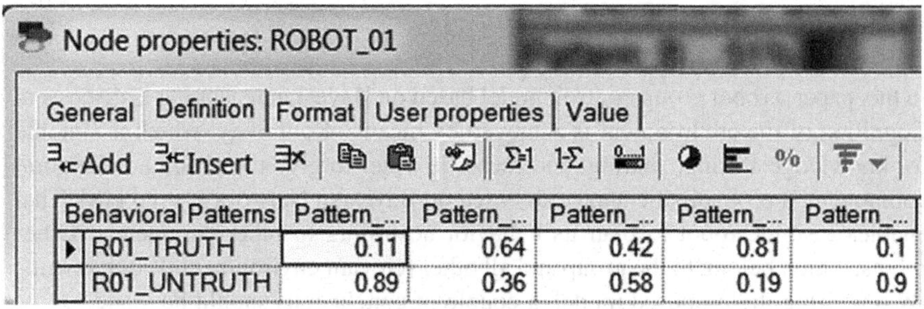


| Behavioral Patterns | Pattern_... | Pattern_... | Pattern_... | Pattern_... | Pattern_... |
|---|---|---|---|---|---|
| R01_TRUTH | 0.11 | 0.64 | 0.42 | 0.81 | 0.1 |
| R01_UNTRUTH | 0.89 | 0.36 | 0.58 | 0.19 | 0.9 |

**Fig. 9.7** CPT table shows related conditional probabilities between robot 01 and BP

```
Capture ontology generated outputs
Call Bayesian Reasoning Mechanism
Check for pattern number
if (pattern_number > 0) then
    do while not goal
    Use the pattern with highest probability
endif
if (pattern_dettected == true and pattern_number > 0) then
  do while not goal
  Set robot to execute the pattern
  Update Bayesian Network
endif
```

**Fig. 9.8** Algorithm—the working procedure for a single robot joining the group

The probability of any node belonging to BN is described using condition probability table (CPT). Probabilities on a particular node are affected by the state of other nodes depending on prior information about the relationships. Figure 9.7 shows the CPT table together with related conditional probabilities between Robot 01 and BP.

Information about conditional probabilities for each robot has to be defined in advance. By altering this information, the system designer gets the opportunity to define system priorities and to achieve certain goals of assembly processes. By using this knowledge, each robot should be able to predict its next actions. The last robot joining the group has the most information about conditional probabilities, because it depends on choices of all other robots.

Figure 9.8 shows the algorithm that is used for describing a procedure that a single robot has to take while determining its further actions respecting the proposed and previously utilized behavioral patterns.

## 3  Conclusions

In this paper a robot group control model based on Bayesian reasoning is discussed. Cognition of the environment is achieved by means of ontology, which is suitable for knowledge storing, sharing, and reuse. In order to give the group a behavioral component, a probabilistic approach based on Bayesian Network is used [18]. That enables a single robot to plan its behavior according to other parameters: Other robots, which belong to the group and all other relevant environmental components. By observing the proposed model, a couple of conclusions should be made.

Deterministic chaos inevitably obstructs absolute expectations, always producing slightly changed situations. To alter uncertain situation, conventional automation methods tend to create technical systems as almost perfect constructions. It seems that such efforts are definitely hopeless and result in expensive and inefficient systems. Such an approach raises more issues that affect almost every contemporary industrial factory in the world: a lack of space and rigidity of the production systems. These problems are even more prominent in Europe. Making industrial systems adaptive, small, cheap, and competitive with the rest of the world is a question that appeared many years ago. Indeed, the system based on contextual perception of the environment can be converted to work on other similar tasks relatively easily compared to classical industrial production lines. A robot with such properties must be able to interpret and understand the context of the environment in order to adapt its strategies to effectively work in a group. The Cognitive-behavioral model for robot group control enables some degree of contextual understanding and provides a way for the robot to plan its actions by observing other robots. Probabilistic reasoning based on BN can be used to increase the level of security by learning and anticipating behavioral patterns of all other robots. Such an approach could also increase the overall security level of the system.

The development of the proposed model has shown the contribution especially in the BN Reasoning part of the model. The model is still in the development phase, and some difficulties are expected to be found. A problem connected with ontology and BN mapping is obvious and has to be solved in the future.

There are a few more things that can be analyzed and implemented into the original model. It is possible to identify certain uncertainties in other parts of the Core Ontology. For example, by altering the assembly process domain ontology it is possible to gain certain production goals by trying to optimize particular production parameters.

Deterministic chaos should be accepted as a natural phenomenon and the development philosophy changes toward the development of intelligent machines capable of adapting their behavior according to the natural imperfect world where nothing is absolutely ideal or accurate.

**Acknowledgment** Authors would like to acknowledge the support of Croatian Ministry of Science, Education and Sports, through projects No.: 01201201948-1941, Multiagent Automated Assembly and joint technological project TP-E-46, with EGO-Elektrokontakt d.d.

# References

1. Wikipedia [online] (2012) Available: http://en.wikipedia.org/wiki/ubiguitous_computing
2. Mark weiser's Home Page [online] (2012) Available: http:/www.ubi2.com/hypertext/weiser/ubihome.htm
3. Curkovic P, Jerbic B, Stipancic T (2009) Swarm-based approach to path planning using honey-bees Mating algorithm and ART neural network. Solid state phenomena, vol 147-149, pp 74–79
4. Curkovic P, Jerbic B, Stipancic T (2008) Hybridization of adaptive genetic algorithm and ART neural architecture for efficient path planning of a mobile robot. Trans FAMENA 32(2):11–21
5. Stipancic T, Jerbic B (2010) Self-adaptable vision system. Emerging Trends in Technological Immovation. Springer, Heidelberg, pp 195-202
6. Stipancic T, Curkovic P, Jerbic B (2008) Robust autonomous assembly in environment with relatively high level of uncertainty. MITIP 2008. Czech Republic Bily slon, s.r.o. Press, pp 193–198
7. Svaco M, Sekoranja B, Jerbic B (2011) Autonomous planning framework for distributed multiagent robotic systems. Technological innovation for sustainability. Springer, Heidelberg, pp 147–154
8. Gruber TR (1994) Toward principles for the design of ontologies used for knowledge sharing. Int J Hum Comput Stud 43:907–928
9. Costa Paulo C G, Laskey Kathryn B, Alghamdi G (2006) Bayesian ontologies in AI systems. In: Proceedings of the fourth Bayesian modelling applications workshop, held at the twenty second conference on uncertainty in artificial intelligence. Cambridge, MA, USA
10. Lemaignan S, Siadat A, Dantan J Y, Semenenko A (2006) MASON: a proposal for an ontology of manufacturing domain. Distributed intelligent systems: collective intelligence and its applications, IEEE workshop. pp 195–200
11. Lohse N (2006) Towards an ontology framework for the integrated design of modular assembly systems. PhD thesis, university of Nottingham.
12. Rampersad HK (1994) Integrated and simultaneous design for robotic assembly. Wiley, Chichester. ISBN 0-471-95018-1
13. Al-Safi Y, Vyatkin V (2007) An ontology-based reconfiguration agent for intelligent mechatronic systems. Holonic and multi-agent systems for manufacturing. Lecture notes in computer science, vol 4659/2007, pp 114–126
14. World wide web consortium (W3C) [Online] (2012) Available: http://www.w3.org/standards/techs/ow/
15. Larik A, Haider S (2010) Efforts to blend ontology with Bayesian networks: an overview. In: Proceedings of 3rd international conference on advanced computer theory and engineering. Chendgu, China
16. Thurn S, Burgard W, Fox D (2006) Probabilistic Robotics. MIT, Cambridge
17. Wikipedia [online] (2012) Available: http://en.wikipedia.org/wiki/A_priori_and_a_posteriori
18. Stipancic T, Jerbic B, Curkovic P (2011) A robot group control based on Bayesian reasoning. In: Proceedings of the world congress on engineering 2011, WCE 2011. Lecture notes in engineering and computer science. London, UK, 6–8 July 2011, pp 1056–1060

# Chapter 10
# A Palmprint Recognition System Based on Spatial Features

**Madasu Hanmandlu, Neha Mittal, and Ritu Vijay**

## 1 Introduction

Any human can be recognized based on his/her physiological or behavioral characteristics [1]. Each characteristic should be universal, distinct, time invariant, and easily collectable. The palmprint as a biometric modality is gaining acceptance in the field of biometrics. As compared to other biometric modalities, it is bestowed with enormous information (i.e., Wrinkles, ridges, and principal lines) serving as its discriminating power. We will discuss a few important contributions made on this modality.

A new method of extracting features from palmprints is proposed in [2], using the competitive coding scheme and angular matching. The competitive coding scheme uses multiple 2-D Gabor filters to extract the orientation information from palm lines. The competitive code constitutes the feature vector. In [3] a low-cost multispectral palmprint system capable of acquiring high-quality images is developed to operate in real time. The palmprint images in the visible and NIR spectra are collected. The information provided by the multispectral palmprint images are fused at the score level. The results are improved after fusion.

M. Hanmandlu (✉)
Electrical Engineering Deparment, Indian Institute of Technology, New Delhi, India
e-mail: mhmandlu@gmail.com

N. Mittal
Electronics & Communication Engineering Department, J.P. Institute
of Engineering & Technology, P.O. Rajpura, Mawana Road, Meerut 250001, India
e-mail: nehamittal2008@gmail.com

R. Vijay
Electronics and Communication Engineering Department, Banasthali Vidyapeeth,
Banasthali University, Rajasthan, India
e-mail: rituvijay1975@yahoo.co.in

S.-I. Ao and L. Gelman (eds.), *Electrical Engineering and Intelligent Systems*,
Lecture Notes in Electrical Engineering 130, DOI 10.1007/978-1-4614-2317-1_10,
© Springer Science+Business Media, LLC 2013

An indexing algorithm is proposed in [4] based on the palmprint classification. This algorithm uses a novel representation involving a two-stage classifier that provides the even-distributed categories. The representation scheme is directly derived from the principal line structures. This scheme does not use wrinkles and singular points and is capable of tolerating poor image quality.

Matching of palmprints dealt in [5] explores the feasibility of identifying a person based on a set of features extracted along the prominent palm lines (and the associated line orientation) from a palmprint. Next a decision is made as to whether two palmprints belong to the same hand by computing a matching score between the corresponding sets of features of the reference and test palmprints. These two sets of features/orientations are matched using point matching technique which takes into account the nonlinear deformations as well as the outliers present in the two sets.

Two new features, viz., fuzzy and wavelet features, are proposed in [6] for the palmprint-based authentication. A palmprint recognition method that uses the phase-based correspondence matching is presented in [7]. To handle nonlinear distortion corresponding points between two images using phase-based correspondence matching are found and also a similarity between local image blocks around the corresponding points is evaluated.

In this chapter, we will explore features such as sigmoid, energy, and entropy from a palmprint. An effort is made to bank upon on two counts: an efficient extraction of region of interest (ROI) and an effective feature selection.

The organization of the chapter is as follows: Sect. 2 presents the extraction of ROI. The extraction of sigmoid, energy, and entropy features is described in Sect. 3. Matching and results of implementation are described in Sects. 4 and 5, respectively. Finally, conclusions are given in Sect. 6.

## 2 ROI Extraction

A coordinate system is formed to align the palm so as to extract ROI in [8]. The central portion of the palm is earmarked as ROI so that the consistent features can be extracted. More or less the same procedure is followed here for ROI extraction.

### 2.1 The Preprocessing for ROI Extraction

The following steps outline the ROI extraction from a palmprint as described in [6]:

1. Take the original image (see Fig. 10.1) and convert it from RGB to gray scale.
2. Crop a fixed section of the image not touching the glass (see Fig. 10.2).
3. Rotate this section based on the type of hand—left or right so that the image is in the specified direction.
4. Find the histogram of the image.

**Fig. 10.1** A typical
palmprint

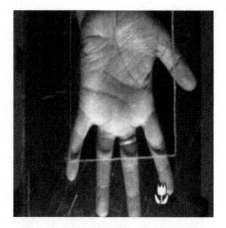

**Fig. 10.2** Cropped section of
image in Fig. 10.1

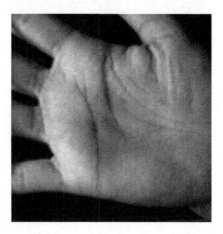

5. Compute the moving average of the histogram, and find the minimum of this
   average. The point of minimum provides us with the threshold for binarizing
   the image (Fig. 10.3). The binary conversion [9] is such that it inverts the
   image, i.e., all dark regions including the cavities between fingers get bright
   whereas the hand region becomes black.
6. Application of the morphological operators removes very small connected
   regions including any holes in the white connected regions as shown in
   Fig. 10.2.
7. Search for the cavities on the left side of the binary image. Rotating the hand
   the two cavities—one between the little finger and the ring finger and another
   between the middle finger and the index finger are detected.
8. The Laplacian edge detector is applied on the fingers to get the contours of the
   cavities.
9. A tangent is drawn between the two contours of the cavities in Fig. 10.4 such
   that all the points of both the contours lie on one side of the tangent.

**Fig. 10.3** Binarized image

**Fig. 10.4** A common tangent
between two curves

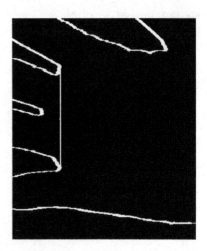

10. The perpendicular bisector of this tangent is taken as the $x$-axis, and the tangent is taken as the $y$-axis to form a coordinate system that facilitates earmarking the ROI.
11. The ROI is rotated and resized to save as a file (See Fig. 10.5).

## 3   Feature Extraction

The ROI is partitioned into a fixed number of nonoverlapping windows. From each of these windows features are extracted [10]. We have selected sigmoid, energy, and entropy as our three feature types for trail on the palmprints. While deriving the three types of features, the average intensity and maximum intensity of subimage are required. We will now elaborate on each feature type.

**Fig. 10.5** The extracted ROI
of palmprint

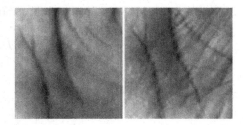

## 3.1   Sigmoid Features

Here, the sigmoid function is used to generate a feature from each window of ROI
requiring the following three steps:

1. The average intensity $I_{avg}$ is calculated from

$$I_{avg} = \frac{\sum_{i=1}^{m}\sum_{j=1}^{n} I(i,j)}{m \times n}, \tag{10.1}$$

   where $(m \times n)$ is the total number of pixels in a window.
2. The maximum intensity, $I_{max}$, is found.
3. The sigmoid feature denoted by sig is obtained as

$$\text{Sig} = \frac{I_{avg}}{1 + \exp\left(-\frac{I_{avg}}{I_{max}}\right)}. \tag{10.2}$$

## 3.2   Energy Features

This feature gives the distribution of energy in a window of ROI. Energy features
are derived from the windows using the steps:

1. The average intensity $I_{avg}$ is now computed by removing the center pixel
   intensity:

$$I_{avg} = \frac{\left(\sum_{i=1}^{m}\sum_{j=1}^{n} I(i,j)\right) - C}{(mxn) - 1} \tag{10.3}$$

   Where C is the gray level of the center pixel.

2. The fuzzy membership function μ is computed using

$$\mu_{ij} = 1 - \frac{|I(i,j) - I_{\text{avg}}|}{I_{\text{max}}}. \tag{10.4}$$

3. The Energy is computed from

$$E_g = \frac{1}{mxn} \sum_{i=1}^{m} \sum_{j=1}^{n} \mu_{ij}^2 \tag{10.5}$$

## 3.3  *Entropy Features*

Entropy is a measure of uncertainty. Hence we use the membership function values of gray levels $\mu_{ij}$ in the windows of palmprints to compute the Shannon Entropy from:

$$E_n = \frac{-1}{(mxn)\log 2} \sum_{i=1}^{m} \sum_{j=1}^{n} \left( \mu_{ij} \log \mu_{ij} + (1 - \mu_{ij}) \log(1 - \mu_{ij}) \right) \tag{10.6}$$

## 4  Matching

### 4.1  *Euclidean Distance Measure*

Having extracted the two sets of features from the training and testing samples, a matching algorithm ascertains the degree of similarity between the training sample features and the test sample features. The Euclidean distance is adopted as a measure of dissimilarity for matching the palmprints. Genuine scores are derived from the Euclidean distances between the sample features of the same user whereas the imposter scores are calculated from the Euclidean distances between the sample features of two different users. The receiver operating characteristic (ROC) is a plot between GAR (genuine acceptance rate) and FAR (false acceptance rate) [8].

### 4.2  *Support Vector Machines*

Support vector machine (SVM) operates on the principle of structural risk minimization (SRM) [11, 12]. It constructs a hyper-plane or a set of hyper-planes on a high-dimensional space for the classification of the input features. Considering a two-class problem to be solved by an SVM, we start with a training sample described by a set of features $x_i \in R^n$; $n$ being the number of features belonging to one of the two classes designated by the label $y_i \in \{+1,-1\}$. The data to be classified by the SVM may not be linearly separable in the original feature space. In the linearly non-separable case the data are projected onto a higher dimensional feature space using Kernel function [13]. The adjustable parameters are the slope ($\alpha$), the constants $a$, $b$, $c$, and the degree of polynomial $d$ is varied in the polynomial Kernel function. For the classification of the feature data, LIBSVM [14] has been used. The values of the parameters are selected as $a = 1$, $c = 0$, $T = 1$, and $d = 1$, 2, and 3.

## 5   Results of Implementation

We have used two databases: one IIT database consisting of 125 users with 5 samples per user totaling $125 \times 5 = 625$ images and another PolyU database [15] consisting of 386 users with 5 samples per user, totaling $386 \times 5 = 1930$ images. Several experiments are conducted for different ratios of training to testing samples. The size of windows is also varied and the recognition rates due to sigmoid, energy, and entropy features are obtained. The entropy features yield GAR of 92.8% at FAR of $10^{-3}$ on PolyU database and GAR of 98.4% at FAR of $10^{-2}$ on IITD database for the training to test ratio of 4:1 at a window size of $9 \times 9$ with the Euclidean distance classifier. The corresponding figures for the training to test ratio of 3:2 are: GAR of 86.8% at FAR of $10^{-3}$ on PolyU and GAR of 91.6% at a FAR of $10^{-2}$ on IITD database for the same window size. The sigmoid features yield GAR of 96.9% at FAR of $10^{-3}$% on PolyU, and GAR of 100% at FAR of $10^{-2}$% on IITD database for the window size of $9 \times 9$ and the ratio of 4:1. The corresponding figures are: GARs of 94.8%, 92.2% on PolyU and IITD database, respectively, for the same window size and the ratio of 3:2.

The usefulness of the features can be judged from the comparative differences. For this purpose, we discuss the plots of entropy features in Fig. 10.6 belonging to samples of two different users and in Fig. 10.7 those belonging to the same user to get an idea of how they differ. The differences between the features of two different users and the same users are depicted in Fig. 10.8. It can be clearly seen that the differences are prominent when the users are different as against the differences between the features of the same users.

These plots reveal the fact that the sigmoid and entropy features yield better authentication results over the energy feature on both the classifiers. However, it is not possible to compare both the classifiers because SVM gives only the classification accuracy unlike the Euclidean classifier that gives the error needed to draw ROC.

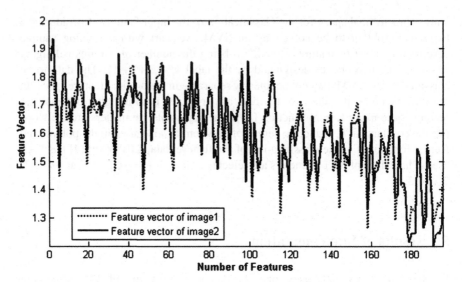

**Fig. 10.6** Sigmoid feature plot of two different image from same user on PolyU database

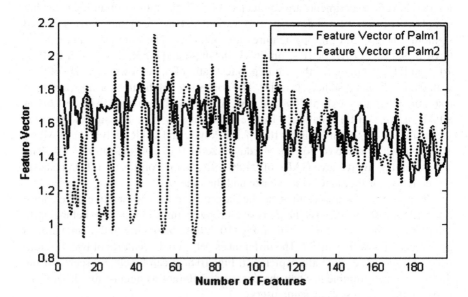

**Fig. 10.7** Sigmoid feature plot of two different users on PolyU database

The performance of Euclidean distance and SVM classifiers on sigmoid, energy and entropy features obtained using three window sizes ($5 \times 5$, $7 \times 7$, $9 \times 9$) is shown in Tables 10.1–10.6 on both databases. Both entropy and energy features have neck to neck competition. The sigmoid features have slightly superior performance.

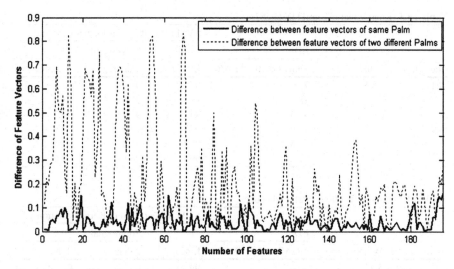

**Fig. 10.8** The difference of sigmoid feature vectors in Figs. 10.6 and 10.7 on PolyU database

**Table 10.1** Sigmoid feature on IITD database

| (train: test) | Window size | ED %GAR at $10^{-2}$ FAR | SVM using different Kernel function | | |
|---|---|---|---|---|---|
| | | | Linear | Polynomial | |
| | | | | 2 degree | 3 degree |
| (4:1) | 5×5 | 100 | 100 | 100 | 100 |
| | 7×7 | 100 | 100 | 100 | 100 |
| | 9×9 | 100 | 100 | 96 | 96.8 |
| (3:2) | 5×5 | 92.5 | 100 | 99.6 | 99.2 |
| | 7×7 | 93.2 | 100 | 96 | 100 |
| | 9×9 | 92.2 | 92.8 | 92.8 | 91.6 |

**Table 10.2** Sigmoid feature on PolyU database

| (train: test) | Window size | ED %GAR at $10^{-3}$ FAR | SVM using different Kernel function | | |
|---|---|---|---|---|---|
| | | | Linear | Polynomial | |
| | | | | 2 degree | 3 degree |
| (4:1) | 5×5 | 96.4 | 99.74 | 100 | 100 |
| | 7×7 | 96.4 | 99.74 | 99.74 | 100 |
| | 9×9 | 96.9 | 99.74 | 97.66 | 98.7 |
| (3:2) | 5×5 | 94.4 | 99.22 | 98.96 | 98.86 |
| | 7×7 | 94.8 | 99.09 | 98.96 | 98.96 |
| | 9×9 | 94.8 | 99.09 | 97.27 | 97.02 |

**Table 10.3** Energy feature on IITD database

| (train: test) | Window size | ED %GAR at $10^{-2}$ FAR | SVM using different Kernel function | | |
| | | | Linear | Polynomial 2 degree | 3 degree |
|---|---|---|---|---|---|
| (4:1) | 5×5 | 76 | 98.4 | 98.4 | 98.4 |
| | 7×7 | 88 | 99.2 | 97.6 | 98.4 |
| | 9×9 | 93.5 | 99.2 | 98.4 | 99.2 |
| (3:2) | 5×5 | 65.2 | 97.6 | 97.2 | 97.2 |
| | 7×7 | 80.5 | 99.6 | 96.8 | 97.2 |
| | 9×9 | 87.2 | 99.6 | 96.8 | 99.6 |

**Table 10.4** Energy feature on PolyU database

| (train: test) | Window size | ED %GAR at $10^{-3}$ FAR | SVM using different Kernel function | | |
| | | | Linear | Polynomial 2 degree | 3 degree |
|---|---|---|---|---|---|
| (4:1) | 5×5 | 83.6 | 99.22 | 99.22 | 99.22 |
| | 7×7 | 92.5 | 99.74 | 99.22 | 99.22 |
| | 9×9 | 93.2 | 99.48 | 98.96 | 99.74 |
| (3:2) | 5×5 | 76.1 | 98.96 | 98.96 | 98.96 |
| | 7×7 | 87.5 | 99.09 | 98.83 | 98.83 |
| | 9×9 | 88.5 | 98.96 | 98.83 | 99.22 |

**Table 10.5** Entropy feature on IITD database

| (train: test) | Window size | ED %GAR at $10^{-2}$ FAR | SVM using different Kernel function | | |
| | | | Linear | Polynomial 2 degree | 3 degree |
|---|---|---|---|---|---|
| (4:1) | 5×5 | 88 | 99.2 | 98.4 | 96.8 |
| | 7×7 | 94.4 | 100 | 98.4 | 96.8 |
| | 9×9 | 98.4 | 100 | 98.4 | 96 |
| (3:2) | 5×5 | 76.5 | 99.6 | 95.6 | 95.2 |
| | 7×7 | 86.4 | 100 | 96.8 | 95.2 |
| | 9×9 | 91.6 | 100 | 96.8 | 95.6 |

**Table 10.6** Entropy feature on PolyU database

| (train: test) | Window size | ED %GAR at $10^{-3}$ FAR | SVM using different Kernel function | | |
| | | | Linear | Polynomial 2 degree | 3 degree |
|---|---|---|---|---|---|
| (4:1) | 5×5 | 79.5 | 98.96 | 98.96 | 98.86 |
| | 7×7 | 90.6 | 98.70 | 98.70 | 98.96 |
| | 9×9 | 92.8 | 98.44 | 98.44 | 98.44 |
| (3:2) | 5×5 | 72 | 98.70 | 98.70 | 98.83 |
| | 7×7 | 85.8 | 98.31 | 98.05 | 98.05 |
| | 9×9 | 86.8 | 97.92 | 98.18 | 98.31 |

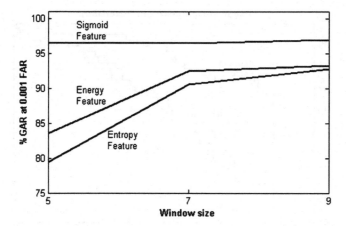

**Fig. 10.9** Comparative plots of sigmoid, entropy, and energy features with window sizes (5×5, 7×7, 9×9) pixels with PolyU database (4:1) with Euclidean classifier

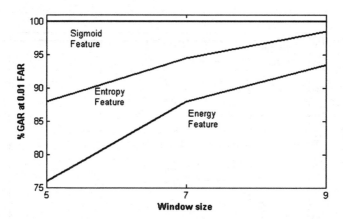

**Fig. 10.10** Comparative plots of sigmoid, entropy, and energy features with window sizes (5×5, 7×7, 9×9) pixels with IITD database (4:1) with Euclidean classifier

The results of authentication on PolyU and IITD database arising out of the use of the three feature types are depicted in Figs. 10.9 and 10.10, respectively, for the training to testing ratio of (4:1). A comparison of the performance of the three feature types appears in the form of ROC plots in Figs. 10.11 and 10.12 on PolyU and IITD databases, respectively. These ROCs correspond to the window size of 9×9.

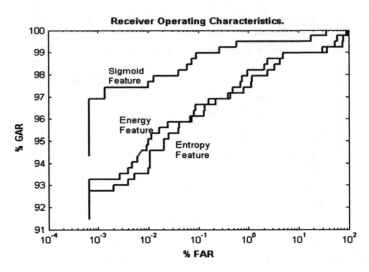

**Fig. 10.11** Comparative ROC plots of sigmoid, entropy, and energy features with window sizes (9×9) pixels with PolyU database (4:1)

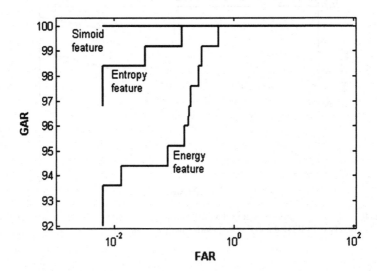

**Fig. 10.12** Comparative ROC plots of sigmoid, entropy, and energy features with window sizes (9×9) pixels with IITD database (4:1)

## 6  Conclusion and Future Work

A palmprint-based biometric authentication system has been developed. Three feature types are experimented on the palmprints for the suitability and usefulness of the palmprint-based authentication. Of these the sigmoid features are found to be most suitable in terms of accuracy among all the feature types tested on the two databases. The energy and entropy features are lagging behind sigmoid features.

The recognition rates of 100% are achieved with both sigmoid and entropy features on both the databases using SVM classifier. Irrespective of the type of features used, there is a marked difference in the results obtained using the Euclidean distance classifier and those with SVM classifier. SVM classifier with a linear Kernel function and two polynomial kernels of degrees 2 and 3 perform with an accuracy of 100% recognition score on PolyU using sigmoid features.

The main problem for achieving very good authentication rates lies in the choice of the ratio of the samples of both the training and the testing. We have made several experiments by varying these ratios. It is observed that as the training samples increases the matching scores increase but as the number of testing samples increase the matching scores decrease correspondingly. The *cross validation* is also done and the results are almost the same.

The future work will be concerned with developing new features and classifiers.

# References

1. Jain Anil K, Ross Arun, Prabhakar Salil (2004) An introduction to biometric recognition. IEEE Trans Circ Syst Video Tech (Special issue on image and video-based biometrics) 14(1):4–20
2. Kong A, Zhang D (2004) Competitive coding scheme for palmprint verification. In: Proceedings of the 17th international conference on pattern recognition, Pattern recognition, 2004 ICPR 2004, vol 1. pp 520–523
3. Zhang D, Guo Z, G Lu, Zhang L, Zuo W (2010) An online system of multi-spectral palmprint verification. IEEE Trans Instrum Meas 59(2):480–490
4. Fang L, Leung MKH, Shikhare T, Chan V, Choon KF (2006) Palmprint classification. IEEE Int Conf Syst Man Cybern 4(8–11):2965–2969
5. Duta N, Jain AK, Mardia KV (2001) Matching of palmprint. Pattern Recogn Lett 23 (4):477–485
6. Hanmandlu M, Gupta HM, Mittal N, Vasikarla S (2009) An authentication system based on palmprint. In: Proceedings ITNG, IEEE Computer Society, pp 399–404
7. Ito K, Litsuka S, Aoki T (2009) A palmprint recognition algorithm using phase based correspondence matching. ICIP 2009, pp 1977–1980
8. Zhang D, Kong W-K, You J, Wong M (2003) Online palmprint identification. IEEE Trans Pattern Anal Mach Intell 25(9):1041–1050
9. Gonzalez RC, Woods RE (1993) Digital image processing. Addison Wesley publishers, MA
10. Hanmandlu M, Vijay R, Mittal N (2011) A study of some new features for palmprint authentication. In: Proceedings of the world congress on engineering 2011. Lecture notes in engineering and computer science, WCE 2011. 6–8 July, London, UK, pp 1623–1628
11. Vapnik VN (1998) Statistical learning theory. Wiley-Interscience, New York
12. Scholkopf B, Burges CJC, Smola AJ (1998) Advances in kemel methods-support vector learning. MIT, Cambridge, MA
13. Li H, Liang Y, Xu Q (2009) Support vector machines and its applications in chemistry. Chemometr Intell Lab Syst 95:188–198
14. Chang C-C, Lin C-J (2001) LIBSVM: a library for support vector machines. http://www.csie. ntu.edu.tw/~cjlin/libsvm
15. H K Polytechnic University (2005) Palmprint database. Biometric research center website. http://www.comp.polyu.edu.hk/~biometrics/

# Chapter 11
# An Interactive Colour Video Segmentation: A Granular Computing Approach

Abhijeet Vijay Nandedkar

## 1 Introduction

A central and prerequisite step in colour image and video understanding is segmentation. In this paper, a novel interactive colour video segmentation (CVS) approach using granular reflex fuzzy min–max neural network (GrRFMN) [1] is presented. It is observed that most of the image and video segmentation techniques are pixel based [2–6]. In the proposed segmentation technique instead of pixels, data granules of an image or a video frame are processed. A data granule represents a bunch or group of pixels in the form of hyperbox [1].

Many successful applications of computer vision to image or video manipulation are interactive by nature. However, parameters of such systems are often trained neglecting the user [7]. It is observed that the conventional image segmentation is carried out in an unsupervised mode, i.e. pixels of an image are classified into different categories using some homogeneity criteria. Unsupervised segmentation is attractive in real-time systems. However, unsupervised segmentation may fail in many cases [8]. In the proposed interactive segmentation, user interaction is added to the segmentation process. With the help of user interaction a semantic object can be defined easily and some uncertain borders can be decided [9]. User interaction happens usually at the initial stage of the segmentation.

In general, colour image segmentation techniques can be classified as histogram thresholding based, neighbourhood based, clustering based and neuro-fuzzy based. Histogram thresholding is one of the simple and widely used techniques for image segmentation. The underlying assumption of histogram thresholding is that objects in the scene give rise to explicit peaks in image histogram. Thus, the segmentation

A.V. Nandedkar (✉)
Department of Electronics and Telecommunication Engineering, S.G.G.S. Institute of Engineering & Technology, Vishnupuri, Nanded, Maharashtra 431606, India
e-mail: avnandedkar@yahoo.com

S.-I. Ao and L. Gelman (eds.), *Electrical Engineering and Intelligent Systems*,
Lecture Notes in Electrical Engineering 130, DOI 10.1007/978-1-4614-2317-1_11,
© Springer Science+Business Media, LLC 2013

task is reduced to find thresholds dissecting the image histogram [2]. An adaptive multi-thresholding approach for CIS can be found in [3]. However, a major drawback in the histogram thresholding techniques is the lack of use of spatial relationship amongst the pixels [4]. The neighbourhood-based approach (e.g. region growing) generally uses the uniformity criteria to segment regions in the image. These methods are better than histogram thresholding since they consider spatial relationship amongst. However, problem with these methods is the selection of initial seed points and the order in which pixels and regions are examined [5]. Clustering-based approaches generally use fuzzy logic to define membership of the pixels [6]. Regions are created by inspecting the membership values of pixels using partition method (e.g. fuzzy c-means (FCM) clustering algorithm) [6].

This work proposes use of image granules for video segmentation. Granulation of information is an inherent and omnipresent activity of human beings carried out with intent of better understanding of the problem [10]. In fact, information granulation supports conversion of clouds of numeric data into more tangible information granules. The concept of information granulation within the frame work of fuzzy set theory was formalized by Zadeh in his pioneering work [11]. He observed that humans mostly employ words in computing and reasoning, and information granulation is a part of human cognition and proposed theory of fuzzy information granulation (TFIG) and fuzzy logic as a tool for computing with words (CW) [11]. This paper is a small step towards implementation of this concept for segmentation of colour video sequences. The major advantage of this technique is that it incorporates user's knowledge for segmentation. In addition, since granules of image are used for segmentation, execution of algorithm is fast. The rest paper is organized as follows. Section 2 briefly discusses GrRFMN architecture and its learning. Section 3 elaborates the proposed method. Section 4 presents experimental results and conclusions.

## 2   GrRFMN Architecture

The segmentation task in the proposed method is carried out using GrRFMN (Fig. 11.1). GrRFMN is a fuzzy hyperbox set based neural network. A hyperbox is a simple geometrical object which can be defined by stating its min and max points (e.g. is shown in Fig. 11.1). GrRFMN learns different classes by aggregating hyperbox fuzzy sets [1]. It accepts input in the form of min–max point of a granule, i.e. in the form of hyperbox. Advantage of GrRFMN is that it is capable to handle data granules of different sizes efficiently, online in a single pass through data.

GrRFMN architecture is divided into two sections as (1) classifying neurons section and (2) reflex section. The main task of classification of input data is carried out by Classifying neurons section. It computes membership with the learned classes. Reflex section, which is based on reflex mechanism of human brain, adds compensation to the output if input sample belongs to class overlap region. Reflex section is further subdivided as overlap compensation and containment compensation sections.

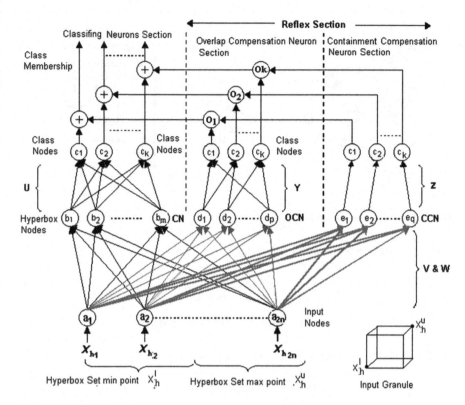

**Fig. 11.1** GrRFMN architecture

Neurons in these sections get activated only if the input sample or granule belongs to the overlap region. This action is very similar to the reflex action of human brain which takes over the control unconsciously in hazardous conditions.

Here, it is assumed that all input features are scaled in the range [0–1]. An n-dimensional input granule is represented by, $X_h = [X_h^u, X_h^l]$ where $X_h^l$, $X_h^u = (x_{h1}, x_{h2}, \ldots, x_{hn})$ are min and max point vectors of the input granule, respectively. A point data is a special case with $X_h^l = X_h^u$. Appending min and max point vectors, the input is connected to the nodes $x_{h1}$–$x_{h2n}$.

## 2.1   Classifying Neurons

The neurons $b_1$–$b_m$ are classifying neurons. Outputs of classifying neurons (CNs) belonging to a class are collected at a class node $C_i$ in the output layer. The activation function of the *classifying neuron* $b_j$ is given by (11.1)–(11.6),

$$\mu_{b_j}(X_h, V_j, W_j, \gamma) = \frac{1}{n}\sum_{i=1}^{n} A_{ji} + B_{ji} \tag{11.1}$$

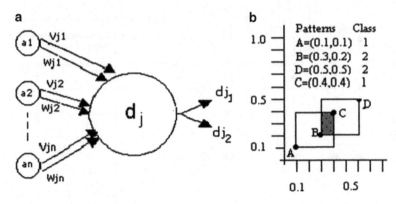

**Fig. 11.2** Overlap compensatory neuron

where $\gamma$ = Fuzzyness parameter, $y = 1/\gamma$

$$A_{ji} = \frac{(r_{ji} + l_{ji} + c_{ji})}{a_i - b_i + \varepsilon} \tag{11.2}$$

$$c_{ji} = \max(\min(b, w_{ji}) - \max(v_{ji}, a), 0) \tag{11.3}$$

$$l_{ji} = \frac{1}{2y}(\max((\min(v_{ji}, b_i) - \max(v_{ji} - y, a_i)), 0)$$
$$\times (\max(a_i, v_{ji} - y) + \min(b_i, v_{ji}) - 2 \times (v_{ji} - y))) \tag{11.4}$$

$$r_{ji} = \frac{1}{2y}(\max((\min(b_i, w_{ji} + y) - \max(w_{ji}, a_i)), 0)$$
$$\times (-(\max(a_i, w_{ji}) + \min(b_i, w_{ji} + y) + 2 \times (w_{ji} + y)))) \tag{11.5}$$

$$B_{ji} = U(a_i - b_i) \times \max\left(\min\left(\frac{a_i - (v_i - y)}{y}, \frac{(w_i + y) - a_i}{y}, 1\right), 0\right) \tag{11.6}$$

$[a, b]$, $[v, w]$: min–max points of input and hyperbox fuzzy set (HBF set), respectively. $\varepsilon = 10^{-10}$ avoids division by zero error in case data is in point form, $U$: unit step function.

## 2.2 Reflex Section: Overlap and Containment Compensation Neurons

While training GrRFMN, situation depicted in Figs. 11.2b and 11.3b where hyperboxes of different classes are overlapping is bound to occur. Overlap compensation neurons (OCN) and containment compensation neurons (CCN) are trained to handle these situations. Nodes $d_1$–$d_p$ and $e_1$–$e_q$ represent OCNs and CCNs, respectively. Outputs of these neurons are collected at the respective class nodes.

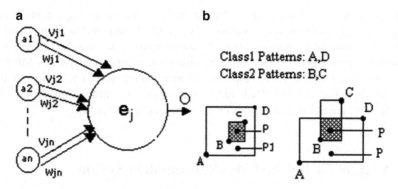

**Fig. 11.3** Containment compensatory neuron

Figure 11.2a shows details of an OCN, which represents a hyperbox of size equal to the overlap region between two hyperboxes. Outputs of OCN are connected to the respective nodes of classes facing the overlap problem. OCN activation function is given by (11.7, 11.8).

$$d_{j_p} = U(b_j(X_h, V, W, \gamma) - 1)) \times \left(-1 + \max\left(b_{j1}\left(X_h^u, V_p, W_p\right), b_{j1}\left(X_h^l, V_p, W_p\right)\right)\right)$$ (11.7)

where

$$b_{j1}\left(X_h^k, V_p, W_p\right) = \left(\frac{1}{n}\sum_{i=1}^{n}\max\left(\frac{x_{hi}^k}{w_{pji}}, \frac{v_{pji}}{x_{hi}^k}\right)\right)$$ (11.8)

$p = 1, 2$. $d_{j1}$ and $d_{j2}$ are outputs for Class 1 and Class 2, $V$, $W$ the min–max point of OCN, $V_1$, $W_1$, $V_2$, $W_2$ the min–max point of overlapping hyperboxes, $U(x)$ a unit step function and $b_j()$ is same as (11.1).

A CCN is shown in Fig. 11.3a. This neuron represents a hyperbox of size equal to the overlap between two classes as shown in Fig. 11.3b. The activation function of CCN is:

$$O_{c_j} = -1 \times U(b_j(X_h, V, W, \gamma) - 1)$$ (11.9)

where $Oc_j$ is the output, $V,W$ the min–max point of CCN, $U(x)$ the unit step function and $bj(\ )$ is same as (11.1).

The output of CCN is connected to the class that contains the hyperbox of other class. The number of output layer nodes in CL section is same as number of classes learned. The number of class nodes in reflex section depends on the nature of overlap network faces during the training process. Final membership calculation is given by

$$\mu_i = \max_{j=1...m} (b_j u_{ji}) + \min\left(\min_{j=1...p} (d_j y_{ji}), \min_{j=1...q} (e_j z_{ji})\right)$$ (11.10)

where $u, y, z$ are connections for the neurons in three sections. $m, p, q$ are number of neurons in respective sections. Training of GrRFMN begins by presenting training granules sequentially. Network tries to accommodate the training samples in hyperboxes for the given class. During training if hyperboxes belonging to different class are found overlapping, respective compensation neuron is added to the network. Training of GrRFMN is online and single pass through the data. More details about the training algorithm for GrRFMN are given in [1].

## 3   A Granular Colour Video Segmentation System

CVS is an important issue for understanding a scene in a video. Here, the problem of CVS is tackled using granular computing. The proposed method utilizes capability of GrRFMN to acquire knowledge through granules of data. The main endeavour behind development of this technique is to demonstrate that instead of processing individual colour pixels (a 3D vector), a group of pixels (granules) can be processed very easily using GrRFMN. Obviously this reduces lot of computational cost required to process individual pixels. The proposed CVS technique is shown in Fig. 11.4.

This CVS method uses CIE (L-a-b) colour space. The reason is that CIE space can control colour and intensity information more independently than RGB colour space. This colour space is especially efficient in the measurement of small colour changes, as a result direct colour comparison can be performed based on geometric separation within this colour space [12].

To perform CVS, the proposed CVS system is trained with a user intervention using the first video frame. Subsequently, proposed CVS system segments the frames based on the earlier acquired knowledge. The detailed scheme is explained as follows.

The main steps in CVS implementation are (a) granulation of user interaction through samples of an image or a video frame presented for training, (b) training of GrRFMN, (c) to segment subsequent frames, a difference with respect to previous frame is computed, (d) granulation of pixels from current frame belonging to difference region and (e) classification of data granules contributing to difference region.

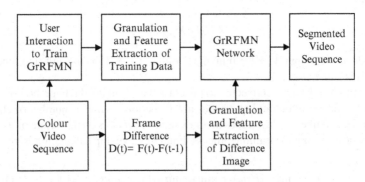

**Fig. 11.4** Proposed interactive colour video segmentation (CVS) technique

In the CVS system, in the beginning GrRFMN is trained with labelled granules (seed granules) constructed from the seed images of different objects/parts in the first training video frame. These seed images are sub-divided into grids of size (k×k). A granule for each grid is represented by a hyperbox which can be represented by simply stating its min and max vertices. Thus, a hyperbox is computed by finding min–max values of the pixels in that grid

$$\text{i.e., } \begin{aligned} V &= [L_{min}, a_{min}, b_{min}] \\ W &= [L_{max}, a_{max}, b_{max}] \end{aligned}. \tag{11.11}$$

Along with this information, a mean value of the grid for the three planes is also extracted and added to the min–max vector as

$$\text{i.e., } \begin{aligned} V &= [L_{min}, a_{min}, b_{min}, L_{mean}, a_{mean}, b_{mean}] \\ W &= [L_{max}, a_{max}, b_{max}, L_{mean}, a_{mean}, b_{mean}] \end{aligned} \tag{11.12}$$

Such granules are then used to train GrRFMN.

To segment a given frame F(t), it is subtracted from previous frame (i.e. D(t) = F(t)−F(t−1)) and a difference frame D(t) is computed. Then F(t) which is to be segmented and its corresponding difference frame D(t) are sub-divided into small grids of size (n×n). The granulation of pixels for the grids in F(t) is computed if corresponding grid in D(t) is non-zero. These granules are then fed to GrRFMN for classification. The final segmentation for F(t) is computed based on segmentation results of previous frame F(t−1) and classification results of the granules representing difference region in D(t). Segmentation labels of F(t−1) are continued for the grids in F(t) whose corresponding grid in D(t) is zero.

An example of colour image segmentation using proposed method is shown in Fig. 11.5. Here, four sub-images (Fig. 11.5b) belonging to two different classes (i.e. house and sky) are used to train GrRFMN. During training seed images are divided into a grid (e.g. size 3×3, 5×5 and 10×10) and granules (hypeboxes) are formed. These granules are then used for training GrRFMN. Here, GrRFMN is trained with an expansion coefficient [1] (θ) equal to 0.2. In the test phase, the given image is granulated with various grid sizes (5×5, 10×10, 15×15). The segmentation results are shown in Fig. 11.5. Note that for various training and test grid sizes, the performance of GrRFMN is almost consistent.

The following section discusses few more experimental results of colour image and video segmentation.

# 4   Experimental Results

Here, aim is to test proposed system on some real images and video sequences. The results are as follows:

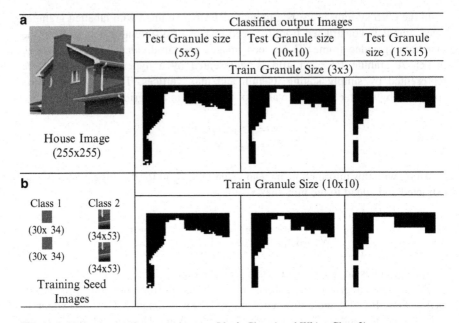

**Fig. 11.5** House image (for output images: *Black*: Class 1 and *White*: Class 2)

## 4.1 Colour Image Segmentation

In this case, training granule size here is kept as $5 \times 5$ and test granule size is $3 \times 3$ and $5 \times 5$, expansion coefficient of GrRFMN is ($\theta = 0.2$). Note that GrRFMN training is done on samples of classes and are represented by a pseudo colours in segmented output image (Fig. 11.6).

Note that in Fig. 11.6d, the image consists of two classes: (1) lady and (2) background. Observe that Class 1 consists of three different colours. To segment the lady properly, the training sample is chosen such that it consists of both coloured body parts of lady. It may be observed that the two classes in the image, i.e. lady and background are segmented properly. This shows the capability of proposed system to group different colours in a class, if required. From results demonstrated in Fig. 11.6, one can note that proposed is capable to classify image granules efficiently.

## 4.2 Colour Video Segmentation

The performance of the proposed CVS system on video sequence is tested in this section. In this experiment, with the help of user interaction, few image/frame samples of each class are selected. These samples are granulated using grid size ($5 \times 5$).

These granules are used to train GrRFMN. Expansion coefficient ($\theta$) of GrRFMN was kept equal to 0.2. To segment subsequent frames, test granules with size ($5\times5$) are formed. The results depicted in Fig. 11.7 demonstrate the capability of proposed CVS system. One may add more user interaction to improve resultant segmentation.

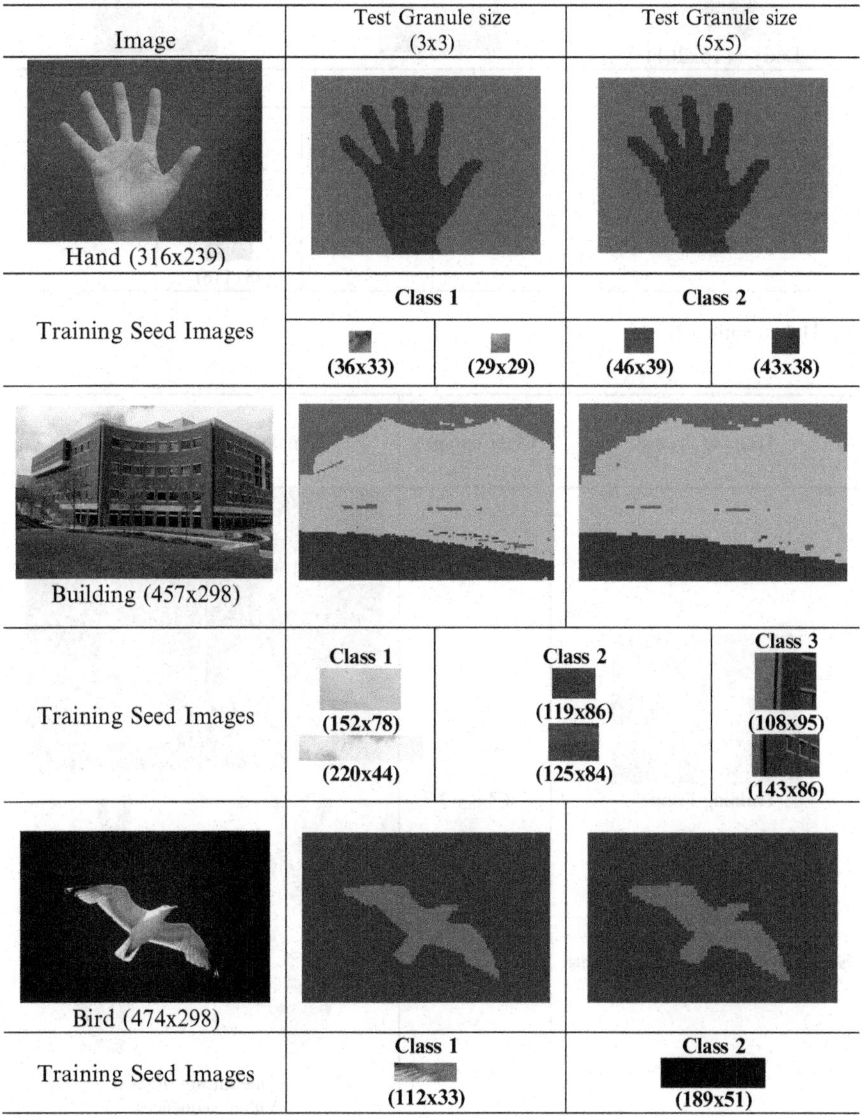

| Image | Test Granule size (3x3) | Test Granule size (5x5) |
|---|---|---|
| Hand (316x239) | | |
| Training Seed Images | Class 1 | Class 2 |
| | (36x33)   (29x29) | (46x39)   (43x38) |
| Building (457x298) | | |
| Training Seed Images | Class 1 (152x78) (220x44) | Class 2 (119x86) (125x84) | Class 3 (108x95) (143x86) |
| Bird (474x298) | | |
| Training Seed Images | Class 1 (112x33) | Class 2 (189x51) |

Fig. 11.6 Segmentation results on different real images

Lady (116x261)
Class 1
Class 2
Training Seed Images
(27x56)
(16x56)
(48x116)
(47x31)


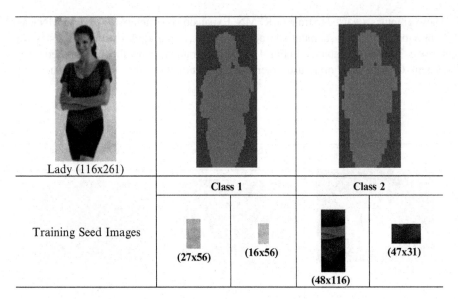

**Fig. 11.6** (continued)

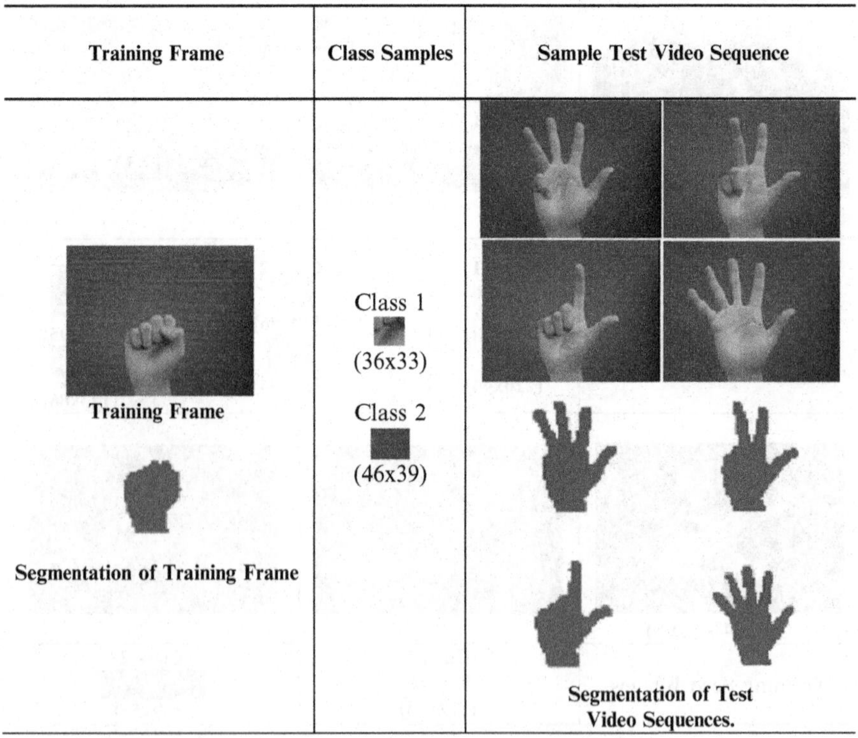

Training Frame
Class Samples
Sample Test Video Sequence
Training Frame
Segmentation of Training Frame
Class 1
(36x33)
Class 2
(46x39)
Segmentation of Test Video Sequences.


**Fig. 11.7** Result for colour video sequences

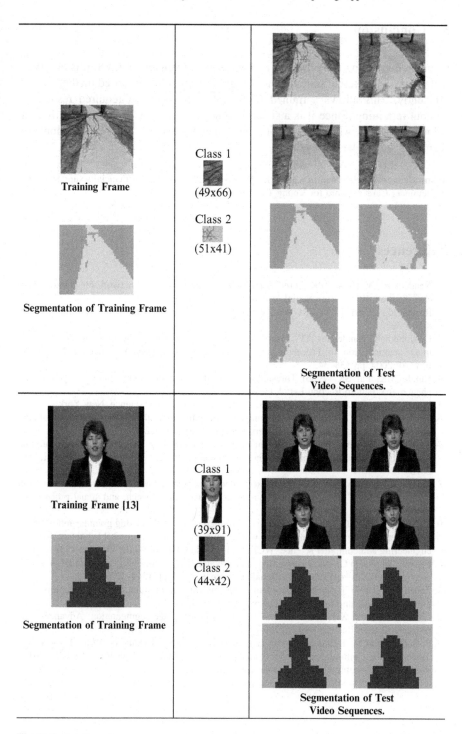

**Fig. 11.7** (continued)

# 5  Conclusion

Granular computing is a powerful tool and is found suitable in CVS. It is observed that even with different training and test granule sizes the proposed method works efficiently. This allows a trained CVS system to work on required resolution without retraining. Since it is a supervised segmentation, we can assign different coloured regions which belong to one category easily. Hence, an improved segmentation can be achieved.

**Acknowledgement**  The author would like thank AICTE New Delhi for financially supporting the project under Career Award for Young Teachers Scheme.

# References

1. Nandedkar AV, Biswas PK (2009) A reflex fuzzy min max neural network for granular data classification. IEEE Trans Neural Network 20(7):1117–1134
2. Rosenfeld A, Davis LS (1978) Iterative histogram modification. IEEE Trans Syst Man Cybern 8:300–302
3. Deshmukh K, Nandedkar AV, Joshi YV, Shinde GN (2004) Multilevel approach for color image segmentation. In: Proceedings of Indian conference on computer vision and graphics, ICVGIP 2004. pp 338–342
4. Kundu MK, Pal SK (1986) Thresholding for edge detection using human psychovisual phenomena. Pattern Recogn Lett 4:433–441
5. Pratt WK (1991) Digital image processing, 2nd edn. Wiley-Inter science, New York
6. Lim YW, Lee SU (1990) On the color image segmentation algorithm based on the thresholding and the fuzzy c-means technique. Pattern Recogn 23(9):935–952
7. Nickisch H, Rother C, Kohli P, Rhemann C (2010) Learning an interactive segmentation system. In: Proceedings of Indian conference on computer vision and graphics. Chennai, Dec 2010
8. Chen H, Qi F, Zhang S (2003) Supervised video object segmentation using a small number of interactions. In: IEEE international conference on acoustics, speech, and signal processing, (ICASSP '03), vol 3(3). pp 365–368
9. Nandedkar A V (2011) An interactive colour video segmentation using granular reflex fuzzy neural network. In: Proceedings of the world congress on engineering 2011. Lecture notes in engineering and computer science. WCE 2011, London, UK, 6–8 July 2011. pp 1688–1693
10. Pedrycz W (2001) Granular computing: an introduction. In: Proceedings of joint IFSA world congress and 20th NAFIPS international conference, vol 3. pp 1349–1354
11. Zadeh LA (1997) Towards a theory of fuzzy information granulation and its centrality in human reasoning and fuzzy logic. Fuzzy Set Syst 90:111–127
12. Cheng HD, Jiang XH, Sun Y, Wang J (2001) Colour image segmentation: Advances and propects. Pattern Recogn 34(12):2259–2281
13. P. Seeling and M. Reisslein in print (2012) Video Transport Evaluation With H.264 Video Traces. IEEE Communications Surveys and Tutorials. Online: DOI 10.1109/SURV.2011. 082911.00067: weblink: http://trace.eas.asu.edu.

# Chapter 12
# A Novel Approach for Heart Murmurs Detection and Classification

Maamar Ahfir and Izzet Kale

## 1 Introduction

Physical examination is a fundamental skill in primary medicine and is often used as the tool with which to determine whether a referral of patients to a specialist is necessary or not. In all over the world, auscultation of the human heart by the use of a stethoscope is still the primary diagnostic tool for detecting and judging the class (innocent or pathological) of cardiac murmurs. The delayed recognition of a pathological heart murmur may have a serious impact on the long-term outcome of the affected patient. Although experienced cardiologists can usually evaluate heart murmurs with a high sensitivity and specificity, non-specialists with less clinical experience (primary care physicians) may have more difficulty. Also the ability of the human ear to distinguish defects from the sound of a heartbeat (single heart beat) is very limited. Therefore, an automated system for an additional recording and analysis of the cardiac sounds could enable primary care physicians to make the initial diagnosis objectively.

The decision to determine whether a cardiac murmur exists or not is purely subjective, and is not always shared by primary care physicians. Physicians range their listening findings from grade 1 to grade 6 (subjective score of audibility). Although murmurs ranged from 4 to 6 are all easily detected by a traditional stethoscope, those ranged from 3 down to 1 may escape to the primary care physicians listening. But even if they are detected, the human ear is not able to determine if the detected murmur is systolic or diastolic in timing, with respect to

M. Ahfir (✉)
Department of Informatics, University of Laghouat, BP: 37G, Laghouat, Algeria
e-mail: m.ahfir@mail.lagh-univ.dz

I. Kale
Applied DSP and VLSI Research Group, Department of Electronic Systems,
University of Westminster, 115 New Cavendish Street, London, UK
e-mail: kalei@westminster.ac.uk

S.-I. Ao and L. Gelman (eds.), *Electrical Engineering and Intelligent Systems*,
Lecture Notes in Electrical Engineering 130, DOI 10.1007/978-1-4614-2317-1_12,
© Springer Science+Business Media, LLC 2013

the first heart and the second heart sounds of the cardiac cycle of a normal functioning heart. Because murmurs timing is a key characteristic to determine their classes and types, then the representation technique of digitized auscultation heart sounds as a phonocardiogram (PCG) for visualisation and diagnostic aid, additional to the historical and physical examination findings can help physicians to make difference between innocent and pathological murmurs.

Although there exist many papers in which different methods have been investigated to develop a PCG-based interface for cardiac ausculatation interpretation, for example the recent one [1], however this technique (PCG-based interface) is not practical in the auscultation context of a large number of patients, for example: for school children and/or waiting lists for the echocardiography analysis of the cardiology services. Also it requires that primary care physicians must be highly educated in signal processing techniques in order for them to read and interpret phonocardiograms. The development of an automated algorithm can then serve as a rapid and low cost device to screen numericaly *objective criteria* for every single heart sound picked up from the ausculation areas corresponding to the four cardiac valves. The objective criteria values can serve as indictors for detection and classification of possible heart murmurs.

In the previous work reported in [2], three automated algorithms that show significant potential promising in their use as an alternative diagnostic tool to the traditional stethoscope for the classification of heart sounds into normal/innocent and pathological classes were developed. These are based on three different methods: The direct ratio which is applied to a single heart beat signal represented in time domain, wavelet processing applied to the equivalent frequency domain representation of the signal and artificial knowledge based neural networks (AKBNN). The later method (AKBNN) was adopted to overcome the common limitations of the two previous methods, which are [2]:

- The only systolic murmurs identification, because of the problem of locating the position of the second heart sound S2 for some pathological cases, and then the diastolic timing of the cardiac cycle is excluded for classification.
- The amplitude of the first heart sound S1 (S2 in case of diastolic murmur identification) which is liable to the indicators of pathology (objective criteria) is not constant for every heart sound recording. This can lead to wrong classification.
- At least 4 objective criteria (algorithm's outputs) are needed for screening only systolic events, which may be not desirable for the auscultation of a large number of patients.

The AKBNN method operates on some periods of the cardiac cycle to analyze a given individual heart sound recording, which makes it able to detect murmurs of both classes (systolic and diastolic) and all possible types. The obtained results in the context of the common limited database (only congenital murmurs captured from children) indicated that AKBNN produced the best performance as an automated murmurs classifier with an optimum sensitivity and specificity of 92.9 and 92 % respectively, with respect to the two evaluated previous methods. But its Hardware implementation requires a very large memory space because of its training DATA-set which must represent all possible types of murmurs which exist within both systolic and diastolic timing classes. However, capturing all possible different murmurs types for both classes is not guaranteed during DATA collection. This is because pathological murmurs are not only congenital which can be captured from a certain age limit (childhood), but there exist also other types which can affect patients at the adult's age because of some bacteria (acquired pathology), and they may be captured only if there are sufficiently associated adult patients who accept to voluntarily participate in DATA collection.

The performance of an implemented automated algorithm based on such method remains strongly dependent to the answers of the two following questions [2]:

- What is the correlation between a specific murmurs type recorded from different patients for the same conditions (gender, age, weight, height, position during auscultation ,..., ect)?
- How many different murmurs types are there, that must be represented in training DATA-set for a network to have a sensitivity and specificity of 100% for all possible heart sound recordings?

This means, an under-representation of all possible murmurs types in the training DATA-set can result in misclassification of some of them. Only very large clinical observations of different categories of patients (children, young adults and adults) with the collaboration of clinicians and cardiologists can help for answers to the two above questions. In the next section, a proposed approach that overcomes the main limitations of the three previous methods will be described. The paper is organized as follows: Section 2 describes the proposed approach which is based on a technique traditionally used in room acoustics for characterization. Preliminary results and comments in the context of restricted database collected from only ten adult patients are given in Sect. 3, and finally Sect. 4 concludes the paper.

## 2   Novel Approach Description

In this approach, the measurement technique of the normalized time dependent sound energy decay used in room acoustics for characterization [3] can be applied for heart sounds analysis. Given a single heart beat signal which can be extracted by an automated algorithm from a clean individual auscultation recording (filtering system output), and after separating the systolic and diastolic timing segments of the signal, normalized time dependent energy decay for both segments can be calculated by a discrete version of the following formula:

$$E(t) = \frac{\int_{t}^{+\infty} h^2(t)\mathrm{d}t}{\int_{0}^{+\infty} h^2(t)\mathrm{d}t}, \qquad (12.1)$$

**Table 12.1** Advantages and disadvantages recapitulation for the previous methods compared to that of the proposed approach

Murmurs identification

| Methods | Systolic | Diastolic | Murmurs types representation in database | Outputs screening |
|---|---|---|---|---|
| The direct ratio | Possible but dependent of S1 | Possible but dependent of S2 | Murmurs of all types not required | At least 4 systolic + 4 diastolic outputs |
| Wavelet processing | Possible but dependent of S1 | Possible but dependent of S2 | Murmurs of all types not required | At least 4 systolic + 4 diastolic outputs |
| AKBNN | Possible not dependent of S1 | Possible not dependent of S2 | Murmurs of all types required | Only 1 logical output 1 or 0 |
| Novel approach | Possible not dependent of S1 | Possible not dependent of S2 | Murmurs of all types not required | Only 1 systolic + 1 diastolic output |

where $h(t)$ represents the time dependent signal (systolic/diastolic timing segment), and $E(t)$ its normalized time dependent energy decay.

If this technique is applied to a systolic timing signal for example (similarly to diastolic segment), it allows overcoming the problem of the first heart sound S1 (S2) where its variable amplitude is used as reference for all the indicators of pathology in the two previous methods (the direct ratio and wavelet processing). Furthermore, a new indicator of pathology independent of S1/S2 is proposed, that is, the early time decay measured once the energy decay, $E(t)$ reaches certain threshold in dB. In the proposed approach an early time decay value above certain threshold represents an indication of pathology. The threshold value which corresponds to healthy heart will be determined after extensive clinical observation and trial for different categories of patients.

Finally, to screen the objective criteria for pathology indication, only one output per timing segment is needed in the novel approach, rather than at least four outputs each in the two previous methods. Although, this novel approach doesn't require all types of pathology in the database for study and analysis, as it is the case for the AKBNN method. Because any timing segment associated to a heart sound with murmur, regardless its type and grade must have a total energy higher than that associated to a heart sound with no murmur (first hypothesis formulated in the previous work). Then in the first case (pathology), the energy decay can be later than that of the second case (normal). In the Table 12.1 below advantages and disadvantages of the three previous methods are recapitulated and compared against that of the proposed approach.

## 3   Results

The novel approach was applied to a restricted database of heart sounds captured individually from only ten different adults patients. These were recorded as wave format files (duration: 6 s, resolution: 16 bits and sampling frequency: 11,025 Hz), [4]. Only three examples of cardiac signals analysis corresponding to three patients are displayed in this section [5]. Two examples correspond to patients with different types of pathology (Aortic Stenosis and Mitral Stenosis) and one example corresponds to a patient with normal heart. In Fig. 12.1, we can see one clean signal period of the cardiac cycle, representing a normal heart. A second example of one clean signal period of the cardiac cycle, representing a pathological heart is shown in Fig. 12.2. Both different periods were extracted by an automated algorithm [2, 3]. After extracting one signal period (one single heart beat) of the cardiac cycle and separating the systolic and diastolic segments of every individual recorded heart sound (assumed to be clean), normalized energy decay for both segments was calculated in order to measure corresponding Early Times Decays which represent the indicators of pathology within the systolic and/or diastolic timing. Figures 12.3 and 12.4 show corresponding curves, where we can see late energy decay in the systole and diastole of the two different pathology types with respect to early energy decay of the normal case. In the Table 12.2 corresponding early time decay T20 in ms for the three cases were calculated for an arbitrary value of 20 dB of energy decay. A great value of T20 as compared to that of the normal case can be considered as indication of pathology for the associated valves. These are in the two cases: aortic and mitral valves.

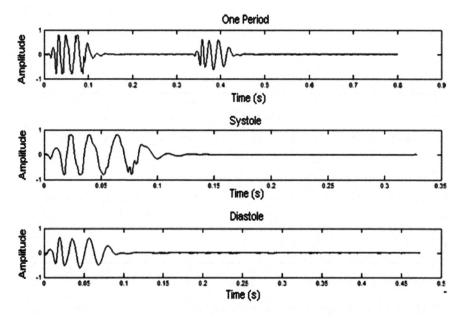

**Fig. 12.1** One signal period of the cardiac cycle representing a normal heart

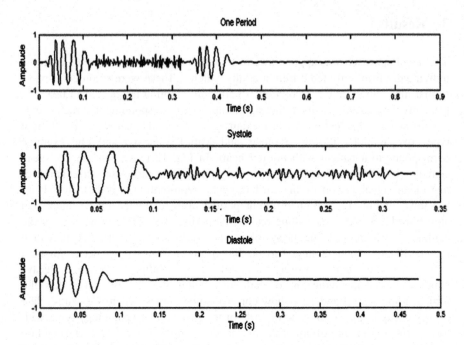

**Fig. 12.2** One signal period of the cardiac cycle representing a pathological heart

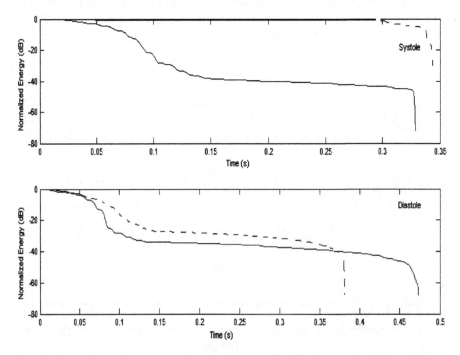

**Fig. 12.3** Normalized energy decay curve for systole (*up*) and diastole (*down*) – Pathology type: aortic stenosis (*discontinued line*) as compared to normal heart (*continued line*)

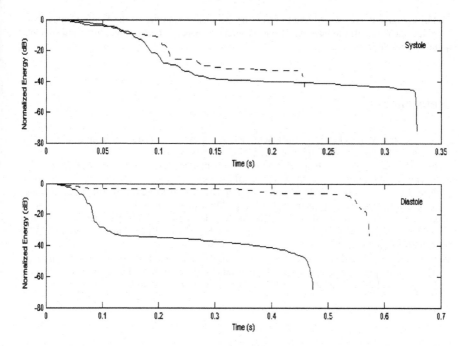

**Fig. 12.4** Normalized energy decay curve for systole (*up*) and diastole (*down*) – Pathology type: mitral stenosis (*discontinued line*) as compared to normal heart (*continued line*)

**Table 12.2** Early time decay T20 for the 3 cases

| T20 (ms) | Systolic | Diastolic |
|---|---|---|
| Normal heart | 92 | 83 |
| Aortic stenosis | 340 | 111 |
| Mitral stenosis | 108 | 568 |

# 4   Conclusion

In this paper, a novel approach for an automated algorithm design for cardiac murmurs detection and classification has been developed. This approach as compared to the previous methods overcomes their main limitations, particularly those recapitulated in the Table 12.1. Although, the initial algorithm stage for extracting one signal period of the cardiac cycle and separating the systolic and diastolic segments was easily designed for some signals (10), associated to the pathology types, where all have the amplitudes of the first and the second heart sounds S1/S2 higher than that of the associated murmurs in the cardiac cycle (clinical grade lower than 5). However, more DATA collection is needed for visualization and observation in order to make the algorithm applicable even for murmurs with high clinical grade. But, murmurs with grade 5 and 6 may be excluded from the database, since they can be easily audible even with the edge of the stethoscope on or off the chest wall of the patient.

**Acknowledgment**   We would like to thank the members committee of the World Congress on Engineering WCE 2011 to have selected our conference paper as book chapter in the edited book published by Springer.

# References

1. Alajarin JM, Merino RR (2005) Efficient method for events detection in phonocardiographic signals. SPIE proceeding 2005. University of Cartagena, Spain.
2. de Vos JP (2005) Automated pediatric cardiac auscultation. Master of Science in Engineering, University of Stellenbosh, South Africa, April 2005.
3. Ahfir M (1997) Mesure et Modélisation des Réponses Impulsionnelles Acoustiques. Thèse de Magister, Octobre 1997, Université de BLIDA, ALGERIA (in French).
4. Gazzam M, Bougrine D, Ahfir M (2010) Analyse des Signaux Cardiaques par le Calcul des Critères Objectifs. Mémoire d'"Ingénieur, Octobre 2010, Université de LAGHOUAT, ALGERIA (in French).
5. Ahfir M, Kale I (2011) Cardiac auscultation improvement by objective criteria computing. The 2011 International Conference of Signal and Image Engineering, Proceeding of the World Congress on Engineering 2011, WCE 2011, 6–8 July 2011, London, UK, 1606–1609

# Chapter 13
# A Fuzzy Logic Approach to Indoor Location Using Fingerprinting

Carlos Serodio, Luis Coutinho, Hugo Pinto, Joao Matias, and Pedro Mestre

## 1 Introduction

From emergency location systems based on mobile networks, such as E911 (Enhanced 911), to the latest concept of applications that are adapted for the end-user and are dynamically delivery based on the user's location [2], all the Location Based Services (LBS) depend on the correct estimation of the users' location.

While in outdoor environments technologies such as GPS (Global Positioning System) can be successfully used, the same is not true when the operating scenarios are indoor environments. In such scenarios alternative location technologies and methodologies must therefore be used, making this a very challenging research area, where in the last years several different types of solutions have been developed.

Some of the most used technologies for indoor location include the use of infra-red [15], ultrasonic waves [16], pressure sensors [8], RFID (Radio Frequency Identification) [13] and wireless communications networks [1, 9, 10]. In what concerns to the methodologies used to obtain the location, they can be divided into three main areas [3]: Triangulation, Proximity and Scene Analysis.

This work is focused on a particular location technique, which uses wireless communications networks as location technology, and a methodology based on scene analysis: location using fingerprinting.

C. Serodio • P. Mestre (✉)
CITAB-UTAD, Vila Real, Portugal
e-mail: cserodio@utad.pt; pmestre@utad.pt

L. Coutinho • H. Pinto
UTAD, Vila Real, Portugal
e-mail: luis_coutinho_86@hotmail.com; htpinto@gmail.com

J. Matias
Centre for Mathematics - UTAD, Vila Real, Portugal
e-mail: j_matias@utad.pt

S.-I. Ao and L. Gelman (eds.), *Electrical Engineering and Intelligent Systems*,
Lecture Notes in Electrical Engineering 130, DOI 10.1007/978-1-4614-2317-1_13,
© Springer Science+Business Media, LLC 2013

One of the objectives of this work is to study methods that can be used to locate standard mobile devices, such as PDAs (Personal Digital Assistant) and mobile phones. These devices usually have wireless communications based on Bluetooth, WiFi (IEEE802.11) and, in the case of mobile phones, have access to the GSM (Global System for Mobile Communications) or 3G communications networks. Bluetooth was not considered for this work mainly due to two reasons: Bluetooth fixed stations are not as ubiquitous as IEEE802.11, GSM or 3G technologies; Bluetooth has a long association time which makes it very difficult to use it in the location of moving terminals. Since a better accuracy is obtained with the lower range technologies [6, 7], when choosing between mobile communications (GSM and 3G) and IEEE802.11, the last has been chosen.

Data to generate the Fingerprint Map and do the location tests were collected using IEEE802.11g Access Points and a laptop computer. These tests were made in classrooms at the University of Trás-os-Montes and Alto Douro, using two Android-based mobile phones. Location estimation was made using two algorithms based on Fuzzy Logic and their performance was compared with the performance of some of the classic methods (Nearest Neighbour, k-Nearest Neighbour and Weighted k-Nearest Neighbour).

## 2 Location Using Wireless Networks

Location using wireless networks is based on the properties of wireless signals. Any property of a wireless signal can be used in location systems, as long as there is a relation between it and the current location of the mobile terminal. The signal properties that usually are used in location systems are the Time-of-Flight (time needed by the information to travel from the transmitter to the receiver) and the Received Signal Strength (RSS).

### 2.1 Location Using Triangulation

Triangulation uses the geometric properties of triangles to determine the location of the mobile node [3]. It can be divided into Lateration and Angulation.

Lateration uses the distances to determine the location. It takes into account the distances between the mobile and the references. It can be made using Circular Triangulation or Hyperbolic Triangulation, with time as independent variable. The distance between the wireless node and the references can be determined based on ToA (Time of Arrival) TDoA (Time Difference of Arrival), or the attenuation value.

To use angulation the angle of incidence of a signal must be known. By analysing the Angle of Arrival of a wireless signal relatively to a given reference, it is possible to determine the location of the mobile node. One example of the application of this method is VOR (VHF Omnidirectional Range).

## 2.2    Localization Using Proximity

Location using this methodology consists in discovering the nearest reference to the mobile terminal, therefore its spatial resolution is dependent on the number of used references.

Cell of Origin is one of the techniques based on Proximity that can be implemented using wireless networks. It is based on the RF cell concept [17]. To determine the location of a mobile node it is only needed to know the cell on which the node is (e.g. the WiFi Access Point to which the node is associated or the GSM cell in which the mobile phone is registered). If the location of the cell is known, then the location of the mobile node is also known. Mobile operators use the Cell of Origin method to determine the localization of their subscribers (as a first approach).

Another technology that uses the Proximity concept to determine the location of objects is RFID. Two different concepts can be used with this technology: Readers are spread along the scenario and detect the presence of the tags carried by the users, when they are within the reading distance as in [13]. Tags are embedded in the environment and users carry small portable RFID readers that detect the presence of the tags as in [11]. In the first case, the location of the user is the location of the reader while in the second is the location of the tag.

## 2.3    Location Using Fingerprinting

Fingerprinting is a scene analysis technique. In scene analysis a scene is "observed" and its patterns and variations along the time are observed. The information about a scene in the case of fingerprinting is obtained from one or more properties of electromagnetic signals generated by the references.

This location methodology consists in reading a given parameter of an electromagnetic signal in real time, typically the value of the Received Signal Strength, and compare it with a set of previously stored values, called the Fingerprinting Map (FM) [1, 4, 14].

Two different concepts of domain are used in fingerprinting: the spatial domain and the signal domain. The first is related with the physical space, i.e., this is the domain where the object to be located is. The second domain is an $N$-dimensional space (each reference is a dimension), which has several values of RSS for each reference. To do the location of a mobile node using fingerprinting, it is first made a search in the signal domain and then it is made the mapping of the information to the spatial domain.

Location using fingerprinting comprises two different phases:

- Calibration phase, which is an off-line phase, i.e., no location is made. It is in this phase that the FM is generated and the mapping between the spatial and the signal domains is made.

- Online phase, on which the location of the mobile node is made. In this phase the value of the RSS signal is acquired from the wireless interface, it is processed by the Location Estimation Algorithm (LEA) and the coordinates, in the spatial domain, are calculated.

# 3   Location Using Fingerprinting

In this section some details about the procedures used to determine the location of a mobile node using fingerprinting are presented. Although any property of the wireless signal can be used in this type of analysis, in this work the fingerprinting analysis will be made based on the RSS values.

## 3.1   Building the Fingerprint Map

Prior to the acquisition of the data to build the fingerprint map, the location of the points (spatial domain) where the RSS readings are going to be made must be established. Data are then acquired for each one of the previously defined points.

Data collected and stored in this phase includes the value of the RSS for each reference found by the mobile terminal, for each point of the spatial domain. The number of references (and therefore dimensions) might be different for each point in the spatial domain.

After collecting all data, the FM is generated. This is made by calculating the average value of the RSS for every reference at each point. In the database it is then stored for each point and reference the corresponding values of the average RSS.

## 3.2   Location Using the Nearest Neighbour Based Algorithms

When using the Nearest Neighbour, the k-Nearest Neighbour and Weighted k-Nearest Neighbour, the first step after acquiring the current value of the received power is to determine the distance between the current point (in the signal domain) and all the points that make part of the FM. This distance can be calculated using the Euclidean distance (13.1):

$$d_j = \sqrt{\sum_{i=0}^{n} (P_{ri} - P_{FMj,i})^2},\tag{13.1}$$

where $d_j$ is the distance to the point $j$, $n$ the number of dimensions, $P_{ri}$ the power received from reference $i$ and $P_{FMj,\,i}$ is the value of the power of reference $i$ registered in the FM for point $j$.

This distance is calculated for all points that belong to the FM and that contain the reference $i$. The mapping of the current location between the signal and the spatial domain is then made using one of the following algorithms:

- Nearest Neighbour—The coordinates of the point in the spatial domain which has the shortest distance (in the signal domain) to the current point are considered as the coordinates of the current location.
- k-Nearest Neighbour—The k nearest points to the current point (in the signal domain) are selected and it is calculated the average of their coordinates in the spatial domain. This value is considered as the coordinates of the current location.
- Weighted k-Nearest Neighbour—In this case, the procedure is similar to the k-Nearest Neighbour. The only difference is that the average of the coordinates is a weighted average.

# 4   Proposed Location Estimation Algorithms Based on Fuzzy Logic

Two distinct LEA based on Fuzzy Logic are presented in this section. The first algorithm is based on the concept of distance where the Fuzzy Logic Algorithm determines which FM points that are considered in the estimation of the current location [12]. The second algorithm is based on pattern search, and it determines the location based on the pattern formed by the RSS values of the detected references [5].

## 4.1   Fuzzy Logic Algorithm Based on the Concept of Distance

In the procedure here described the coordinates for the current location are also estimated by calculating the average of the coordinates of several points of the FM (as in the classic methods). Fuzzy Logic is used to select which points are the most important to calculate the final coordinates of the current location and to assess their corresponding weight in the average.

As for the other algorithms, the first step, after acquiring the current value of the received power, is to determine the distance in the signal domain between the current location and all the points that make part of the fingerprint map. The next step is to transform these distance values into grades of membership, i.e., it is made the fuzzification. For this phase a set of membership functions, such as the one presented in Fig. 13.1, must be used.

In the above presented example the distance, $d$ (in dB), between the current location and the point of the FM (in the signal domain) can be classified as "Very Close" if $d < d_2$, "Near" if $d_1 < d < d_4$ or "Far" if $d > d_3$.

**Fig. 13.1** Membership
functions used to classify
the distance

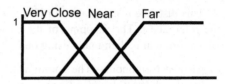

In this case the distance cannot be calculated using the Euclidean distance (13.1) because the number of dimensions might not be the same for all points under test. Since the number of dimensions influences the value of the distance, it would require the definition of several sets of membership functions, one per possible number of dimensions. The number of dimensions cannot be predicted beforehand because it will depend on the maximum number of references, the distance to the references and the sensitivity of the mobile receiver. To cope with this, the distance will be calculated using (13.2):

$$d_j = \sqrt{\frac{1}{n}\sum_{i=0}^{n}(P_{ri} - P_{FMj,i})^2}. \tag{13.2}$$

Based on the distance between the point under test and the reference, a weight for that reference will be chosen. This is done after calculating the grades of membership of the point of the FM under test using the fuzzy inference. In this stage the following simple IF THEN rules are used:

- IF the distance is "Very Close" THEN the point weight is "high."
- IF the distance is "Near" THEN point the weight is "medium."
- IF the distance is "Far" THEN point the weight is "low."

To the values for "high," "medium" and "low" are assigned different weights, $W_1$, $W_2$ and $W_3$, such that $W_1 > W_2 > W_3$.

After the defuzzification, the weight of the point of the FM under test ($W_{FMj}$) is known, and the current coordinates ($C_p$) can be calculated using (13.3):

$$C_p = \frac{\sum_{j=0}^{n}(W_{FMj} \times C_i)}{\sum_{i=0}^{n}(W_{FMi})}, \tag{13.3}$$

where $W_{FMj}$ is the weight of point $j$ of the FM and $C_j$ represents the point coordinates in the spatial domain.

## 4.2  Fuzzy Logic Algorithm Based on the Pattern Search

Although the classic methods are based on the concept of distance, this LEA uses pattern search (based on Fuzzy Logic) to determine the location. This algorithm

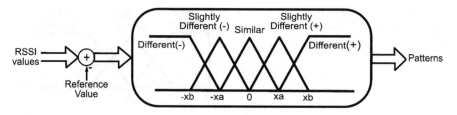

**Fig. 13.2** Generation of the patterns for the points to be used in the FM

was developed to overcome some of the problems associated with the use of RSS values, which include the negative effect of attenuation due to (dynamic) obstacles and the different RSS values obtained with different types of terminals (even under the same operating conditions). The value of the signal power that arrives to receiver is given by (13.4):

$$P_r = P_t - PL + G_t + G_r \tag{13.4}$$

where $P_r$ is the received power in dBm, $P_t$ the transmitted power (dBm), $PL$ the total Path Loss (db) and $G_t$ and $G_r$ are the gains of the transmitting and receiving antenna (in dB or dBi).

As a consequence, different types of terminal will receive different power values because different terminals have different antenna gains. Another issue found in these systems is related with the attenuation due to obstacles. The $PL$ term in (13.4) is dependent both on the free space losses and the attenuation due to obstacles between the transmitting and the receiving antennas. This means that for the same terminal different RSS values for the same references can be obtained at the same point, e.g. a cell phone will have different RSS values when placed on a table or when it is inside a pocket.

### 4.2.1 Building the Patterns

Let us consider the coordinates of point $i$ as a function of the received power ($Pr$) from the $n$ references used in the scenario: $P_i = f(Pr_1, Pr_2, \ldots, Pr_n)$. According to (13.4) different points have different received power values therefore each point of the spatial domain has a different pattern $P_i$. To locate a node, it is then only needed to find similar patterns in the FM.

Due to the above mentioned issues, the values collected from the network interface cannot be directly stored in the FM as a pattern. Instead, a new pattern is generated and stored, as depicted in Fig. 13.2. To each RSS value obtained at a given point it is subtracted a reference value and then the result of this operation is

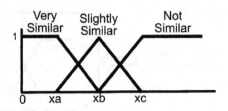

**Fig. 13.3** Membership functions used in the online phase

matched against a set of membership functions which will classify it as "Similar," "Slightly Different" (positive or negative) and "Different" (positive or negative).

This step assigns a weight to each value, according to the following set of IF-THEN rules:

- IF the value is "Different (Negative)" THEN the weight is "high negative."
- IF the value is "Slightly Different (Negative)" THEN the weight is "medium negative."
- IF the value is "Similar" THEN the weight is "low."
- IF the value is "Slightly Different (Positive)" THEN the weight is "medium positive."
- IF the value is "Different (Positive)" THEN the weight is "high positive."

### 4.2.2 Locating the Mobile Node

During the online phase, the location is made by searching in the FM for patterns similar to the one being received by the wireless interface. Since the probability of finding in the FM a pattern that matches perfectly the current pattern is very low, several points must be considered in the location estimation. It is a task of Fuzzy Logic based reasoning to determine which points must be included. For each point of the FM the degree of similarity with the pattern obtained at the current location is determined. The degree of similarity is calculated using (13.5):

$$S_j = \sum_{i=1}^{n} \left( |P_{j,i} - P_{FMi}| \right), \tag{13.5}$$

where $n$ is the number of references, $P_{j,i}$ the value of the pattern of point $j$ for the reference $i$ and $P_{FMi}$ is the value of the pattern in the FM for the point $i$.

These values are then classified as "Very Similar," "Slightly Similar" and "Not Similar." The weight of point $j$ in the final average, used to determine the current location, is dependant on the grade of membership of $S_j$ to the membership functions presented in Fig. 13.3. After determining the grade of membership, the following set of IF THEN rules are used:

- IF the value is "Very Similar" THEN the point weight is "high."
- IF the value is "Slightly Similar" THEN the point weight is "medium."
- IF the value is "Not Similar" THEN the point weight is "low."

After calculating the weight for each point of the FM, the final weighted average of all the coordinates in the spatial domain of the points belonging to FM are considered the coordinates of the current location.

## 5  Tests and Results

In this section the results for the tests made in two classrooms in the University of Trás-os-Montes and Alto Douro are presented. The map of the scenario where the tests where made is presented in Fig. 13.4, where all the points that were considered in this test are marked. These are the points where the data to generate the FM and to do LEA tests were collected. The distance between the points is 2.5 m. In Fig. 13.4 also the location of the Access Points in the testing scenario is presented.

In this test five IEEE802.11g Cisco Aironet 1,200 Access Points, with 6.5 dBi patch antennas, were used. These are the same Access Points and antennas used in our University in the "eduroam" wireless network. In these tests the APsnot under service.

The first step was the acquisition of the data to build the FM. It was made using an application developed using the Java-based Framework presented in [7], using a laptop computer (ASUS Notebook K50IN with an Atheros AR9285 Wireless

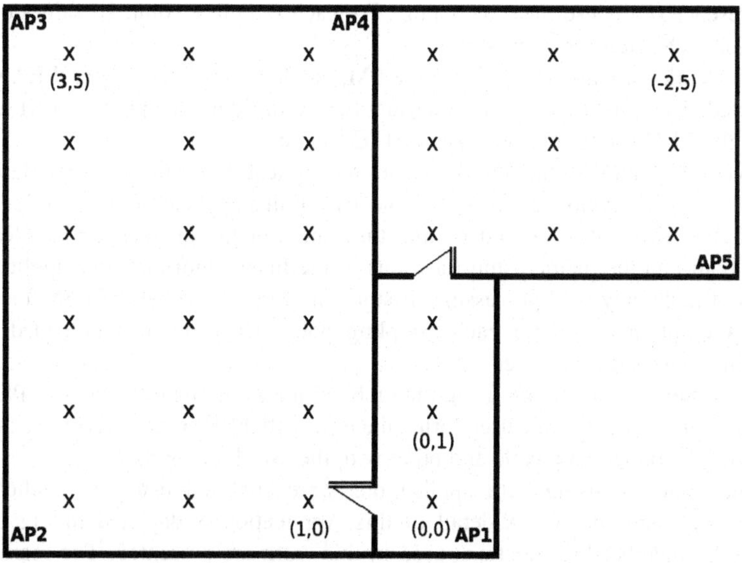

**Fig. 13.4** Map of the area where the location tests were made, the marks represent the location of the points considered for the fingerprint map

**Fig. 13.5** Data collection application developed for the Android platform which was used to collect data using the mobile phones

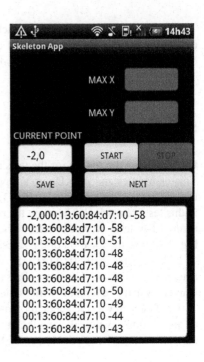

Network Adapter) running Ubuntu Linux. This data collection was made without the presence of the user near the laptop. For each sampling point, 20 values of the RSS for each reference were stored.

After collecting the data to build the FM, the data to test the chosen LEA were collected. These data were collected using two Android mobile phones, a XPERIA X10 mini PRO (Sony Ericsson) and a HTC Desire.

To collect data with the mobile phones an application for the Android platform (Fig. 13.5) was developed. Since the objective of this application is to collect data to test the different LEA and to fine tune the weights of Weighted k-Nearest Neighbour and the membership functions of the fuzzy algorithm, this application does not make any data processing. It stores in a file the values of RSS for each Access Point in range, for each sampling point. Data are then uploaded to a computer where it is analysed.

To do the data collection using the mobile phones, in the case of the XPERIA X10 mini it was held by the user during the test, with the HTC two tests were made, one with the phone on a table and other with the user holding it.

Data collected using both applications were analysed using an application developed using Matlab. A detail of this Application is depicted in Fig. 13.6, where the membership functions used in this work are presented. The marks on the plot represent the different membership degrees for the different points.

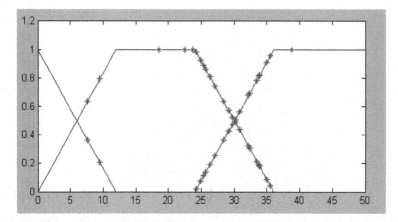

**Fig. 13.6** Detail of the output window of the application developed in Matlab where the used membership functions are presented

**Table 13.1** Comparison of the various methods using three different mobile terminals

|  | Mobile1 (on table) | | Mobile1 (hand) | | Mobile2 (hand) | |
|---|---|---|---|---|---|---|
| LEA | Prec(m) | SDev(m) | Prec(m) | SDev(m) | Prec(m) | SDev(m) |
| N. Neighbour | 6.185 | 1.492 | 7.327 | 1.625 | 5.001 | 2.036 |
| k-N. Neighbour | 4.857 | 0.920 | 6.287 | 0.952 | 3.814 | 1.473 |
| W. k-N. Neighbour | 5.561 | 1.024 | 6.850 | 1.193 | 4.318 | 1.638 |
| Fuzzy logic (distance) | 4.481 | 0.620 | 5.507 | 0.674 | 3.719 | 0.810 |
| Fuzzy logic (pattern) | 3.133 | 1.498 | 3.974 | 1.877 | 3.416 | 1.619 |

In Table 13.1 a comparison of the results obtained with the used Location Estimation Algorithms using the data collected with the two mobile terminals is presented. To be noticed that for the HTC mobile phone all five references were detected and for the other mobile phone only four references were detected at each sampling point. In this test a value of $k = 3$ was used and the weights used for the Weighted k-Nearest Neighbour were 0. 7, 0. 2 and 0. 1, as in [7].

For the tested mobile phones the best method was always the method based on Fuzzy Logic, either considering the standard deviation or the precision. The best value obtained for the precision was 3.133 m and the lower standard deviation was 0.620 m, both using the HTC mobile phone. Also to be noticed that, as it was expected, the worse values for the precision were obtained when the mobile device was being held by the user (comparing the values for the same mobile phone).

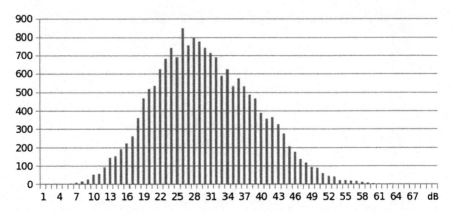

**Fig. 13.7** Histogram of the distances in the Signal Domain for all data samples using the HTC mobile phone on a table

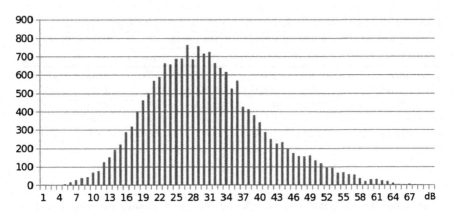

**Fig. 13.8** Histogram of the distances in the Signal Domain for all data samples using the HTC mobile phone held by the user

Observing the histogram for the distances (in the signal domain) of all data samples collected with the three terminals (Figs. 13.7, 13.8 and 13.9), the best performance of the Sonny Ericsson in what concerns to the precision, using the first Fuzzy Logic Algorithm, since it has more points that will be classified as "Very Close" was already expected.

To test the influence of the distribution of the distances, the histogram for the HCT held by the user (Fig. 13.8) was shifted to the left (Fig. 13.10). The new value for the precision was 5.0693 m, which is better than the previous value.

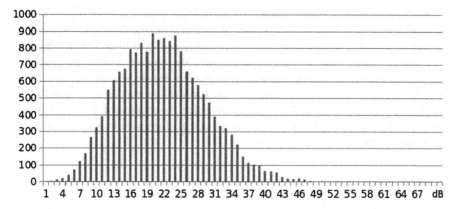

**Fig. 13.9** Histogram of the distances in the Signal Domain for all data samples using the Sony Ericsson mobile phone held by the user

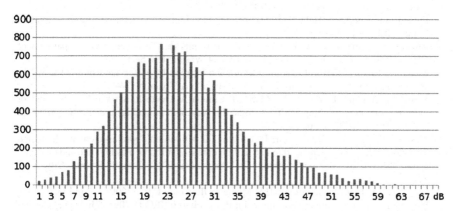

**Fig. 13.10** New Histogram for the distances in the Signal Domain for all data samples using the HTC mobile phone help by the user

## 6   Conclusion

In this work two LEA based on Fuzzy Logic were presented and compared with some of the classic algorithms for location using wireless networks. As a case study, location using IEEE802.11 fingerprinting was presented. Although this work was centred in a particular standard, other wireless network technologies (e.g. IEEE802.15.4) can also be used. In fact in previous works [7] and [6], it has been demonstrated that different types of technologies can even be used together to improve the accuracy.

The best performance was obtained by the methods based on Fuzzy Logic, either considering the precision or the standard deviation as metric. With the LEA based on pattern search a precision of 3.133 m was obtained. It is to be noticed that the FM was generated with another terminal. Also the type of mobile terminal has influence on the performance of the location system. However it can be concluded that, for the tested mobile terminals, the used algorithms can be used, and Fuzzy Logic is a feasible solution. Not being so sensitive to variation of the RSS values, the algorithm based on pattern search had the best performance in what concerns to the values of precision.

# References

1. Bahl P, Padmanabhan VN (2000) RADAR: An in-building RF-based user location and tracking system. INFOCOM 2000. Nineteenth annual joint conference of the IEEE computer and communications societies. Proc IEEE 2:775–784
2. Bellavista P, Kupper A, Helal S (2008) Location-based services: back to the future. Pervasive Comput IEEE 7(2):85–89
3. Hightower J, Borriello G (2001) Location sensing techniques. Technical report, Department of Computer Science and Engineering, University of Washington, Seattle
4. Komar C, Ersoy C (2004) Location tracking and location based service using IEEE 802.11 WLAN infrastructure. In: European wireless. Barcelona, Spain, pp 24–27
5. Mestre P, Coutinho L, Reigoto L, Matias J, Correia A, Couto P, Serodio C (2011) Indoor location using fingerprinting and fuzzy logic. In: Advances in intelligent and soft computing, vol 107. Springer, Berlin, pp 363–374
6. Mestre P, Pinto H, Moura J, Oliveira P, Serôdio C (2010) Multiple wireless technologies fusion for indoor location estimation. International conference on Indoor Position and Indoor Navigation (IPIN2010). Abstract volume, 15–17 September 2010, Campus Science City, ETH Zurich, pp 69–70
7. Silva PMMA, Pinto H, Serodio CMJA, Monteiro JL, Couto CAC (2009) A multi-technology framework for LBS using fingerprinting. In: Proceedings of IECON '09, 35th Annual Conference of IEEE Industrial Electronics, 3–5 November 2009, Porto, Portugal, pp 2693–2698
8. Orr RJ, Abowd GD (2000) The smart floor: A mechanism for natural user identification and tracking. In: CHI '00: CHI '00 extended abstracts on Human factors in computing systems, pp 275–276, New York, NY, USA, ACM
9. Otsason V, Varshavsky A, LaMarca A, de Lara E (2005) Accurate GSM indoor localization. In: Proceedings of ubiquitous computing, pp 141–158
10. Prasithsangaree P, Krishnamurthy P, Chrysanthis P (2002) On indoor position location with wireless LANs. The 13th IEEE international symposium on personal, indoor and mobile radio communications, 2002, vol 2, pp 720–724
11. Seco F, Plagemann C, Jiménez AR, Burgard W (2010) Improving RFID-based indoor positioning accuracy using gaussian processes. In: Proceeding of IPIN 2010, International conference on Indoor Position and Indoor Navigation. 15–17 September 2010, Campus Science City, ETH Zurich, pp 1–8
12. Serodio C, Coutinho L, Pinto H, Mestre P (2011) A comparison of multiple algorithms for fingerprinting using IEEE802.11. In: Lecture notes in engineering and computer science: proceedings of the world congress on engineering 2011 (WCE 2011), 6–8 July, 2011, pp 1710–1715. London, UK

13. Silva PM, Paralta M, Caldeirinha R, Rodrigues J, Serodio C (2009) TraceMe – indoor real-time location system. In: Proceedings of IECON '09, 35th Annual Conference of IEEE Industrial Electronics, 3–5 November 2009, Porto, Portugal, pp 2721–2725
14. Taheri A, Singh A, Agu E (2004) Location fingerprinting on infrastructure 802.11 wireless local area networks. In: LCN'04: Proceedings of the 29th annual IEEE international conference on Local Computer Networks, IEEE Computer Society. pp 676–683. Washington, DC, USA
15. Want R, Hopper A, Falcao V, Gibbons J (1992) The active badge location system. ACM Trans Inf Syst 10(1):91–102
16. Ward A, Jones A, Hopper A (1997) A new location technique for the active office. IEEE Pers Comm 4(5):42–47
17. Zeimpekis V, Giaglis, GM, Lekakos G (2003) A taxonomy of indoor and outdoor positioning techniques for mobile location services. SIGecom Exch 3(4):19–27

# Chapter 14
# Wi-Fi Point-to-Point Links

## Extended Performance Studies of IEEE 802.11 b,g Laboratory Links Under Security Encryption

J.A.R. Pacheco de Carvalho, H. Veiga, N. Marques,
C.F. Ribeiro Pacheco, and A.D. Reis

## 1 Introduction

Contactless communication techniques have been developed using mainly electromagnetic waves in several frequency ranges, propagating in the air. Examples of wireless communications technologies are Wi-Fi and FSO, whose importance and utilization have been growing.

Wi-Fi is a microwave-based technology providing for versatility, mobility and favourable prices. The importance and utilization of Wi-Fi have been growing for complementing traditional wired networks. It has been used both in ad hoc mode and in infrastructure mode. In this case an access point, AP, permits communications of Wi-Fi devices with a wired-based LAN through a switch/router. In this way a WLAN, based on the AP, is formed. Wi-Fi has reached the personal home, forming a WPAN, allowing personal devices to communicate. Point-to-point and point-to-multipoint configurations are used both indoors and outdoors, requiring specific directional and omnidirectional antennas. Wi-Fi uses microwaves in the 2.4 and 5 GHz frequency bands and IEEE 802.11a, 802.11b, 802.11g and 802.11n standards [1]. As the 2.4 GHz band becomes increasingly used and interferences increase, the 5 GHz band has received considerable attention, although absorption increases and ranges are shorter.

Nominal transfer rates up to 11 (802.11b), 54 Mbps (802.11 a, g) and 600 Mbps (802.11n) are specified. CSMA/CA is the medium access control. Wireless communications, wave propagation [2, 3] and practical implementations of WLANs [4] have been studied. Detailed information has been given about the

J.A.R.P. de Carvalho (✉) • C.F.R. Pacheco • A.D. Reis
Unidade de Detecção Remota, Universidade da Beira Interior, 6201-001 Covilhã, Portugal
e-mail: pacheco@ ubi.pt; a17597@ubi.pt; adreis@ubi.pt

H. Veiga • N. Marques
Centro de Informática, Universidade da Beira Interior, 6201-001 Covilhã, Portugal
e-mail: hveiga@ubi.pt; nmarques@ubi.pt

S.-I. Ao and L. Gelman (eds.), *Electrical Engineering and Intelligent Systems*,
Lecture Notes in Electrical Engineering 130, DOI 10.1007/978-1-4614-2317-1_14,
© Springer Science+Business Media, LLC 2013

802.11 architecture, including performance analysis of the effective transfer rate. An optimum factor of 0.42 was presented for 11 Mbps point-to-point links [5]. Wi-Fi (802.11b) performance measurements are available for crowded indoor environments [6].

Performance has been a fundamentally important issue, giving more reliable and efficient communications. In comparison to traditional applications, new telematic applications are specially sensitive to performances. Requirements have been pointed out, such as 1–10 ms jitter and 1–10 Mbps throughput for video on demand/moving images; jitter less than 1 ms and 0.1–1 Mbps throughputs for Hi Fi stereo audio [7].

Wi-Fi security is very important. Microwave radio signals travel through the air and can be easily captured by virtually everyone. Therefore, several security methods have been developed to provide authentication, such as by increasing order of security, WEP, WPA and WPA2. WEP was initially intended to provide confidentiality comparable to that of a traditional wired network. A shared key for data encryption is involved. In WEP, the communicating devices use the same key to encrypt and decrypt radio signals. The CRC32 checksum used in WEP does not provide a great protection. However, in spite of its weaknesses, WEP is still widely used in Wi-Fi communications for security reasons. WPA implements the majority of the IEEE 802.11i standard [1]. It includes a MIC, message integrity check, replacing the CRC used in WEP. WPA2 is compliant with the full IEEE 802.11i standard. It includes CCMP, a new AES-based encryption mode with enhanced security. WPA and WPA2 can be used in either personal or enterprise modes. In this latter case an 802.1x server is required. Both TKIP and AES cipher types are usable and a group key update time interval is specified.

Several performance measurements have been made for 2.4 and 5 GHz Wi-Fi open [8, 9], WEP [10], WPA2 links [11] as well as very high speed FSO [12]. In the present work further Wi-Fi (IEEE 802.11 b,g) results arise, using personal mode WPA2, through OSI levels 4 and 7. Performance is evaluated in laboratory measurements of WPA2 point-to-point links using available equipments. Comparisons are made to corresponding results obtained for open, WPA and WEP links.

The rest of the paper is structured as follows: Chapter 2 presents the experimental details i.e. the measurement setup and procedure. Results and discussion are presented in Chap. 3. Conclusions are drawn in Chap. 4.

## 2  Experimental Details

The measurements used Linksys WRT54GL wireless routers [13], with a Broadcom BCM5352 chip rev0, internal diversity antennas, firmware DD-WRT v24-sp1-10011 [14] and a 100-Base-TX/10-Base-T Allied Telesis AT-8000S/16 level 2 switch [15]. The wireless mode was set to bridged access point. In every type

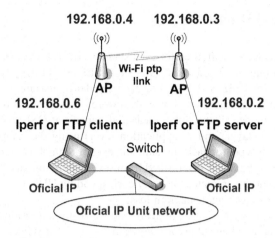

**Fig. 14.1** Experimental laboratory setup scheme

of experiment, interference free communication channels were used. This was checked through a portable computer, equipped with a Wi-Fi 802.11 a/b/g adapter, running NetStumbler software [16]. For WPA2 and WPA, personal encryption was activated in the APs, using AES and a shared key composed of 9 ASCII characters. WEP encryption activation used 128 bit encryption and a shared key composed of 13 ASCII characters. The experiments were made under far-field conditions. No power levels above 30 mW (15 dBm) were required, as the access points were close.

A laboratory setup was planned and implemented for the measurements, as shown in Fig. 14.1. At OSI level 4, measurements were made for TCP connections and UDP communications using Iperf software [17], permitting network performance results to be recorded. For a TCP connection, TCP throughput was obtained. For a UDP communication with a given bandwidth parameter, UDP throughput, jitter and percentage loss of datagrams were determined. TCP packets and UDP datagrams of 1,470 bytes size were used. A window size of 8 kbytes and a buffer size of the same value were used for TCP and UDP, respectively. One PC, with IP 192.168.0.2, was the Iperf server and the other, with IP 192.168.0.6, was the Iperf client. Jitter, which indicates the smooth mean of differences between consecutive transit times, was continuously computed by the server, as specified by RTP in RFC 1889 [18]. The scheme of Fig. 14.1 was also used for FTP measurements, where FTP server and client applications were installed in the PCs with IPs 192.168.0.2 and 192.168.0.6, respectively.

The server and client PCs were HP nx9030 and nx9010 portable computers, respectively, running Windows XP. They were configured to maximize the resources allocated to the present work. Batch command files were written to enable the TCP, UDP and FTP tests. The results were obtained in batch mode and written as data files to the client PC disk. Each PC had a second network adapter, to permit remote control from the official IP Unit network, via switch.

## 3 Results and Discussion

The access points were configured for each standard IEEE 802.11 b, g with typical nominal transfer rates (1, 2, 5.5 and 11 Mbps for IEEE 802.11 b; 6, 9, 12, 18, 24, 36, 48 and 54 Mbps for IEEE 802.11g). For each standard, measurements were made for every fixed transfer rate. In this way, data were obtained for comparison of the laboratory performance of the links, measured namely at OSI levels 1 (physical layer), 4 (transport layer) and 7 (application layer) using the setup of Fig. 14.1. In each experiment type, for each standard and every nominal fixed transfer rate, an average TCP throughput was determined from several experiments. This value was used as the bandwidth parameter for every corresponding UDP test, giving average jitter and average percentage datagram loss.

At OSI level 1, noise levels (N, in dBm) and signal-to-noise ratios (SNR, in dB) were monitored and typical values are shown in Fig. 14.2 for WPA2 and open links. Similar values were obtained for WPA and WEP links.

The main average TCP and UDP results are summarized in Table 14.1, both for WPA2 and open links. In Fig. 14.3 polynomial fits were made to the 802.11b, g TCP throughput data for WPA2 and open links, where $R^2$ is the coefficient of determination. It was found that the best TCP throughputs are for 802.11g. A fairly good agreement was found for the 802.11 b, g data both for WPA2, open, WPA and WEP links. In Figs. 14.4–14.6, the data points representing jitter and percentage datagram loss were joined by smoothed lines. Jitter results are shown in Fig. 14.4 for WPA2 and open links, and in Fig 14.5 for WPA and WEP links. It was found that, on average, the best jitter performances are for 802.11g for all link types. On average, both for 802.11 b and 802.11g, the best jitter performances were found for open links. The best jitter performances were found, by descending order, for open, WEP, WPA and WPA2 links. Increasing security encryption was found to degrade jitter performance. Results for percentage datagram loss are illustrated in Fig. 14.6 for WPA2 and open links. No significant sensitivities were found in the data (1.3 % on average), within the experimental errors, either to standard or link type.

At OSI level 7 we measured FTP transfer rates versus nominal transfer rates configured in the access points for the IEEE 802.11b, g standards. Every measurement was the average for a single FTP transfer, using a binary file size of 100 Mbytes. The average results thus obtained are summarized in Table 14.1, both for WPA2 and open links. In Fig. 14.7 polynomial fits are shown to 802.11 b, g data for WPA2 and open links. It was found that, for each link type, the best FTP performances were for 802.11g. A fairly good agreement was found for the 802.11 b, g data for all link types. These results show the same trends found for TCP throughput.

Generally, except for jitter, the results measured for WPA2 links were found to agree, within the experimental errors, with corresponding data obtained for open, WEP and WPA links. Increasing security encryption was found to degrade jitter performance.

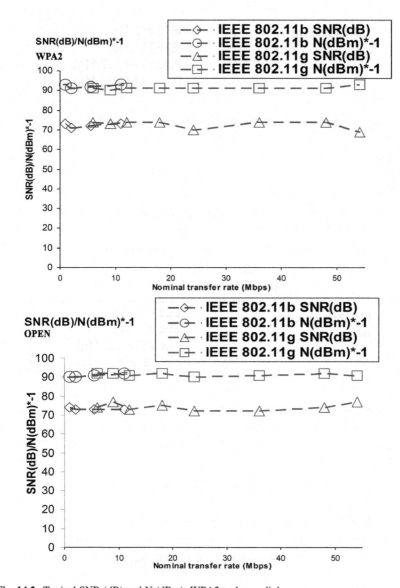

**Fig. 14.2** Typical SNR (dB) and N (dBm); WPA2 and open links

**Table 14.1** Average Wi-Fi (IEEE 802.11 b,g) results; WPA2 and open links

| Link type | WPA2 | | Open | |
|---|---|---|---|---|
| Parameter/IEEE standard | 802.11b | 802.11g | 802.11b | 802.11g |
| TCP throughput (Mbps) | 2.9 ± 0.1 | 13.9 + 0.4 | 2.9 ± 0.1 | 13.9 ± 0.4 |
| UDP-jitter (ms) | 3.6 ± 0.2 | 2.4 ± 0.1 | 2.1 ± 0.2 | 1.2 ± 0.1 |
| UDP-% datagram loss | 1.2 ± 0.2 | 1.3 ± 0.1 | 1.3 ± 0.2 | 1.6 ± 0.1 |
| FTP transfer rate (kbyte/s) | 349.7 ± 10.5 | 1504.1 ± 45.1 | 342.3 ± 10.3 | 1508.3 ± 45.2 |

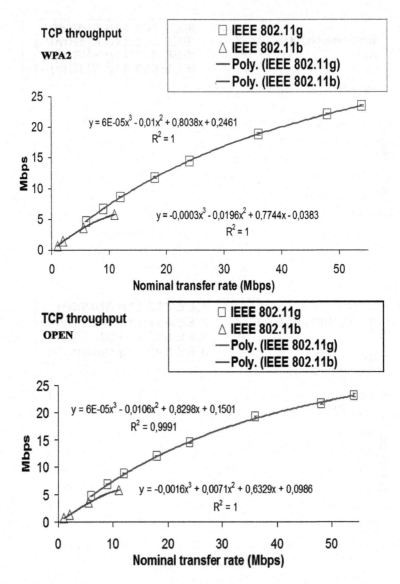

**Fig. 14.3** TCP throughput results (*y*) versus technology and nominal transfer rate (*x*); WPA2 and open links

# 4 Conclusions

In the present work a laboratory setup arrangement was planned and implemented, permitted systematic performance measurements of available wireless equipments (WRT54GL wireless routers from Linksys) for Wi-Fi (IEEE 802.11 b, g) in WPA2

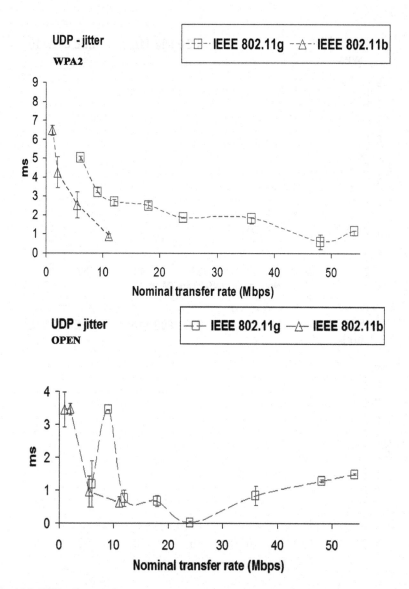

**Fig. 14.4** UDP – jitter results versus technology and nominal transfer rate; WPA2 and open links

point-to-point links. Comparisons were made to corresponding results obtained for
open, WPA and WEP links.

Through OSI layer 4, TCP throughput, jitter and percentage datagram loss were
measured and compared for each standard. The best TCP throughputs were found
for 802.11g. A fairly good agreement was found for the 802.11 b, g data for all link
types. Concerning jitter, it was found that on average the best performances were
for 802.11g for all link types. On average, both for 802.11 b and 802.11g, the best

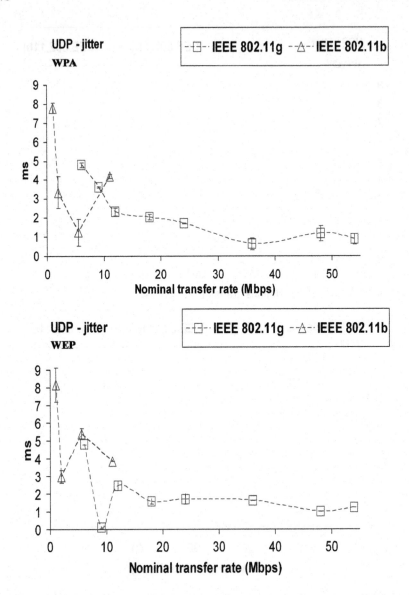

**Fig. 14.5** UDP – jitter results versus technology and nominal transfer rate; WPA and WEP links

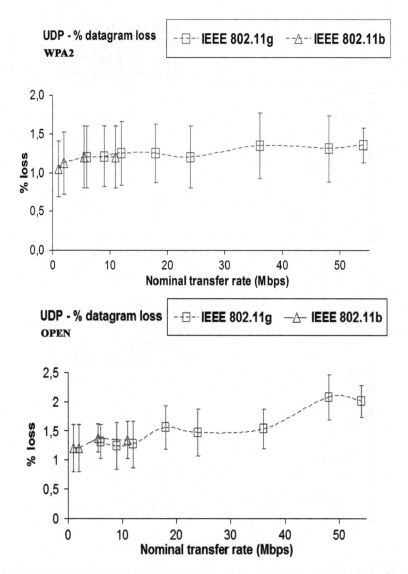

**Fig. 14.6** UDP – percentage datagram loss results versus technology and nominal transfer rate; WPA2 and open links

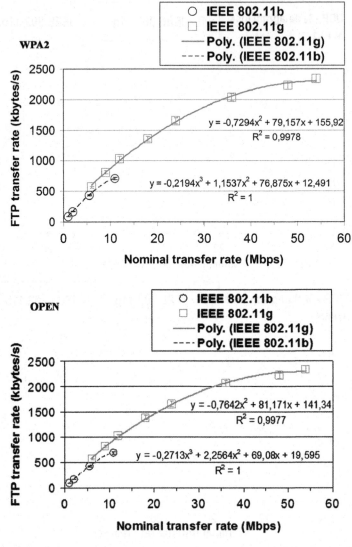

**Fig. 14.7** FTP transfer rate results ($y$) versus technology and nominal transfer rate ($x$); WPA2 and open links

jitter performances were found for open links. The best jitter performances were found, by descending order, for open, WEP, WPA and WPA2 links. Increasing security encryption was found to degrade jitter performance. For percentage datagram loss, no significant sensitivities were found, within the experimental errors, either to standard or link type.

At OSI layer 7, it was found that, for each link type, the best FTP performances are for 802.11g. A fairly good agreement was found for the 802.11 b, g data for all link types. The results show the same trends found for TCP throughput.

Generally, except for jitter, the results measured for WPA2 links were found to agree, within the experimental errors, with corresponding data obtained for open, WEP and WPA links.

Additional performance measurements either started or are planned using several equipments and security settings, not only in laboratory but also in outdoor environments involving, mainly, medium range links.

**Acknowledgement**    Supports from University of Beira Interior and FCT (Fundação para a Ciência e a Tecnologia)/POCI2010 (Programa Operacional Ciência e Inovação) are acknowledged.

# References

1. IEEE 802.11a, 802.11b, 802.11g (2007) 802.11n, 802.11i (2009) standards; http://standards. ieee.org/getieee802. Accessed 10th Dec 2010
2. Mark JW, Zhuang W (2003) Wireless communications and networking. Prentice-Hall, Inc., Upper Saddle River, NJ
3. Rappaport TS (2002) Wireless communications principles and practice, 2nd edn. Prentice-Hall, Inc., Upper Saddle River, NJ
4. Bruce WR III, Gilster R (2002) Wireless LANs end to end. Hungry Minds, Inc., NY
5. Schwartz M (2005) Mobile wireless communications. Cambridge University Press, Cambridge
6. Sarkar NI, Sowerby KW (2006) High performance measurements in the crowded office environment: a case study. Proceedings of ICCT'06-International conference on communication technology. Guilin, China, 27–30 Nov 2006, pp 1–4.
7. Monteiro E, Boavida F (2002) Engineering of informatics networks, 4th edn. FCA-Editor of Informatics Ld, Lisbon
8. Pacheco de Carvalho JAR, Gomes PAJ, Veiga H, Reis AD (2008) Development of a university networking project. In: Putnik GD, Cunha MM (eds) Encyclopedia of networked and virtual organizations. IGI Global, Hershey, PA (Pennsylvania), pp 409–422
9. Pacheco de Carvalho JAR, Veiga H, Gomes PAJ, Ribeiro Pacheco CF, Marques N, Reis AD (2010) Wi-Fi point-to-point links- performance aspects of IEEE 802.11 a,b,g laboratory links.

In: Ao SI, Gelman L (eds) Electronic engineering and computing technology. Series: lecture notes in electrical engineering, vol. 60. Springer, Netherlands, pp 507–514.
10. Pacheco de Carvalho JAR, Veiga H, Marques N, Ribeiro Pacheco CF, Reis AD (2011) Wi-Fi WEP point-to-point links- performance studies of IEEE 802.11 a,b,g laboratory links. In: Ao SI, Gelman L (eds) Electronic engineering and computing technology. Series: lecture notes in electrical engineering, vol. 90. Springer, Netherlands, pp 105–114.
11. Pacheco de Carvalho JAR, Veiga H, Marques N, Ribeiro Pacheco CF, Reis AD (2011) Laboratory performance of Wi-Fi IEEE 802.11 b,g WPA2 point-to-point links: a case study. In: Proceedings of the WCE 2011 – world congress on engineering 2011, vol. II. Imperial College London, London, England, pp 1770–1774, 6–8 July 2011.
12. Pacheco de Carvalho JAR, Veiga H, Gomes PAJ, Cláudia FFP, Pacheco R, Reis AD (2008) Experimental performance study of very high speed free space optics link at the university of beira interior campus: a case study. In: Proceedings of the ISSPIT 2008-8th IEEE international symposium on signal processing and information technology. Sarajevo, Bosnia and Herzegovina, pp 154–157, 16–19, Dec 2008.
13. Linksys (2005) WRT54GL wireless router technical data; http://www.linksys.com. Accessed 13th Jan 2011
14. DD-WRT (2008) DD-WRT firmware; http://www.dd-wrt.com. Accessed 13th Jan 2011
15. Allied Telesis (2008) AT-8000S/16 Layer 2 managed fast ethernet switch. http://www.alliedtelesis.com. Accessed 20 Dec 2008.
16. NetStumbler software (2005); http://www.netstumbler.com. Accessed 21st March 2011
17. Iperf software (2003) NLANR, http://dast.nlanr.net. Accessed 10th Jan 2008
18. Network Working Group (1996) RFC 1889-RTP: a transport protocol for real time applications; http://www.rfc-archive.org. accessed 10th Feb 2011

# Chapter 15
# Wi-Fi IEEE 802.11 B,G WEP Links

## Performance Studies of Laboratory Point-to-Point Links

**J.A.R. Pacheco de Carvalho, H. Veiga, N. Marques, C.F. Ribeiro Pacheco, and A.D. Reis**

## 1 Introduction

Contactless communication techniques have been developed using mainly electromagnetic waves in several frequency ranges, propagating in the air. Examples of wireless communications technologies are Wi-Fi and FSO, whose importance and utilization have been growing.

Wi-Fi is a microwave-based technology providing for versatility, mobility and favourable prices. The importance and utilization of Wi-Fi have been growing for complementing traditional wired networks. It has been used both in ad hoc mode and in infrastructure mode. In this case an access point, AP, permits communications of Wi-Fi devices (such as a personal computer, a wireless sensor, a PDA, a smartphone, video game console and a digital audio player) with a wired-based LAN through a switch/router. In this way a WLAN, based on the AP, is formed. Wi-Fi has reached the personal home, forming a WPAN, allowing personal devices to communicate. Point-to-point and point-to-multipoint configurations are used both indoors and outdoors, requiring specific directional and omnidirectional antennas. Wi-Fi uses microwaves in the 2.4 and 5 GHz frequency bands and IEEE 802.11a, 802.11b, 802.11g and 802.11n standards [1]. The 2.4 GHz band has been increasingly used, leading to higher interferences. Therefore, the 5 GHz band has received considerable attention, although absorption increases and ranges are shorter.

J.A.R.P. de Carvalho (✉) • C.F.R. Pacheco,
• A.D. Reis
Unidade de Detecção RemotaUniversidade da Beira Interior, 6201-001 Covilhã, Portugal
e-mail: pacheco@ubi.pt; a17597@ubi.pt; adreis@ubi.pt

H. Veiga • N. Marques
Centro de Informática, Universidade da Beira Interior, 6201-001 Covilhã, Portugal
e-mail: hveiga@ubi.pt; nmarques@ubi.pt

S.-I. Ao and L. Gelman (eds.), *Electrical Engineering and Intelligent Systems*,
Lecture Notes in Electrical Engineering 130, DOI 10.1007/978-1-4614-2317-1_15,
© Springer Science+Business Media, LLC 2013

Nominal transfer rates up to 11 (802.11b), 54 Mbps (802.11 a, g) and 600 Mbps (802.11n) are specified. The medium access control is CSMA/CA. Wireless communications, wave propagation [2, 3] and practical implementations of WLANs [4] have been studied. Details about the 802.11 architecture have been given, including performance analysis of the effective transfer rate for 802.11b point-to-point links [5]. Wi-Fi (802.11b) performance measurements are available for crowded indoor environments [6].

Performance has been a most relevant issue, resulting in more reliable and efficient communications. New telematic applications are specially sensitive to performances, when compared to traditional applications. Requirements have been presented, such as 1–10 ms jitter and 1–10 Mbps throughput for video on demand/moving images; jitter less than 1 ms and 0.1–1 Mbps throughputs for Hi Fi stereo audio [7].

Wi-Fi security is very important as microwave radio signals travel through the air and can be very easily captured. WEP was initially intended to provide confidentiality comparable to that of a traditional wired network. A shared key for data encryption is involved. In WEP, the communicating devices use the same key to encrypt and decrypt radio signals. The CRC32 checksum used in WEP does not provide a great protection. In spite of presenting weaknesses, WEP is still widely used in Wi-Fi networks for security reasons, mainly in point-to-point links.

Several performance measurements have been made for 2.4 and 5 GHz Wi-Fi open [8, 9], WEP [10, 11] links as well as very high speed FSO [12]. Following, in the present work further Wi-Fi (IEEE 802.11 b,g) results are presented, using WEP encryption, namely through OSI levels 1, 4 and 7. Performance is evaluated in laboratory measurements of WEP point-to-point links using available equipments. Detailed comparisons are made about the performances of the links.

The rest of the paper is structured as follows: Chapter 2 presents the experimental details, i.e. the measurement setup and procedure. Results and discussion are presented in Chap. 3. Conclusions are drawn in Chap. 4.

## 2 Experimental Details

Two types of experiments were carried out, which are referred as Expb and Expc. The measurements of Expb used Enterasys RoamAbout RBT-4102 level 2/3/4 access points (mentioned as APb), equipped with 16–20 dBm IEEE 802.11 a/b/g transceivers and internal dual-band diversity antennas [13], and 100-Base-TX/10-Base-T Allied Telesis AT-8000S/16 level 2 switches [14]. The access points had transceivers based on the Atheros 5213A chipset, and firmware version 1.1.51. The configuration was for minimum transmitted power and equivalent to point to point, LAN to LAN mode, using the internal antenna. Expc used Linksys WRT54GL wireless routers [15] (mentioned as APc), with a Broadcom BCM5352

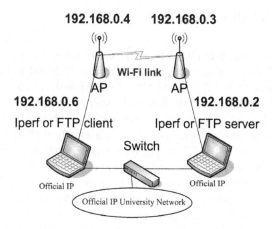

**Fig. 15.1** Experimental laboratory setup scheme

chip rev0, internal diversity antennas, firmware DD-WRT v24-sp1-10011 [16] and the same type of level 2 switch [14]. The firmware from the manufacturer did not provide for point-to-point links. The wireless mode was set to bridged access point. In every type of experiment, interference-free communication channels were used. This was checked through a portable computer, equipped with a Wi-Fi 802.11 a/b/g adapter, running NetStumbler software [17]. WEP encryption was activated in the APs, using 128 bit encryption and a shared key for data encryption composed of 13 ASCII characters. The experiments were made under far-field conditions. No power levels above 30 mW (15 dBm) were required, as the access points were close.

A laboratory setup has been planned and implemented for the measurements, as shown in Fig. 15.1. At OSI level 4 , measurements were made for TCP connections and UDP communications using Iperf software [18], permitting network perfor-mance results to be recorded. For a TCP connection, TCP throughput was obtained. For a UDP communication with a given bandwidth parameter, UDP throughput, jitter and percentage loss of datagrams were determined. TCP packets and UDP datagrams of 1,470 bytes size were used. A window size of 8 kbytes and a buffer size of the same value were used for TCP and UDP, respectively. One PC, with IP 192.168.0.2, was the Iperf server and the other, with IP 192.168.0.6, was the Iperf client. Jitter, which represents the smooth mean of differences between consecutive transit times, was continuously computed by the server, as specified by RTP in RFC 1889 [19]. The scheme of Fig. 15.1 was also used for FTP measurements, where FTP server and client applications were installed in the PCs with IPs 192.168.0.2 and 192.168.0.6, respectively.

The server and client PCs were HP portable computers running Windows XP. They were configured to maximize the resources available to the present work. Batch command files have been written to enable the TCP, UDP and FTP tests. The results were obtained in batch mode and written as data files to the client PC disk. Each PC had a second network adapter, to permit remote control from the official IP University network, via switch.

## 3   Results and Discussion

In each type of experiment Expb and Expc, the corresponding access points APb and APc, respectively, were configured for each standard IEEE 802.11 b, g with typical nominal transfer rates (1, 2, 5.5 and 11 Mbps for IEEE 802.11 b; 6, 9, 12, 18, 24, 36, 48 and 54 Mbps for IEEE 802.11g). For each experiment type, measurements were made for every fixed transfer rate. In this way, data were obtained for comparison of the laboratory performance of the links, measured namely at OSI levels 1 (physical layer), 4 (transport layer), and 7 (application layer) using the setup of Fig. 15.1. In each experiment type, for each standard and every nominal fixed transfer rate, an average TCP throughput was determined from several experiments. This value was used as the bandwidth parameter for every corresponding UDP test, giving average jitter and average percentage datagram loss.

At OSI level 1, noise levels (N, in dBm) and signal-to-noise ratios (SNR, in dB) were monitored and typical values are shown in Fig. 15.2 for Expb and Expc.

The main average TCP and UDP results are summarized in Table 15.1. In Fig. 15.3, both for Expb and Expc, polynomial fits were made to the 802.11b, g TCP throughput data, where $R^2$ is the coefficient of determination. It is seen that, for each AP type, the best TCP throughputs are for 802.11g. APc shows, on average, a better TCP throughput performance than APb, both for 802.11g (+49.0 %) and 802.11 b (+11.5 %). In Figs. 15.4 and 15.5, the data points representing jitter and percentage datagram loss were joined by smoothed lines. It follows that, on average, APb and APc present similar jitter performances for 802.11 b, within the experimental error. For 802.11g APc shows a better average jitter performance (1.8 ± 0.1 ms) than APb (2.6 ± 0.1 ms). Concerning percentage datagram loss data (1.2% on average) there is a fairly good agreement for both APb and APc, and for both standards.

At OSI level 7 we measured FTP transfer rates versus nominal transfer rates configured in the access points for the IEEE 802.11b, g standards. Every measurement was the average for a single FTP transfer, using a binary file size of 100 Mbytes. The average results thus obtained are summarized in Table 15.1 and represented in Fig. 15.6 both for Expb and Expc. Polynomial fits to data were made for the implementation of each standard. It was found that, for each AP type, the best FTP performances are for 802.11g. APc shows, on average, a better FTP performance than APb, both for 802.11g and 802.11b. These results show the same trends found for TCP throughput.

Generally, the results measured for WEP links were not found as significantly different, within the experimental errors, from corresponding data obtained for open links, except for jitter, where the best performances were found for open links. WEP encryption contributed to degradation of jitter performance.

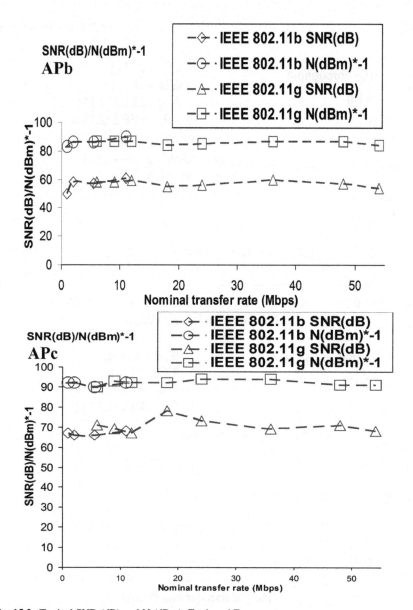

**Fig. 15.2** Typical SNR (dB) and N (dBm); Expb and Expc

**Table 15.1** Average Wi-Fi (IEEE 802.11 b,g) results; Expb and Expc

| Experiment type | Expb | | Expc | |
|---|---|---|---|---|
| Parameter/IEEE standard | 802.11b | 802.11g | 802.11b | 802.11g |
| TCP throughput (Mbps) | 2.6 ± 0.1 | 9.6 ± 0.3 | 2.9 ± 0.1 | 14.3 ± 0.4 |
| UDP-jitter (ms) | 4.8 ± 0.3 | 2.6 ± 0.1 | 5.1 ± 0.3 | 1.8 ± 0.1 |
| UDP-% datagram loss | 1.2 ± 0.2 | 1.3 ± 0.1 | 1.1 ± 0.2 | 1.3 ± 0.1 |
| FTP transfer rate (kbyte/s) | 300.5 ± 9.0 | 1106.4 ± 55.3 | 347.9 ± 10.4 | 1557.7 ± 46.7 |

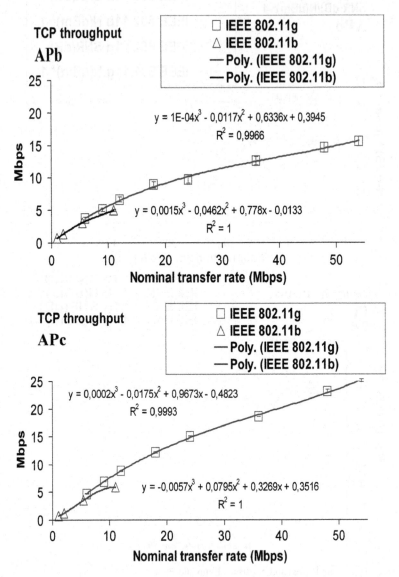

**Fig. 15.3** TCP throughput results (*y*) versus technology and nominal transfer rate (*x*); Expb and Expc

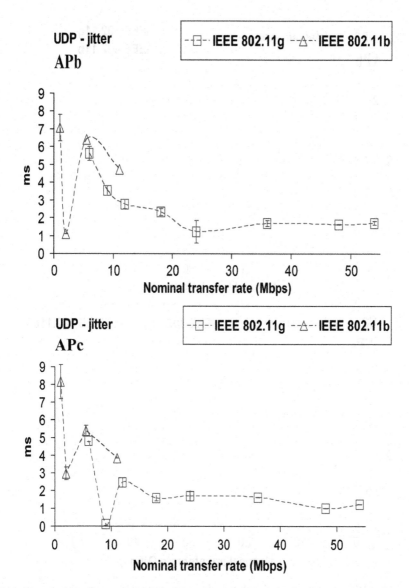

**Fig. 15.4** UDP – jitter results versus technology and nominal transfer rate; Expb and Expc

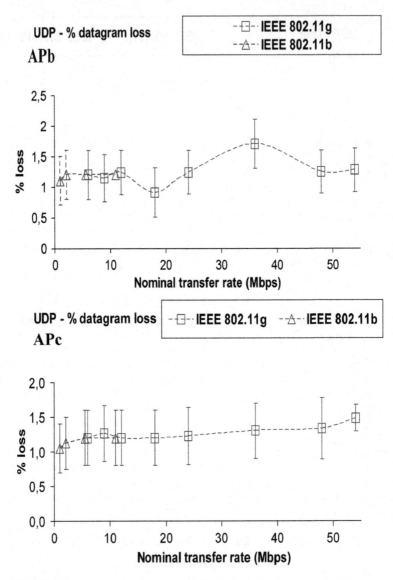

**Fig. 15.5** UDP – percentage datagram loss results versus technology and nominal transfer rate; Expb and Expc

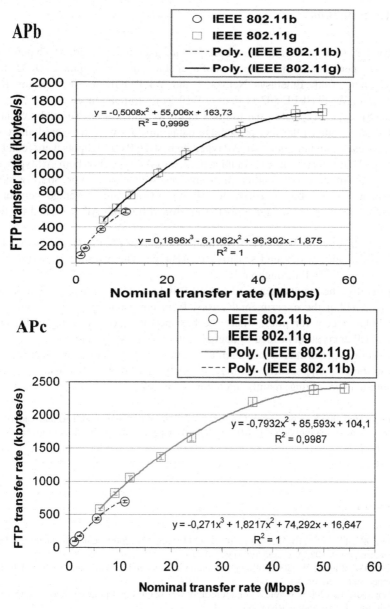

**Fig. 15.6** FTP transfer rate results ($y$) versus technology and nominal transfer rate ($x$); Expb and Expc

192 J.A.R.P. de Carvalho et al.

# 4 Conclusions

A laboratory setup arrangement has been planned and implemented, which permitted systematic performance measurements of available wireless equipments (RBT-4102 access points from Enterasys Networks and WRT54GL wireless routers from Linksys) for Wi-Fi (IEEE 802.11 b, g) in WEP point-to-point links.

At OSI level 1, typical signal-to-noise ratios were measured. Through OSI layer 4, TCP throughput, jitter and percentage datagram loss were measured and compared for each standard. For each AP type, the best TCP throughputs were found for 802.11g. TCP throughput was found sensitive to AP type. For both AP types and 802.11 b, similar average jitter performances were found. For 802.11g, jitter performance was found sensitive to AP type. Concerning average percentage datagram loss, fairly good agreements were found for both AP types and for both standards.

At OSI layer 7, the best FTP performances were, for each AP type, for 802.11g. FTP performance was found sensitive to AP type. These results show the same trends found for TCP throughput.

Generally, the results measured for WEP links were not found as significantly different, within the experimental errors, from corresponding data obtained for open links, except for jitter, where the best performances were found for open links. WEP encryption has led to degradation of jitter performance.

Additional performance measurements either started or are planned using several equipments and security settings, not only in laboratory but also in outdoor environments involving, mainly, medium range links.

**Acknowledgement** Supports from University of Beira Interior and FCT (Fundação para a Ciência e a Tecnologia)/POCI2010 (Programa Operacional Ciência e Inovação) are acknowledged.

# References

1. IEEE 802.11a, 802.11b, 802.11g (2007), 802.11n, 802.11i (2009) standards. http://standards. ieee.org/getieee802. Accessed 10th Dec 2010
2. Mark JW, Zhuang W (2003) Wireless communications and networking. Prentice-Hall, Inc., Upper Saddle River, NJ
3. Rappaport TS (2002) Wireless communications principles and practice, 2nd edn. Prentice-Hall, Inc., Upper Saddle River, NJ
4. Bruce WR III, Gilster R (2002) Wireless LANs end to end. Hungry Minds, Inc., NY
5. Schwartz M (2005) Mobile wireless communications. Cambridge University Press, Cambridge
6. Sarkar NI, Sowerby KW (2006) High performance measurements in the crowded office environment: a case study. In: Proceedings of ICCT'06-International conference on communication technology. Guilin, China, 27–30 Nov 2006, pp 1–4.
7. Monteiro E, Boavida F (2002) Engineering of informatics networks, 4th edn. FCA-Editor of Informatics Ld, Lisbon

8. Pacheco de Carvalho JAR, Gomes PAJ, Veiga H, Reis AD (2008) Development of a university networking project. In: Putnik GD, Cunha MM (eds) Encyclopedia of networked and virtual organizations. IGI Global, Hershey, PA (Pennsylvania), pp 409–422
9. Pacheco de Carvalho JAR, Veiga H, Gomes PAJ, Ribeiro Pacheco CF, Marques N, Reis AD (2010) Wi-Fi point-to-point links- performance aspects of IEEE 802.11 a,b,g laboratory links. In: Ao SI, Gelman L (eds) Electronic engineering and computing technology. Series: lecture notes in electrical engineering, vol. 60. Springer, Netherlands, pp 507–514.
10. Pacheco de Carvalho JAR, Veiga H, Marques N, Ribeiro Pacheco CF, Reis AD (2011) Wi-Fi WEP point-to-point links- performance studies of IEEE 802.11 a,b,g laboratory links. In: Ao SI, Gelman L (eds) Electronic engineering and computing technology. Series: lecture notes in electrical engineering, vol. 90. Springer, Netherlands, pp 105–114.
11. Pacheco de Carvalho JAR, Veiga H, Marques N, Ribeiro Pacheco CF, Reis AD (2011) Comparative performance evaluation of Wi-Fi IEEE 802.11 b,g WEP point-to-point links. In: Proceedings of the WCE 2011 – world congress on engineering 2011, vol. II. Imperial College London, London, England, 6–8 July 2011, pp 1745–1750
12. Pacheco de Carvalho JAR, Marques N, Veiga H, Ribeiro Pacheco CF, Reis AD (2011) A medium range GBPS FSO link- extended field performance measurements. In: Ao SI, Gelman L (eds) Electronic engineering and computing technology. Series: lecture notes in electrical engineering, vol. 90. Springer, Netherlands, pp 125–134.
13. Enterasys (2007) RoamAbout RBT-4102 access point technical data. http://www.enterasys.com. Accessed 10th Jan 2010
14. Allied Telesis (2008) AT-8000S/16 layer 2 managed fast ethernet switch. http://www.alliedtelesis.com. Accessed 20 Dec 2008
15. Linksys (2005) WRT54GL wireless router technical data. http://www.linksys.com. Accessed 13th Jan 2011
16. DD-WRT (2008) DD-WRT firmware. http://www.dd-wrt.com. Accessed 13th Jan 2011
17. NetStumbler software (2005). http://www.netstumbler.com. Accessed 21st March 2011
18. Iperf software, NLANR (2003). http://dast.nlanr.net. Accessed 10th Jan 2008
19. Network Working Group (1996). RFC 1889-RTP: a transport protocol for real time applications. http://www.rfc-archive.org. Accessed 10th Feb 2011

# Chapter 16
# Integrating Mobility, Quality-of-service and Security in Future Mobile Networks

**Mahdi Aiash, Glenford Mapp, Aboubaker Lasebae, Raphael Phan, and Jonathan Loo**

## 1 Introduction

The world is experiencing the development and large-scale deployment of several wireless technologies; these range from next generation cellular networks to personal/home networks such as WLANs and metropolitan ones such as WiMax. Users will expect to connected to several networks at the same time and ubiquitous communication will be achieved by seamless switching between available networks using handover techniques.

As explained in [15], handover is defined as the changing of the network point of attachment of a mobile device. When the device moves to a new point of attachment which is technologically identical to the previous point of attachment, this is called horizontal handover. An example of horizontal handover occurs when the calls on a 3G phone are moved from one 3G Base-station to another 3G Base-station. Vertical handover is defined as a handover where the new point of attachment comprises a different technology when compared with the previous point of attachment. In this paper, we are primarily concerned with vertical handovers.

However, since these networks have different characteristics in terms of speed, latency, and reliability, vertical handover will therefore have an impact on the network service experienced by ongoing applications and services as mobile nodes move around. Some applications such as multimedia may be able to adapt, while others may need support to deal with varying quality-of-service (QoS). This situation is also

M. Aiash (✉) • G. Mapp • A. Lasebae • J. Loo
School of Engineering and Information Sciences, Middlesex University, Middlesex, UK
e-mail: m.aiash@mdx.ac.uk; g.mapp@mdx.ac.uk; a.lasebae@mdx.ac.uk

R. Phan,
Department of Electrical and Electronic Engineering,
Loughborough University, Loughborough, UK
e-mail: r.phan@lbro.ac.uk

S.-I. Ao and L. Gelman (eds.), *Electrical Engineering and Intelligent Systems*,
Lecture Notes in Electrical Engineering 130, DOI 10.1007/978-1-4614-2317-1_16,
© Springer Science+Business Media, LLC 2013

reflected on the server side which must be ready to adapt its service delivery when the QoS or security parameters change. So service level agreements (SLAs) must be adapted to handle changing network conditions.

As pointed out in [14], vertical handover can cause radical changes in QoS. Hence, it is important that as much control as possible is exercised by mobile devices to achieve optimum vertical handover. It is therefore necessary to develop new techniques which could make other layers of the protocol stack aware of impending handover decisions and, thus, allow them to take steps to minimize the effects. So a new QoS Framework is required which must be integrated with mobility mechanisms such as handover.

Another key observation is that heterogeneous networking will mean that the core of the network will have to be based on an open architecture, for example, an IP backbone. This core will be very fast and will be surrounded by slower peripheral wireless networks. However, the openness of the core network will give rise to a number of security threats such as denial-of-service (DoS) attacks that we see in the Internet today. So security must also be integrated into future mobile networks to deal with these threats. Thus, future mobile systems must encompass communications, mobility, QoS, and security. A new framework is therefore required to build such systems.

Examples of such future frameworks are the Mobile Ethernet as described in [13], Ambient Networks explained in [8] and the Y-Comm framework highlighted in [12]. The Y-Comm group has been working on introducing novel approaches to support a seamless and secure vertical handover in heterogeneous environments. This paper will consider the mechanisms introduced by the Y-Comm group to address mobility, QoS, and security issues and explains how these mechanisms could be integrated in the context of the Y-Comm framework.

The rest of this paper is organized as follows: Sect. 2 examines the structure of future networks while Sect. 3 looks at how handover is classified. Section 4 describes the Y-Comm Architecture and outlines how proactive handover is achieved in Y-Comm and Sect. 5 details how QoS is integrated with proactive handover mechanisms. Section 6 specifies how security in Y-Comm is integrated with proactive handover. The paper concludes in Sect. 7.

## 2   The Structure of Future Networks

In current closed systems, such as second-generation and third-generation cellular environments, the core network is owned by a sole operator, who is responsible for managing all aspects of the system including security and QoS provision. However, as previously explained, the heterogeneity of future networks leads to a new open architecture for the core network, where the infrastructure is not controlled by a single operator. Rather, multiple operators will have to coexist in the core network. To deal with the interoperability issue between the different operators, the Y-Comm group in [1] and the Daidalos II Project in [6], adopted and

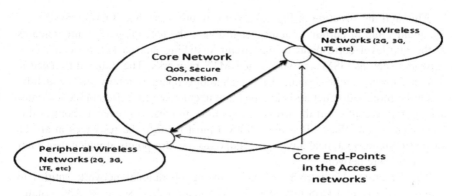

**Fig. 16.1**  The future internet structure

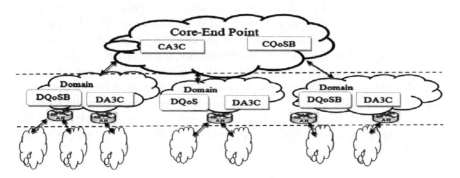

**Fig. 16.2**  The network structure at the core endpoint

enhanced the concept of a central management entity, as proposed by the ITU [9], to control the different operators. Hence, the concept of the core endpoint (CEP) as a central administrative domain that controls the operation of different network operators was introduced.

As shown in Fig. 16.1, the future Internet could be viewed as composed of several CEPs, interconnected over a super-fast core. Each CEP is responsible for managing multiple, wireless peripheral networks such as Wimax, WiFi, or cellular technologies. A detailed view of this arrangement along with its components are explained in [1] and shown in Fig. 16.2. It is a hierarchical network structure composed of three levels. At the top level, the CEP acts as a gateway to the Internet and is responsible for managing multiple, mid-level domains. Each domain is technology specific and is controlled by a single operator. For instance, two domains might be connected to the same CEP, each controlled by a different technology operator such as WiMAX and GSM. The bottom level comprises individual peripheral wireless networks, controlled by access routers or Base-stations through which the mobile terminal has access to the wider network.

Although the structure in Fig. 16.2 is for future networks, it has evolved from the architecture of current systems; for instance, the technology-specific domains at the mid-level correspond to the circuit switching and packet switching core networks such as GSM and GPRS or UMTS networks. The major difference is that in current mobile systems each technology operates totally independently, while the proposed structure is an open architecture, where different technologies and operators could join the network at core endpoints. However, to control this open architecture, the CEP at the top level must manage the resources in all the various domains in a local area. Core endpoints are also points at which data enter or exit the core network.

The structure also introduces the following operational network entities to support security and QoS-related tasks: The central A3C server (CA3C) shares the SLA information with the subscriber. The central QoS broker (CQoSB) negotiates QoS parameters in case of inter CEPs connection and handover. The domain A3C server (DA3C) deals with the security aspects in terms of Authentication, Authorization, and Access control at the domain level. The domain QoS broker (DQoSB) manages the resources of the attached peripheral networks with respect to user preferences and network availability; it also makes per-flow admission control decisions. The access router (AR) facilitates communication between the mobile terminal and the network. It also enforces admission control decisions, taken by the DQoSB. Since the AR acts as a relay between the mobile terminal (MT) in the peripheral network and the DA3C, using security terminology, the AR will be referred to as the Authenticator (Auth). A detailed explanation of the structure of these components is found in [1].

## 3  Handover Classification

Given the future structure of the network as detailed in Figs. 16.1 and 16.2, vertical handover will occur between networks connected to the same core endpoint (called an intra-CEP handover) or between networks connected to adjacent core endpoints (called an inter CEP handover). In order to optimize the handover process, it is necessary to examine handover in detail. This advanced classification is shown in Fig. 16.3.

At the top level, handover can be divided into two types: imperative and alternative. Imperative handover occurs because the signal from the current network is fading, hence network service will degrade if handover does not occur. Alternative handover occurs for reasons other than technical ones, such as user-preferences, pricing, etc. Imperative handover is further divided into two main types: reactive and proactive. With reactive handover, the mobile terminal is reacting to signals from its network interfaces about the changes of network availability as users move around. These signals are referred to as Level 2 (L2) triggers or events. Reactive handovers are further divided into two types: anticipated and unanticipated handovers. In anticipated handover, there are

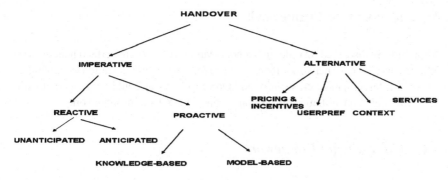

**Fig. 16.3**  Classification of handovers

alternative Base-stations to which the mobile terminal may handover. In unantic-ipated handover, the signal from the current network is fading and there is no other point of attachment to which the mobile node can handover. In this situation, the mobile terminal is forced to temporarily shut down its ongoing connections and only resume them when a new network point of attachment is found. The actions needed to be taken by the transport layer with regard to handover have been detailed in [12] while this paper looks at the QoS and security responses.

Proactive handover attempts to find the best time and place to handover before the user actually reaches that point. Proactive handovers can also be divided into two classes: knowledge-based and model-based handovers. Knowledge-based handovers involve finding out network coverage and signal strengths in a given region by physically measuring these parameters using a suitably equipped vehicle such as a Sentient Van. This is explored in [7]. In model-based handover, signal strengths and network coverage are modelled using propagation models. In this context, handover is initiated when the signal strength drops below a given threshold. Given that the mobile terminal knows its location as well as its velocity, via GPS say, as well as the position of the Access Point and the power being radiated at the transmitter of the Access Point, it is possible to calculate when handover should occur. This technique is explained in [16].

Proactive handover will also allow mobile terminals to calculate the time before vertical handover (TBVH) which can be used to signal to the upper layers that handover is about to occur. This would allow different parts of the protocol stack to take appropriate action to minimize the effects of handover. For example, reliable protocols such as TCP may begin to buffer packets and store acknowledgements ahead of handover so that connections can be quickly resumed once handover occurs. Knowing TBVH will also allow QoS and security issues to be negotiated before handover occurs. This minimizes handover latency and, therefore, signifi-cantly increases good user experience. The authors believe that proactive handovers will be a key requirement of future mobile systems. The Y-Comm framework has been designed to support proactive handovers.

# 4 The Y-Comm Framework

As previously discussed, in order to support seamless and secure vertical handover in future, heterogeneous networks, there is a need for novel communication frameworks, wherein different handover, security and QoS mechanisms could be integrated with the communication system. This section outlines the Y-Comm framework.

## 4.1 The Y-Comm Framework

Y-Comm is a communication architecture to support a heterogeneous networking environment. The architecture has two frameworks:

- The Peripheral Framework deals with operations needed on the mobile terminal to support a heterogeneous environment where there are several networks attached to a single Core endpoint. The Peripheral Framework therefore runs on the mobile terminal.
- The Core Framework deals with functions in the core network to support different peripheral networks. These functions are distributed throughout the network infrastructure of the core network.

As shown in Fig. 16.4, both frameworks diverge in terms of functionality but the corresponding layers interact to provide support for heterogeneous environments. The framework also uses a multilayered security system which joins the two systems. An explanation of the complete Y-Comm Architecture is given in [11].

## 4.2 Proactive Handover in Y-Comm

In this section we look at how proactive handover is done in Y-Comm. This is detailed in Fig. 16.5. Vertical handover is facilitated in Y-Comm using the three

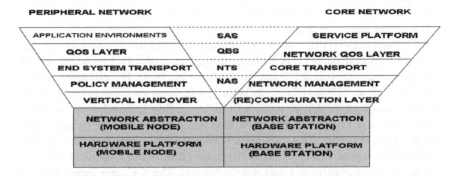

**Fig. 16.4** The complete Y-Comm framework

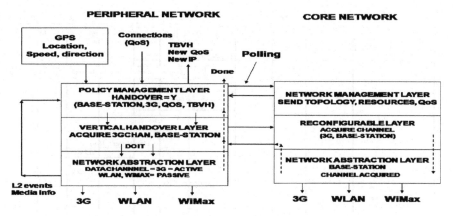

**Fig. 16.5**  Proactive handover in Y-Comm

layers in both the Peripheral and Core Frameworks. In the Peripheral network, the layers are the network abstraction layer (NAL) which allows the mobile node to control its wireless interfaces; the vertical handover layer (VHL) which executes the handover and the policy management layer (PML) which decides when and where a handover will occur. In the Core Framework, the NAL controls the resources on an individual Base-station, the re-configuration layer (RAL) controls the resources of several Base-stations and access routers and the network management layer (NML), knows about all the different networks connected to the core endpoint.

The handover scenario is described as follows: the PML polls the NML about networks in the local area. The NML sends this information about local networks and their topologies back to the PML. The PML uses this information to decide when, where, and to which network to handover and then gives that information to VHL to execute the handover at the chosen time. The VHL contacts the Reconfiguration layer in the core network to request a channel on the target Base-station. The RAL will acquire a channel on behalf of the mobile terminal. When the TBVH period expires, the VHL then instructs its NAL to handover. It is very important to see that the PML also informs the higher layers of impending handover, giving the new QoS, TBVH, and the IP address on the new network. This is used by the Handover Module in the QoS Framework to initiate QoS and security mechanisms to ensure a smooth vertical handover.

## 5   Quality of Service and Handover Integration in Y-Comm

This section describes different mechanisms for supporting an end-to-end QoS signalling and cross-operator security provision. These mechanisms along with other enhancements to the current network infrastructure will enable secure vertical

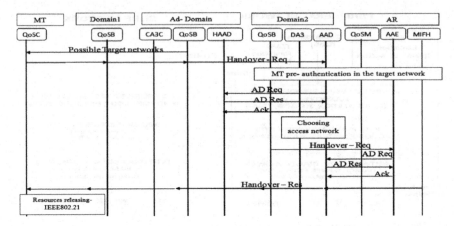

**Fig. 16.6** The QoS intra-CEP handover model

handover in heterogeneous environments. Y-Comm uses three QoS signalling models: the Registration model which describes the procedure that is followed when the MT first attaches to the peripheral network, the Connection Model which deals with the case when the MT starts a connection to a server, and the Handover Model which explains how QoS and security parameters are transferred to the new network in the case of a handover. This model is explored further by looking at how this QoS model implements an intra core endpoint handover. More details on the complete QoS Framework are found in [1].

## 5.1 The QoS Handover Model

In this model, after being signalled by the PML layer about the impending handover, the QoS-context is transferred and used by the access control mechanism in the new network to enforce the right access admission policy. After configuring the access policy in the target access router, a successful handover response message is sent back to the MT. A signal is then sent to the VHL that the QoS parameters have been transferred and the actual handover may begin.

As shown Fig. 16.6, in the case of intra-CEP handover, where the MT moves between different domains within the same CEP, this model deploys the Pre-Authentication (Pre-AKA) protocol to achieve Pre-Authentication and Key Agreement as well as launching the security materials in the target network before the actual handover takes place and thus, reducing the handover disruption to the handover caused by the security mechanisms. Also in this step, the QoS-context is transferred and used by the access control mechanism in the new network to enforce the right access admission policy. After configuring the access policy in the target access router, a successful handover response message is sent back to the MT. A signal is then sent to the VHL to indicate that the actual handover may begin.

In the case of the inter-CEP handover, the Old CEP provides the target CEP with the user's security and QoS requirements. Thus, the MT-related information is made available to the target network. The remaining steps are similar to the intra-CEP handover model.

# 6   Security and Handover Integration in Y-Comm

This section describes the security features of the Y-Comm architecture; it starts by explaining its a four-layer integrated security module (ISM) which provides security at different levels of the communication stack.

## 6.1   The Integrated Security Module

To deal with security in heterogeneous networks, Y-Comm employs a multilayer security module which is described in [3]. As shown in Fig. 16.4, the security layers in Y-Comm work together across both frameworks in order to be fully integrated with the new architecture. The highest layer of security is at layer seven and is called Service and Application Security or SAS. In the Peripheral Framework, SAS layer defines the A3C functions at the end-device and is used to authenticate users and applications. SAS in the Core Framework provides A3C functions for services on the Service Platform in the core network. The next security layer is called QoS-Based Security or QBS and is concerned with QoS issues and the changing QoS demands of the mobile environment as users move around, the QBS layer also attempts to block QoS related attacks, such as DoS attacks on networks and servers.

The next security layer is at layer five and is called Network Transport Security or NTS. In the Peripheral Framework, NTS is concerned with access to and from end-devices and the visibility of these devices and services on the Internet. In the Core Framework, NTS is used to set up secure connections through the core network. Finally, the fourth and last level of security is defined at layer four. It is called Network Architecture Security or NAS. In the Peripheral Framework, it attempts to address security issues involved in using particular networking technologies and the security threats that occur from using a given wireless technology. So when a mobile device wishes to use any given network, NAS is invoked to ensure that the user is authorized to do so. NAS also ensures that the local LAN environment is as secure as possible. In the Core Framework, NAS is used to secure access to the programmable infrastructure. NAS in this context determines which resources may be used by the network management system.

## 6.2   The Authentication and Key Agreement Framework

In order to build a real security framework it is necessary to develop security protocols which implement the respective functionalities. A number of authentication and key agreement (AKA) protocols were developed which provide the functionality of some of the layers of the ISM. The PL-AKA protocols could be used in the SAS Layer. It authenticates users to use mobile terminals, etc. The NL-AKA and SL-AKA protocols could be used in the NAS and NTS security layers. There are two network level protocols: the first, called Initial NL-AKA, is used to connect users to a network. There is another NL-AKA protocol concerned with handover called the Handover NL-AKA. Similarly, the SL-AKA level which is concerned with the interaction between the user and a given service also has initial and mobile variants. However, the functionality of the QBS layer is defined by integrating the AKA framework with the QoS signalling models described in the previous section.

## 6.3   The Targeted Security Models

In addition to the Integrated Security module, Y-Comm also uses targeted security models (TSMs). These models were developed so as to protect different entities that use the open infrastructure. In such a system, it is not only necessary to protect data that are being exchanged between different entities but also necessary to protect entities from attacking each other. Today, we see these attacks in the Internet in terms of DoS attacks, SPAM, etc. TSMs, therefore, were developed to protect entities in an open environment. As explained in [10], Y-Comm has defined three TSMs. The Connection Security Model controls how users are connected and prevents a user from arbitrarily sending a message to other users, e.g., SPAM. The Ring-based Security Model attempts to protect servers by reducing their visibility to the scope of their functionality. Further details on the Ring-based Model are given in [2]. The final TSM is the Vertical Handover Security Model which provides secure handover by authenticating the node and ensuring that only the resources required by the mobile terminal are actually allocated to it in the new network, so networking resources are not abused during handover. The Connection Security Model and the Vertical Handover model are based on the initial and handover NL-AKA protocols. A detailed explanation of the Connection Security Model is given in [5]. Below we examine the Vertical Handover Security Model.

## 6.4   The Vertical Handover Security Model

This model facilitates secure vertical handover and attempts to prevent network resources from being abused and overloaded. This is done by monitoring resource requests and ensuring access to vulnerable components does not exceed the available QoS.

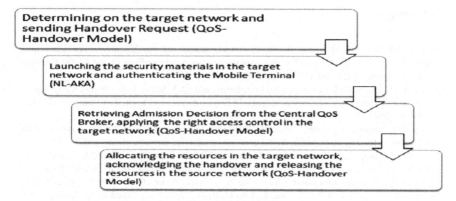

**Fig. 16.7**   The Vertical Handover Security Model

As shown in Fig. 16.7, this model describes the case of proactive vertical handover, where the mobile terminal knows beforehand about all available networks and has decided on a target network to which to handover. This is done by the PML. In order to support secure but seamless handover, the security procedures take place before the actual handover happens, and thus, security is provided with minimum disruption of the handover mechanisms. Therefore, as described in [4], the proposed AKA protocol for securing the handover launches the security parameters and performs the negotiation stage before the actual handover. The next step is to retrieve the Admission Decision in order to configure the right access control on the access routers of the target network. This is accomplished as part of the QoS Handover Model. The last step of this model comprises allocating resources in the target network, acknowledging successful handover and releasing the resources in the source network. This is done by the VHL on the mobile terminal.

# 7   Conclusion

This paper has explored mechanisms for building future networks. These systems will have to integrate mobility, via vertical handover, QoS, and security. The required mechanisms have been explored in the context of the Y-Comm architecture. Proactive handover mechanisms as well as QoS and Security were integrated to provide secure vertical handover. Future work will involve using the mechanisms detailed in this paper to build a prototype system.

# References

1. Aiash M, Mapp G, Lasebae A (2011) A QoS framework for heterogeneous networking. In: Lecture notes in engineering and computer science: proceedings of the World Congress on Engineering 2011, WCE 2011. pp 1765–1769
2. Aiash M, Mapp G, Lasebae A (2011) Security and QoS integration for protecting service providers in heterogeneous environments. IAENG International Journal of Computer Science, 38:4, pp 384–393
3. Aiash M, Mapp G, Lasebae A, Phan R (2010) Providing security in 4G systems: unveiling the challenges. In: The sixth international conference on telecommunications, (ACT 2010)
4. Aiash M, Mapp G, Lasebae A, Phan R, Loo J (2011) A formally verified AKA protocol for vertical handover in heterogeneous environments using Casper/FDR. EURASIP Journal on Wireless Communications and Networking 2012, 2012:57 doi:10.1186/1687-1499-2012-57
5. Aiash M, Mapp G, Lasebae A, Phan R, Loo J (2011) A formally verified initial AKA protocol in heterogeneous environments using Casper/FDR. Int J Inform Secur (submitted to)
6. Aimeida M, Corujo D, Sargento S, Jesus V, Aguiar R (2007) An end-to-end QoS framework for 4G mobile heterogeneous environments. In: Openet workshop. March 27–29, 2007, Diegem, Belgium
7. Cottingham D, Harle RK, Hopper A (2008) Constructing accurate, space-efficient, wireless coverage maps for vehicular contexts. In: Proceedings of the 4th international wireless internet conference (IST WICON)
8. Horn U, Prehofer C, Karl H (2004) Ambient networks - an architecture for communication networks beyond 3g. IEEE Wireless Communications, 11:2, pp 14–22
9. ITU (2008) Global information infrastructure, internet protocol aspects for next generation networks Y.140.1. http://www.itu.int/itu-t/recommendations/index.aspx?ser=Y
10. Mapp G, Aiash M, Lasebae A (2010) Security models for heterogeneous networking. In: International conference on security and cryptography (SECRYPT)
11. Mapp G, Shaikh F, Cottingham D, Crowcroft J, Baliosian J (2007) Y-Comm: a global architecture for heterogeneous networking. In: 3rd annual international wireless (WICON)
12. Mapp G, Shaikh F, Vanni RP, Augusto M, Moreira E (2009) Exploring efficent imperative handover mechanisms for heterogeneous wiresless networks. In: International symposium on emerging ubiquitous and pervasive systems
13. Masahiro K, Mariko Y, Ryoji O, Shinsaku K, Tanaka T (2004) Secure service and network framework for mobile ethernet. Wireless Pers Comm 29:161–190
14. McNair J, Zhu F (2004) Vertical handoffs in fourth-generation multi-network environments. IEEE Wireless Comm 11 pp 8–15
15. Schiller JH (2003) Mobile communications, 2nd edn. Addison Wesley, England
16. Shaikh F, Mapp G, Lasebae A (2007) Proactive policy management using TBVH mechanism in heterogeneous networks. In: Proceedings of the international conference and exhibition on next generation mobile applications, services and technologies (NGMAST'07)

# Chapter 17
# Decision Under Uncertainties of Online Phishing

Ping An Wang

## 1 Introduction

Phishing has been a serious online risk related to privacy, security, and trust and is still a phenomenon of great practical significance for B2C (business-to-consumer) e-commerce [3]. Phishers often try to lure victims into clicking a spoofed universal resource locator (URL) pointing to a rogue Web page to steal sensitive personal and financial information from unsuspecting online consumers [13]. There has been considerable research on online risks and consumer decision making in the B2C e-commerce context [2, 23, 30]. However, existing research in this area of online risks primarily focuses on determinants of subjective probability and value and assumes that consumers judge (1) the subjective probability of a loss and (2) the subjective magnitude of consequences of the loss, and compute an expectation of loss. A significant problem is that neither the probability of occurrence of online risks nor the consequences of risky events are always known to consumers. For example, the likelihood and consequences of a credit card fraud resulting from an online transaction are not known for sure even to experts [21]. Thus, the question arises as to how online consumers judge phishing risks and decide on online purchases under various uncertain knowledge conditions of the risks.

This study argues that consumer decisions in risky online environments are made under uncertain conditions where risk probability information is imprecise, vague, or ambiguous. Based on research in decision under risks and uncertainties, this study categorizes an online consumer's knowledge of the phishing risk as falling under one of four fine-grained uncertainty states: known certainty, known uncertainty, unknown uncertainty, and unknowable uncertainty. An online consumer's risk

P.A. Wang (✉)
Graduate School Cybersecurity Department,
University of Maryland University College, Adelphi, MD 20783, USA
e-mail: pwang@umuc.edu

S.-I. Ao and L. Gelman (eds.), *Electrical Engineering and Intelligent Systems*,     207
Lecture Notes in Electrical Engineering 130, DOI 10.1007/978-1-4614-2317-1_17,
© Springer Science+Business Media, LLC 2013

evaluation and purchase intention and decision are strongly affected by his or her assumption of the variant of uncertainty regarding the extent and severity of the phishing security risk involved in the online transaction.

The primary goal of this research is to investigate how variant degrees of uncertainty of online consumers' knowledge of phishing risks affect their judgment of and behavioral response to the risks. Section 2 reviews relevant information systems (IS) literature on decision under risks and uncertainty. Section 3 discusses the research model and hypotheses proposed. Section 4 introduces the experiment method used for the study. Section 5 reports the data analysis and findings. Section 6 concludes the paper.

## 2  Literature Review

There has been considerable IS research interest in decision under uncertainty and the impacts of online security risks. However, there has not been a systematic model and approach available to address the impacts of variant uncertainties of knowledge of online information security risks on consumer decision making in the B2C e-commerce context. The theoretical basis for prior research on decision under risk and uncertainty primarily falls into three categories: utility theory, attitudinal theories, and the psychometric paradigm.

### 2.1  Risk Studies Based on Utility Theory

The classical notion of risk in decision theory is primarily modeled using utility theory. Utility theory assumes that people are rational and should choose the option that maximizes the expected utility, which is the product of probability and payoff. Utility theory also assumes that all risk probabilities and payoff are known to a point estimate but does not allow ambiguity, or a variant form of uncertainty. In reality, however, uncertainty does occur when risk probabilities or payoff is missing or unknown. The subjective expected utility (SEU) model of utility theory proposed by Savage [26] argues that people's *subjective* preferences and beliefs, rather than objective probabilities, are used in the evaluation of an uncertain prospect for decision making. The SEU model is based on a set of seven axioms designed for consistent and rational behavior.

Subsequent research, however, has shown that people often violate the axioms [27]. Experimental studies by Ellsberg [8] indicated that the choices of many decision makers reveal inconsistent preferences that cannot be explained using the SEU model. Ellsberg's experiment demonstrated that people prefer known and specific probabilities to ambiguity or vagueness, suggesting ambiguity aversion. But Ellsberg did not address the factor of unknowable uncertainty in decision making. In addition, Ellesberg's experimental study was limited to urn tasks and choices for bets.

Another variation of utility theory is the prospect theory proposed by Kahneman and Tversky [18]. The prospect theory views decision under risks from a behavioral economic perspective and recognizes the importance of framing perceptions in risk and outcome evaluation. Kahneman and Tversky argued that attitudes toward risk are jointly determined by perceived values and decision weights of specified prospects or choices. The prospect theory presented a descriptive conceptual model for framing risk perceptions, but it does not address security risks in e-commerce. Also, it is usually a difficult task to determine and measure the reference point for gains and losses. In reality, very few IS research articles use this theory [21].

The maxmin expected utility (MEU) model proposed by Gilboa and Schmeidler [14] argues that a decision maker has a set of prior beliefs and the utility of an act is the minimal expected utility in this set. However, the model did not differentiate uncertainty levels and failed to address the role of subjective beliefs in decision making under uncertainty. The comparative ignorance hypothesis by Fox and Tversky [11] argued that ambiguity aversion is produced by a comparison with less ambiguous events or with more knowledgeable individuals. Like other utility theory approaches, their study neither distinguished different degrees of uncertainty nor studied online phishing risks.

## 2.2   Risk Studies Based on Attitudinal Theories

A large amount of prior research on online risks was based on attitudinal theories involving risk perceptions and behavioral intentions. The conceptual assumption of such models was rooted in the theory of reasoned action (TRA) developed by Fishbein and Ajzen [10]. In TRA, behavioral intentions, determined by attitudes and perceptions, are antecedents to specific individual behaviors. An online customer's perception and attitudes regarding risks, accordingly, will affect his or her behavioral intentions to conduct transactions online. The general assumption of various attitudinal theories is that people's decisions under risks are driven by inconsistent perceptions, beliefs, and emotions.

Hogarth and Kunreuther [16] found uncertainty of risk knowledge an important factor in consumer decision making. The study was one of the few applied to consumer purchase decisions, but it did not involve online security risks. However, they pointed out that the standard lab cases of gambling used in most prior decision studies did not capture the variety of decision choices faced by people in the real world. Roca, Hogarth, and Maule [24] concurred that future decision research should be extended to a broader range of contexts and response modes, such as willingness to pay for uncertainties and risks.

Bhatnagar et al. [2] suggested a negative correlation between knowledge and risk aversion. However, their study focus was not on online security risks but on product risks and financial risks. Miyazaki and Fernandez [20] studied the relationship between consumer perceived privacy and security risks and online purchase behavior. Salisbury et al. [25] studied consumer perceptions of Web security risks in Internet shopping. Pavlou [22] proposed a B2C e-commerce acceptance model of trust, perceived risk, perceived usefulness, and ease of use for predicting e-commerce acceptance and online purchase behavior. Milne et al. [19] studied the online privacy risks from the security perspective and focused on the specific risk of identity theft. However, none of these studies included the consumer risk knowledge factor or uncertainty levels in addressing consumer purchase decisions.

Acquisti and Grossklags [1] recognized the importance of uncertainty in individual decision making in situations that have an impact on privacy. Their concept of privacy risks is relevant to the domain of online information security. However, they did not address the security knowledge factor in e-commerce decision making. Tsai et al. [31] studied the role of privacy policy visibility and privacy protection concerns in online shopping decisions. They found that online consumers value privacy and are willing to pay a premium for privacy protection. However, they did not address knowledge or uncertainty factors for individual decision making.

Dinev and Hu [7] emphasized user awareness as a key determinant of user intention toward adopting protective information technologies. But the awareness construct and its measures were limited to user interest in learning about information security issues and strategies. Jiang et al. [17] studied the relationship between knowledge, trust, and intention to purchase online. However, the concept of knowledge in the study primarily refers to consumers' familiarity with the online shopping environment, such as transaction and payment.

The common assumption of the prior studies from various attitudinal perspectives is that decisions under risks are driven by inconsistent perceptions, beliefs, and emotions. However, they all share two major limitations: (a) no presence of fine-grained degrees of uncertainties and (b) lack of focus on the online phishing risks and e-commerce consumer decision making.

## 2.3 The Psychometric Approach to Risks

Prior IS research based on the psychometric theory suggests that consumers use attributes other than risk probabilities and consequences in their decision making. Fischhoff et al. [9] studied technological risks and benefits using the psychometric paradigm. The study touched upon known and unknown risks but did not address the unknowable risks. Slovic et al. [29] found that risk acceptability is affected by risk attributes, such as familiarity, control, and uncertainty about the level of risk. However, they neither defined the uncertainty concept nor distinguished different degrees of uncertainty of risk knowledge. Slovic [28] further suggested that the level of knowledge attribute seems to influence the relationship between perceived risk, perceived benefit, and risk acceptance. However, he did not distinguish different degrees of uncertainties. Also, his study did not touch upon any online security or phishing risks for e-commerce.

Nyshadham and Ugbaja [21] used psychometric techniques to study how B2C e-commerce consumers organize novel online risks in memory. The study called for further analysis to define the risk dimensions. Using the psychometric paradigm, Gabriel and Nyshadham [12] studied perceptions of online risks that affect online purchase intentions in the B2C e-commerce environment. The focus of the study was to develop a taxonomy of online risks and construct a cognitive map of online consumers' risk perceptions and attitudes. The results suggested that knowledge of risks is an important parameter of online risk perceptions. However, the study did not focus on the variable of knowledge and did not go into fine-grained notion of risk probability.

Glover and Benbasat [15] proposed a "comprehensive model of perceived risk" for e-commerce transactions followed by a field study of online participants. Their study indicated the important role of consumer perceptions of risks in online transactions. Their model of perceived risk is driven by a marketing theory of risk and consists of three dimensions: risk of functionality inefficiency, risk of information misuse, and risk of failure to gain product benefit. Consumers' level of risk knowledge is not one of the dimensions or the focus of the study. Although the information misuse risk dimension seems to be generically inclusive of possible misuse of personal and financial information, the study does not specifically address the online phishing risk.

This research is to address the common limitations of prior studies by focusing on the uncertainty of knowledge of online phishing risks in e-commerce decision making and adopting a fine-grained taxonomy of degrees of uncertainties. The purpose of the study is to measure the effect of knowability of risk on a person's decision making when faced with online phishing risks. Chow and Sarin [6] defined knowability as one's assumption about the availability of information regarding the uncertainty of probability. Decision situations are usually either under certainty or uncertainty. In contrast to known certainty, Chow and Sarin proposed and distinguished three types of uncertainties: known, unknown, and unknowable uncertainties. This fine-grained classification of uncertainties of risk knowledge is the theoretical basis for this study. Accordingly, the uncertainties are broken down into four levels or degrees of conditions: known certainty, known uncertainty, unknowable uncertainty, and unknown uncertainty. Table 17.1 defines the four degrees of uncertainties.

# 3   Research Model

The research model in Fig. 17.1 guides this study. The model was based on the model initially proposed and updated by Wang [32, 33].

The research model is contextualized for the different degrees of uncertainties of risk knowledge. The construct of phishing risk evaluation reflects consumers' subjective beliefs and judgment of online phishing risks and protection mechanisms. Decision behaviors under risks are related to people's degrees of

**Table 17.1** Uncertainties of risk knowledge

| Degree of uncertainty | Definition | Example |
|---|---|---|
| Known certainty | Information on all attributes and alternatives are available | A vendor guarantees that none of its online transactions involves phishing, due to strong online security mechanism |
| Known uncertainty | Risk probability is precisely and officially specified | It is officially confirmed that 3% of online transactions with the vendor involve phishing |
| Unknowable uncertainty | Risk probability is unavailable to all | It is impossible for anyone to know exactly what percentage of online transactions with the vendor involves phishing |
| Unknown uncertainty | Risk probability is missing to one but may be possessed by others | The public is only told that less than 5% online transactions with the vendor involve phishing. But the exact percentage is not disclosed |

**Fig. 17.1** Research model

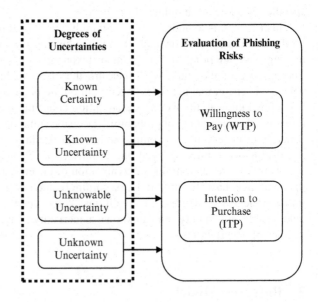

knowledge of the risk probabilities. Hogarth and Kunreuther [16] found that people demonstrate different observable behaviors between situations where they do and do not have knowledge about probabilities and outcomes. Thus, this study proposes that uncertainty levels of risk knowledge affect online shoppers' risk evaluation and their intention and decision to purchase under risks.

Known certainty is obviously the ideal knowledge level for decision making. The constructs of variant uncertainties are based on Chow and Sarin [6]. Chow and Sarin view known uncertainty as the most comfortable uncertainty to people and preferable to vagueness in probability. Unknown uncertainty is less preferable than unknowable uncertainty, and it is the least comfortable level of uncertainty to a decision maker. Unknowable uncertainty, according to Chow and Sarin, is the intermediate comfort level of uncertainty to people and more tolerable than unknown uncertainty. Thus, the following two hypotheses are proposed for this study:

*Hypothesis 1*: Known uncertainty is preferable to unknowable uncertainty in consumer evaluation of online phishing risks.
*Hypothesis 2*: Unknowable uncertainty is preferable to unknown uncertainty in consumer evaluation of online phishing risks.

Consumers' behavioral response to online phishing risks consists of willingness to pay (WTP) to avoid the risks and their intention to purchase (ITP) online under the risks. Prior research in decision theories suggested that individuals are willing to pay a premium to avoid uncertainty of risks [6, 24]. The WTP amount is expected to grow as consumer-perceived phishing risks increase. In addition, according to the theory of reasoned action (TRA), attitudes and perceptions determine behavioral intentions which are antecedents to actual behavior. Thus, this study also proposes that ITP is expected to decrease as the perceived phishing risk level increases.

# 4 Methodology

An experimental study was used to test the research model. Variant degree of uncertainty is the key treatment variable, and WTP and ITP are the primary dependent variables. The design of the experiment and questions were based on the prior model developed and pilot tested by Wang [32].

The experiment for this study was conducted among a total of 120 undergraduate students recruited from a college in northeastern United States. The subjects were randomly divided into four test groups, each receiving a different uncertainty treatment: known certainty (KC), known uncertainty (KU), unknowable uncertainty (UBU), and unknown uncertainty (UNU). The treatment variable was induced among subjects using hypothetical risk scenarios and vignettes of online phishing scenarios adapted from Wang [28]. Each scenario depicts an online phishing risk scenario corresponding to a different uncertainty degree in Table 17.1. Based on the vignette, subjects provided judgments on the amount they are willing to pay (WTP) to avoid the phishing risk and intention to purchase (ITP) online under the risk. An analogy type manipulation check question was also given to check if the treatment variable was properly understood by the subject. Table 17.2 shows the manipulation check used for the four different experiment scenarios. Demographic data were collected from subjects at the end of the experiment.

**Table 17.2** Manipulation checks for experiments

| Treatment variables: | Manipulation check question: |
|---|---|
| KU = known uncertainty,<br>UBU = unknowable uncertainty,<br>UNU = unknown uncertainty,<br>KC = known certainty | If the phishing risk is compared to the chance of randomly drawing a red ball from an urn of 100 red and black balls mixed together, the scenario given resembles which of the following? (Answer choices are: a, b, c, d) |
| **KU treatment vignette:** | **Expected answer:** |
| *A published study concludes that about 3% of online transactions from sites, such as E-WizWire involve phishing risks* | *a. Out of 100 balls in the urn, 3 are red and the rest are black* |
| **UBU treatment vignette:** | **Expected answer:** |
| *Research studies have concluded that, while the probability of phishing occurring due to online transactions with firms like E-WizWire is small, it is not possible to compute a reliable estimate of the rate. Thus, there seems to be no way of knowing the probability of phishing risks arising from a transaction* | *b. Out of 100 balls in the urn, there is no way of knowing how many are red and how many are black* |
| **UNU treatment vignette:** | **Expected answer:** |
| *A published summary of a study says that the estimated rate of transaction from firms like E-WizWire leading to phishing is less than 5%. The study was conducted by a coalition of online vendors and computer security firms. The study was privately funded and thus the details of the study are not made available to the public. The exact rate information may be known only to some insiders but unknown to the public* | *c. Out of 100 balls in the urn, we only know that the number of red balls is below 5. But most people do not know exactly how many are red and how many are black* |
| **KC treatment vignette:** | **Expected answer:** |
| *E-WizWire guarantees in writing and with full guarantee that none of their online transactions will involve phishing risks, due to their strong online security mechanism. Should it happen that a transaction with E-WizWire involves phishing risks, the firm will pay all costs to recover any loss at no expense to the user* | *d. It is officially announced that there are no red balls out of the 100 balls in the urn* |

# 5   Data Analysis and Results

A total of 120 responses were received from the four group experiments. A total of three responses were found to have failed the manipulation check question and were excluded from data analysis.

**Table 17.3** Tukey post hoc tests for WTP

| | (I) Treatment | (J) Treatment | Mean difference (I − J) |
|---|---|---|---|
| Tukey HSD | KC | KU | −2.4503 |
| | | UBU | −4.9466[a] |
| | | UNU | −8.4601[a] |
| | KU | KC | 2.4503[a] |
| | | UBU | −2.4963[a] |
| | | UNU | −6.0098[a] |
| | UBU | KC | 4.9466[a] |
| | | KU | 2.4963[a] |
| | | UNU | −3.5135[a] |
| | UNU | KC | 8.4601[a] |
| | | KU | 6.0098[a] |
| | | UBU | 3.5135[a] |

[a]The mean difference is significant at the 0.05 level

## 5.1  Demographics

Basic data on demographics and relevant online experience were collected from the subjects. The data include age, gender, Internet usage, and experience in online purchase and online credit card payment. The data show that over 90% of the subjects have had prior experience purchasing online and making online payment by credit card. In addition, over 80% of the subjects have used the Internet for four or more years. On average, over 95% of the subjects use the Internet between 1 and 10 hours per day. The age of the subjects for the pilot study falls between 18 and 50. The gender ratio of the subjects (56% female and 44% male) is very close to the gender ratio of the general student population at the sampled college.

## 5.2  ANOVA results

ANOVA was performed on WTP and ITP using the uncertainty treatment level as the independent variable. The ANOVA results suggest that the subjective estimates on willingness to pay to avoid the online phishing risk and on the scale of intention to purchase online are significantly different across the four treatment levels in the experiment. This shows that variant uncertainty levels have a significant effect on online consumer decisions.

Post hoc tests were conducted using SPSS to compare the pairwise differences in WTP and ITP means, shown in Tables 17.3 and 17.4. The test results clearly indicate significant differences across the treatment conditions for both WTP and ITP. Table 17.3 suggests that consumers are willing to pay a statistically significant amount of approximately $2.50 to avoid moving from known uncertainty to unknowable uncertainty and approximately $3.50 to avoid moving from unknowable uncertainty to unknown uncertainty in judging online phishing risk scenarios.

**Table 17.4** Tukey post hoc tests for ITP

|            | (*I*) Treatment | (*J*) Treatment | Mean difference (*I* − *J*) |
|------------|-----------------|-----------------|----------------------------|
| Tukey HSD  | KC              | KU              | 1.2667[a]                  |
|            |                 | UBU             | 2.8333[a]                  |
|            |                 | UNU             | 5.2471[a]                  |
|            | KU              | KC              | −1.2667[a]                 |
|            |                 | UBU             | 1.5667[a]                  |
|            |                 | UNU             | 3.9805[a]                  |
|            | UBU             | KC              | −2.8333[a]                 |
|            |                 | KU              | −1.5667[a]                 |
|            |                 | UNU             | 2.4138[a]                  |
|            | UNU             | KC              | −5.2471[a]                 |
|            |                 | KU              | −3.9805[a]                 |
|            |                 | UBU             | −2.4138[a]                 |

[a]The mean difference is significant at the 0.05 level

In terms of the ITP measure, Table 17.4 suggests that online consumers have statistically greater intentions to purchase online under reduced uncertainty conditions. Table 17.3 shows that the average intention to purchase under the knowable uncertainty condition is 1.5667 greater than that under the unknowable condition. The average ITP under unknowable uncertainty is 2.4138 greater than that under the unknown uncertainty condition.

## 6 Conclusion and Future Work

This study proposed a fine-grained approach to understanding variant degrees of uncertainties of consumer knowledge of online phishing risks. The goal of the study was to investigate the effect of variant levels of uncertainties on B2C e-commerce consumer decision making in online purchase. The experimental results provided empirical support for the research model and the hypotheses of this study. The finding suggests that consumer judgment of online phishing risks and intention to purchase vary systematically with the uncertainty conditions of their risk knowledge. The pairwise differences for WTP and ITP indicate that consumers prefer known uncertainty over unknowable uncertainty over unknown uncertainty in this order in judging online phishing risks. This study can be further extended to future studies of other online security risks involving decision under uncertainty.

A practical implication of the finding of this study is for B2C e-commerce vendors. The research suggests that online vendors may increase consumer intention to purchase by lowering uncertainty and presenting online phishing risks with more precise risk probability and outcome estimates. B2C e-commerce consumers will find this research model and findings helpful to improving their knowledge of online phishing risks and enhancing their online purchase decision process.

There could be promising further research in this area. One valuable research topic could be to develop a more comprehensive model of how B2C e-commerce consumers view and respond to online phishing risks. This model could incorporate not only the dimension of risk knowledge but also attributes of personal characteristics and the decision task. The study by Cai and Xu [4] has found that aesthetic design qualities, such as color, graphics, and the layout of an online shopping site have an important effect on consumers' online shopping value and experiences. The future model could also incorporate the variables of online transaction environment and potentially measure and compare consumers' levels of priorities among various concerns and risks in e-commerce transactions. Chan and Chen [5] developed a driving aptitude test to predicate one's performance for safe and quality driving. Similarly, an anti-phishing aptitude test could be developed to measure B2C e-commerce consumers' knowledge of online phishing risks and predict their performance in online purchase decisions.

# References

1. Acquisti A, Grossklags J (2005) Uncertain, ambiguity, and privacy. In: Proceedings of the 4th annual workshop on economics and information security. pp 1–21
2. Bhatnagar SM, Rao HR (2000) On risk, convenience, and internet shopping behavior. Commun ACM 43(11):98–105
3. Bose I, Leung ACM (2007) Unveiling the masks of phishing: threats, preventive measures, and responsibilities. Commun Assoc Inform Syst 19:544–566
4. Cai S, Xu Y (2011) Designing not just for pleasure: effects of web site aesthetics on consumer shopping value. Int J Electron Commer 15(4):159–187
5. Chan AHS, Chen K (2011) The development of a driving aptitude test for personnel decisions. Eng Lett 19(2):112–118
6. Chow CC, Sarin RK (2002) Known, unknown, and unknowable uncertainties. Theory Decis 52:127–138
7. Dinev T, Hu Q (2007) The centrality of awareness in the formation of user behavioral intention toward protective information technologies. J Assoc Inform Syst 8:386–408
8. Ellsberg D (1961) Risk, ambiguity and the savage axioms. Q J Econom 75:643–669
9. Fischhoff B, Slovic P, Lichtenstein S (1978) How safe is safe enough? A psychometric study of attitudes towards technological risks and benefits. Policy Sci 9(2):127–152
10. Fishbein M, Ajzen I (1975) Belief, attitude, intention, and behavior: an introduction to theory and research. Addison-Wesley, Reading
11. Fox CR, Tversky A (1995) Ambiguity aversion and comparative ignorance. Q J Econom 110 (3):585–603
12. Gabriel IJ, Nyshadham E (2008) A cognitive map of people's online risk perceptions and attitudes: an empirical study. In: Proceedings of the 41st annual Hawaii international conference on systems sciences. Big Island, HI, pp 274–283
13. Garera S, Provos N, Chew M, Rubin AD, November 2, 2007. A framework for detection and measurement of phishing attacks. In: WORM'07, Alexandria, pp 1–8
14. Gilboa I, Schmeidler D (1989) Maxmin expected utility with non-unique prior. J Math Econom 18:141–153
15. Glover S, Benbasat I (2011) A comprehensive model of perceived risk of e-commerce transactions. Int J Electron Commer 15(2):47–78

16. Hogarth RM, Kunreuther H (1995) Decision making under ignorance: arguing with yourself. J Risk Uncertain 10:15–36

17. Jiang J, Chen C, Wang C (2008) Knowledge and trust in e-consumers' online shopping behavior. In: International symposium on electronic commerce and security., pp 652–656

18. Kahneman D, Tversky A (1979) Prospect theory: analysis of decision under risk. Econometrica 47(2):263–292

19. Milne GR, Rohm AJ, Bahl S (2004) Consumers' protection of online privacy and identity. J Consum Aff 38(2):217–232

20. Miyazaki AD, Fernandez A (2001) Consumer perceptions of privacy and security risks for online shopping. J Consum Aff 35(1):27–44

21. Nyshadham EA, Ugbaja M (2006) A study of ecommerce risk perceptions among B2C consumers: a two country study. In: Proceedings of the 19th Bled eConference, Bled, Slovenia

22. Pavlou PA (2003) Consumer acceptance of electronic commerce: integrating trust and risk with the technology acceptance model. Int J Electron Commer 7(3):69–103

23. Pavlou PA, Fygenson M (2006) Understanding and predicting electronic commerce adoption: an extension of the theory of planned behavior. MIS Q 30(1):115–143

24. Roca M, Hogarth RM, Maule AJ (2002) Ambiguity seeking as a result of the status quo bias. J Risk Uncertain 32:175–194

25. Salisbury WD, Pearson RA, Pearson AW, Miller DW (2001) Perceived security and world wide web purchase intention. Ind Manag Data Syst 101(4):165–176

26. Savage LJ (1954) The foundations of statistics. Wiley, New York (Revised and enlarged edition, Dover, New York, 1972)

27. Shafer G (1986) Savage revisited. Stat Sci 1(4):463–501

28. Slovic P (1987) Perception of risk. Science 236:280–285

29. Slovic P, Fischhoff B, Lichtenstein S (1982) Why study risk perception? Risk Anal 2(2):83–93

30. Son J, Kim SS (2008) Internet users' information privacy-protective responses: a taxonomy and a nomological model. MIS Q 32(3):503–529

31. Tsai J, Cranor L, Egelman S, Acqusiti A (2007) The effect of online privacy information on purchasing behavior: an experimental study. In: Proceedings of the twenty eighth international conference on information systems. Montreal, pp 1–17

32. Wang P (2010) The effect of knowledge of online security risks on consumer decision making in B2C e-commerce. Ph.D. Dissertation, Nova Southeastern University, FL (UMI No. 3422425)

33. Wang P (2011) Uncertainties of online phishing risks and consumer decision making in B2C e-commerce. In: Lecture notes in engineering and computer science: proceedings of the world congress on engineering 2011, WCE 2011. London, 6–8 July 2011, pp 469–474

# Chapter 18
# Using Conceptual Graph Matching Methods to Semantically Mediate Ontologies

Gopinath Ganapathy and Ravi Lourdusamy

## 1 Introduction

A requirement common to many knowledge applications is to determine whether two or more knowledge representations, encoded using the same ontology, capture the same knowledge. The task of determining whether two or more representations encode the same knowledge is treated as a graph matching problem. The knowledge representation is encoded using conceptual graph. The representations capture the same knowledge if their corresponding graphs match. The multiple encoding of the same knowledge rarely matches exactly, so a matcher must be flexible to avoid a high rate of false negatives. However, a matcher that is too flexible can suffer from a high rate of false positives [1].

Previous solutions to this problem have produced two types of matchers. *Syntactic matchers* use only the graphical form of the representations, judging their similarity by the amount of common structures shared [2, 3] or the number of edit operations required to transform one graph into the other [4–7]. Approaches that focus on the amount of shared common structures do not handle mismatches. Approaches that use edit operations can handle mismatches but are sensitive to the cost assigned to them. In this approach, handling these parameters such as cost is problematic.

Semantic matchers use this knowledge to determine the match of two representations [1, 8–10]. The knowledge encoded can be equivalently represented as conceptual graphs [11]. The algorithms developed for matching [12] and merging of $m$ conceptual graphs are illustrated using health care domain ontologies [13]. Further with the help of transformations the matching process is improved to provide semantic matching.

G. Ganapathy (✉)
Department of Computer Science, Bharathidasan University, Tiruchirappalli 620 023, India
e-mail: gganapathy@gmail.com

R. Lourdusamy
Department of Computer Science, Sacred Heart College, Tirupattur 635 601, India
e-mail: ravi.lourdusamy@gmail.com

S.-I. Ao and L. Gelman (eds.), *Electrical Engineering and Intelligent Systems*,
Lecture Notes in Electrical Engineering 130, DOI 10.1007/978-1-4614-2317-1_18,
© Springer Science+Business Media, LLC 2013

## 2 Preliminary

**Definition 2.1** Transitive and part ascendant transformations conform to a more general notion called "transfers through [1]." A relation $r$ transfers through another relation $r'$ if

$$X \xrightarrow{r} Y \xrightarrow{r'} Z \Rightarrow X \xrightarrow{r} Z \qquad (18.1)$$

**Definition 2.2** A triple is a 3-tuple of the form (*head, relation, tail*) where head and tail are concepts or instances (i.e., nodes in a conceptual graph) and relation is an edge in the graph. Every two nodes connected by an edge in a conceptual graph can be mechanically converted into a triple and hence a conceptual graph into a set of triples [1].

**Definition 2.3** The $n$ triples $t_1 = (head_1, relation_1, tail_1)$, $t_2 = (head_2, relation_2, tail_2)$, ..., $t_n = (head_n, relation_n, tail_n)$ of graph $G$ align if $head_1 \geq head_2 \geq, ..., \geq head_n$, $uneg(relation_1) \geq uneg(relation_2) \geq, ..., \geq uneg(relation_n)$ and $tail_1 \geq tail_2 \geq, ..., \geq tail_n$. The unneg (relation) unnegates relation if it is negated otherwise returns the relation.

**Definition 2.4** For $\ell = \{(t_{11}, t_{21}, ..., t_{m1}), (t_{12}, t_{22}, ..., t_{m2}), ..., (t_{1n1}, t_{2n}, ..., t_{mnm})\}$, a list of aligned triples of $m$ graphs, the bindings for $\ell$, i.e., $b(\ell) = \{(head_{11}/ head_{21}/, ..., /head_{m1}, tail_{11}/tail_{21}/, ..., /tail_{m1}), (head_{12}/head_{22}/, ..., /head_{m2}, tail_{12}/tail_{22}/, ..., /tail_{m2}), ..., (head_{1m1}/head_{2m2}/, ..., /head_{mnm}, tail_{1m1}/tail_{2n2}/, ..., /tail_{mnm})\}$.

## 3 Matching of $m$-Conceptual Graphs

Given $m$ graphs $G_1 = \{t_{11}, t_{12}, ..., t_{1n1}\}$, $G_2 = \{t_{21}, t_{22}, ..., t_{2n2}\}$, ..., $G_m = \{t_{m1}, t_{m2}, ..., t_{mnm}\}$, where $n_1, n_2, ..., n_m$ are the number of triples of $G_1, G_2, ..., and G_m$, respectively, and a set of $r$ transformation $R$, where $R = \{R_1, R_2, ..., R_r\}$. The aim of the algorithm is to find a common subgraph of $G_1, G_2, ..., and G_m$ called SG. Construct a list $m$ of all possible alignments between the triple of $G_1, G_2, ...,$ and $G_m$. Each element of $m$ is of the form $\ell = \{(t_{11}, t_{21}, ..., t_{m1}), (t_{12}, t_{22}, ..., t_{m2}), (t_{1n1}, t_{2n2}, ..., t_{mnm})\}$. The generalized algorithm for finding a match between $m$ representations is presented as Algorithm 18.1.

The steps for finding match between $m$ representations are illustrated with an example of three graphs, $G_1, G_2, ...,$ and $G_3$ generated by organization structure of three hospitals shown in Figs. 18.1–18.3. For reference, we label each triple in $G_1$ with unique number from 1 to 22, each triple in $G_2$ with unique uppercase letter from A to V and each triple in $G_3$ with a unique lowercase letter from a to z and from aa to hh. We use subscripts to differentiate terms that appear multiple times (e.g., Hospital_Model_1).

**Algorithm 18.1** Outline of the generalization matching algorithm.

1) $M = $ NIL and $\ell = NIL$
   *FOR* each triple $t_{1i1}$ in $G_1$
   *FOR* each triple $t_{2i2}$ in $G_2$

   ............................

   *FOR* each triple $t_{mim}$ in $G_m$
   If $t_{1i1}$, $t_{2i2}$,..., $t_{mim}$ are aligned
   *THEN* add $(t_{1i1}, t_{2i2},..., t_{mim})$ to $\ell$.
   ADD $\ell$ to $M$ and rest $\ell$ to *NIL*.
2) Use $M$ to construct a common subgraph of $G_1$, $G_2$,..., $Gm$
   called SG.
   SG $= \{(t_{11},..., t_{ml}), (t_{12},..., t_{m2}),..., (t_{1nl},..., t_{mnm})\}$ where $(t_{1i},...,$
   $t_{mi})$ are the aligned triples of $G_1$, $G2$,..., $Gm$ respectively.
3) IF SG is inconsistent *THEN* stop and return *NIL*
4) *FOR* each rule $R_i$ in $R$,
   *FOR* each $j = 1, 2,..., m$
   *FOR* each $k = 1, 2,..., m$
   *IF* $R_i$ is applicable to $G_j$ with respect to $G_k$
   *THEN* apply $(R_i, G_j, G_k)$
5) *FOR* each $j = 1, 2,..., m$
   *FOR* each unaligned triples $t_{jij}$ in $G_j$
   *IF* $t_{1i1j}$, $t_{2i2j}$,..., $t_{mimj}$ are aligned and
   $b (\{(t_{1i1j}, t_{2i2j},..., t_{mimj})\})$ is consistent with $b$ (SG),
   *THEN* add $(t_{1i1j}, t_{2i2j},..., t_{mimj})$ to SG and break.
   *UNTIL* SG reaches quiescence go to step 4.
6) *RETURN SG*

**Algorithm 18.2** Outline of the generalised merging algorithm.

(1) Set *SG* as *G*
(2) *FOR* each unmatched triple $t_{1i1}$ in $G_1$
   *FOR* each unmatched triple $t_{2i2}$ in $G_2$
   *FOR* each unmatched triple $t_{mim}$ in $G_m$
   Search *G* for the head of $t_{jij}$, in $n_i$
   If the search is successful add tjij as a subnode
   Else return the resultant graph as G
(3) *RETURN G*

   In step 1, the generalized algorithm compares each triple in $G_1$ with each triple in $G_2, G_3, \ldots, G_m$ to find all possible alignments. In our example, $t_{11}$ aligns with $t_{21}$ and $t_{31}$. Triple $t_{11}$, however, does not align with triple $t_{22}$ and its combinations because the relations differ. The initial match is denoted by $M$. $M = \{\ell_1, \ell_2, \ldots, \ell_p\}$, where $p$ is the total number of possible alignments. Each element of $M$ is a list called $\ell_i$. For example $\{(t_{11}, t_{21}, t_{31})\}$ is called $\ell_1$, $\{(t_{12}, t_{22}, t_{32})\}$ is called $\ell_2$, etc. In step 2, the generalized algorithm uses $M$ to construct common subgraph of $G_1$, $G_2$, ..., $G_m$ called *SG*. The generalized algorithm begins by selecting a member, $\ell_i$, of $M$ to serve as the seed of the construction process (recall that $M = \{\ell_1, \ell_2, \ldots, \ell_s\}$) and $\ell_i = \{(t_{1i11}, t_{2i21}, t_{mim1}), (t_{1i12}, t_{2i22}, t_{mim2}), \ldots, (t_{1i1k}, t_{2i2k}, t_{mimk})\}, I = 1,2, \ldots, s$. This seed is selected based on a heuristic scoring function

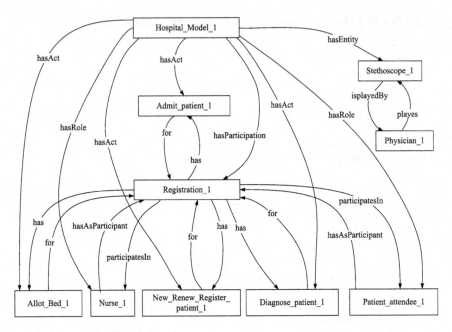

**Fig. 18.1** Graph $G_1$ of hospital model 1

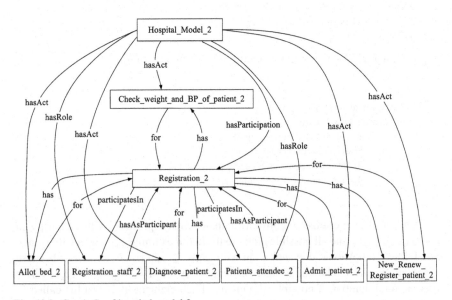

**Fig. 18.2** Graph $G_2$ of hospital model 2

$$h(\ell i) = \frac{1}{K_i} \sum\nolimits_{j=1}^{k_i} n(\text{term1}) + n(\text{term2}) \qquad (18.2)$$

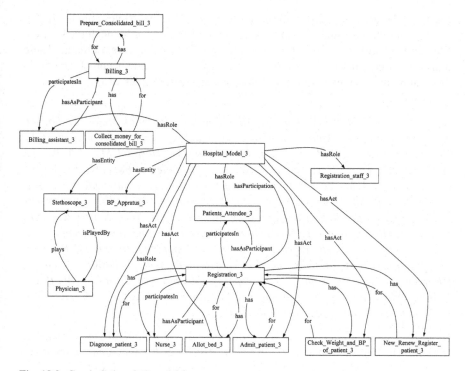

**Fig. 18.3** Graph $G_3$ hospital model 3

where

$$term1 = head_{1i1j}/head_{2i2j}/, \ldots, /head_{mimj}$$
$$term2 = tail_{1i1j}/tail_{2i2j}/, \ldots, /tail_{mimj}$$

are the bindings of $t_{1i1j}$, $t_{2i2j}$, ..., $t_{mimj}$, and $n(b)$ is the number of times the bindings $b$ occur in *bindings* $(M)$. This function h is a heuristic that favors those $\ell_i$ in $M$ with interconnectivity. Bindings that occur frequently indicate high interconnectivity. We want to select these $\ell_i$ as the seeds because they have more potential for allowing larger common subgraphs to be constructed. Therefore, the algorithm selects the $\ell_i$ in $M$ with the highest score, as determined by the function $h$. *SG* is extended with those pairs of aligned triples in $M$ whose bindings intersect the binding of the pairs in *SG*. Pairs in $M$ that extend *SG* are removed from $M$ along with $\ell_j$ they belong to. This process is repeated until *SG* can no longer be extended.

In steps 3–5, the generalized algorithm checks if *SG* is consistent. *SG* is inconsistent if it contains an aligned $m$-tuples of triples $(t_{1i}, t_{2i}, \ldots, t_{mi})$ where the relation of at least one $t_{ji}$ is negated, and the relation of at least one $t_{ki}$ is not negated. If *SG* contains such a $m$-tuples, then generalized algorithm stops and returns *NIL*. Otherwise, the generalized algorithm applies transformations to improve the match. Steps 4 and 5 are repeated until *SG* reaches quiescence. In step 4, the generalized algorithm applies transformations to resolve mismatches among $G_1$, $G_2$, ..., $G_m$.

In step 5, the generalized algorithm tries to align additional triples among $G_1$, $G_2$, ..., $G_m$. Step 5 is similar to step 1 except that it focuses on the unaligned triples. We have identified a set of transformations for health care domain. These transformations are used to improve the matching of $m$-conceptual graphs in the health care domain. Returning to our example, the triples of $G_1$, $G_2$, and $G_3$ are as follows:

$G_1 = \{$(Hospital_Model_1, hasAct, Allot_Bed_1), (Hospital_Model_1, hasAct, New_Renew_Register_patient_1), (Hospital_Model_1, hasAct, Admit_patient_1), (Hospital _Model_1, hasParticipation, Registration_1), (Hospital_Model_1, hasAct, Diagnose_patient_1), (Hospital_Model_1, hasRole, Patients_attendee_1), (Admin_patient_1, for, Registration_1), (Registration_1, has, Admit_patient_1), (Registration_1, has, Allot_Bed_1), (Allot _Bed_1, for, Registration_1), (New_Renew_Register_patient_1, for, Registration_1), (Registration_1, has, New_ Renew_Register_ patient_1), (Registration _1, has, Diagnose_patient_1), (Diagnose_patient_1, for, Registration _1), (Patients_attendee_1, hasAsParticipant, Registration_1), (Registration_1, participatesIn, Patients_attendee_1), (Hospital_Model_1, hasRole, Nurse_1), (Hospital _Model_1, hasEntity, Stethoscope_1), (Stethoscope _1, isPlayedBy, Physician_1), (Physician_1, plays, Stethoscope _1), (Nurse_1, hasAsParticipant, Registration_1), (Registration _1, participatesIn, Nurse_1)$\}$.

$G_2 = \{$(Hospital_Model_2, hasAct, Allot_Bed_2), (Hospital_Model_2, hasAct, New_Renew_Register_patient_2), (Hospital_Model_2, hasAct, Admit_patient_2), (Hospital _Model_2, hasParticipation, Registration_2), (Hospital_Model_2, hasAct, Diagnose_patient_2), (Hospital_Model_2, hasRole, Patients_attendee_ 2), (Admit_patient_2, for, Registration_2), (Registration_2, has, Admit_patient_2), (Registration_2, has, Allot_Bed_2), (Allot_Bed_2, for, Registration_ 2), (New_Renew_Register_patient_2, for, Registration_2), (Registration_2, has, New_Renew_Register_patient_2), (Registration _2, has, Diagnose_patient_2), (Diagnose_patient_2, for, Registration_2), (Patients_attendee_2, hasAsParticipant, Registration_2), (Registration_2, participatesIn, Patients_attendee_ 2), (Hospital_Model_2, hasRole, Registration _Staff_2), (Hospital_Model_2, hasAct, Check_weight_and_BP_of_patient_2), (Check_weight_and_BP_of_ patient_2, for, Registration_2), (Registration_2, has, Check_weight_and_BP_of_ patient_2), (Registration_Staff_2, hasAsParticipant, Registration_2), (Registration_2, ParticipatesIn, Registration_Staff_2)$\}$.

$G_3 = \{$ (Hospital_Model_3, hasAct, Allot_Bed_3), (Hospital_Model_3, hasAct, New_Renew_Register_patient_3), (Hospital_Model_3, hasAct, Admit_patient_ 3), (Hospital_Model_3, hasParticipation, Registration_3), (Hospital_Model_3, hasAct, Diagnose_patient_3), (Hospital_Model_3, hasRole, Patients_attendee_ 3), (Admit_patient_3, for, Registration_3), (Registration_3, has, Admit_patient_3), (Registration_3, has, Allot_Bed_3), (Allot_Bed_3, for, Registration_3), (New_ Renew_Register_patient_3, for, Registration_3), (Registration_3, has, New_Renew_Register_patient_3), (Regiatration_3, has, Diagnose_patient_3), (Diagnose_patient_3, for, Registration_3), (Patients_attendee_3, hasAsParticipant,

Registration_3), (Registration_3, ParticipatesIn, Patients_attendee_3), (Hospital_ Model_3, hasRole, Billing_Assistant_3), (Hospital_Model_3, hasEntity, Stetho-scope_3), (Hospital_Model_3, hasEntity, BP_Apparatus_3), (Hospital_Model_3, hasRole, Nurse_3), (Hospital_Model_3, hasAct, Check_weight_and_BP_of_patient_3), (Hospital_Model_3, hasRole, Registration_Staff_3), (Nurse_3, hasAs Participants, Registration_3), (Registration_3, ParticipatesIn, Nurse_3), (Registration_3, has, Check_weight_and_BP_of_patient_3), (Check_ weight_and_BP_of_patient_3, for, Registration_3), (Physician_3, Plays, Stethoscope _3), (Stetho-scope _3, isPlayedBy, Physician_3), (Collect_money_for_consolidated_bill_3, for, Billing_3), (Billing_3, has, Collect_money_for_consolidated_bill_3), (Billing_Assistant_3, hasAsParticipant, Billing_3), (Billing_3, ParticipatesIn, Billing_assistant_3), (Prepare_consolidated_bill_3, for, Billing_3), (Billing_3, has, Prepare_ consolidated_bill_3)}.

In step 1, the algorithm compares each triple in $G_1$ with each triple in $G_2$ and $G_3$ to find all possible alignments. Triple 1 of $G_1$ aligns with A of $G_2$ and a of $G_3$, Triple 2 of $G_1$ aligns with B of $G_2$ and b of $G_3$, and so on. Hence, the matched triples of $G_1, G_2$, and $G_3$ are as follows: Binding $(m) = \{(1, A, a), (2, B, b), (3, C, c), (4, D, d), (5, E, e), (6, F, f), (7, G, g), (8, H, h), (9, I, i), (10, J, j), (11, K, k), (12, L, l), (13, M, m), (14, N, n), (15, O, o), (16, P, p)\} = \{\ell_1, \ell_2, \ell_3, \ell_4, \ell_5, \ell_6, \ell_7, \ell_8, \ell_9, \ell_{10}, \ell_{11}, \ell_{12}, \ell_{13}, \ell_{14}, \ell_{15}, \ell_{16}\}$.

In step 2, hscore is calculated as follows: $hscore(\ell_1) = hscore(\ell_2) = hscore(\ell_3) = 10$, $hscore(\ell_4) = 18$, $hscore(\ell_5) = hscore(\ell_6) = 10$, $hscore(\ell j) = 14$, where $j = 7, 8, \ldots, 16$. Select $\ell_4$ and remove it from M and hence the subgraph SG is $\{(4, D, d)\}$.

Binding $(M) = \{\ell_1, \ell_2, \ell_3, \ell_5, \ell_6, \ell_7, \ell_8, \ell_9, \ell_{10}, \ell_{11}, \ell_{12}, \ell_{13}, \ell_{14}, \ell_{15}, \ell_{16}\}$. The head of $\ell_4$ is (Hospital_Model_1/Hospital_Model_2/Hospital_Model_3) is intersecting with $\ell_1, \ell_2, \ell_3, \ell_4, \ell_5, \ell_6$. Remove all from M and hence the subgraph of SG is $\{(4, D, d), (1, A, a), (2, B, b), (3, C, c), (5, E, e), (6, F, f)\}$. Binding $(M) = \{\ell_7, \ell_8, \ell_9, \ell_{10}, \ell_{11}, \ell_{12}, \ell_{13}, \ell_{14}, \ell_{15}, \ell_{16}\}$. The tail $\ell_4$ is (Registration_1/Registration_2/Registration_3) is intersecting with $\ell_7, \ell_8, \ell_9, \ell_{10}, \ell_{11}, \ell_{12}, \ell_{13}, \ell_{14}, \ell_{15}, \ell_{16}$. Hence, the subgraph

$$SG = \{(4, D, d), (1, A, a), (2, B, b), (3, C, c), (5, E, e), (6, F, f), (7, G, g), (8, H, h), (9, I, i), (10, J, j), (11, K, k), (12, L, l), (13, M, m), (14, N, n), (15, O, o), (16, P, p)\}.$$

The result is the maximal subgraph of $G_1, G_2$, and $G_3$ which is shown in Fig. 18.4.

Since SG is consistent, we apply transformation to improve the match. For example, transformation $R_1$ (Fig. 18.5) is applied to graph $G_1, G_2$, and $G_3$, and hence, a triple is added to graphs $G_1, G_2$, and $G_3$. Because of the added triple another match is added in SG. The added triple is shown as an edge of dotted line in Fig. 18.4.

Considering the unaligned triples in the matching process, it is observed that the set of unaligned triples of a model represents the unique feature of that model. Returning to our example, the set of unaligned triples of $G_1$ represents the unique feature of the Hospital_Model_1. Similarly, the set of unaligned triples of $G_2$ and $G_3$

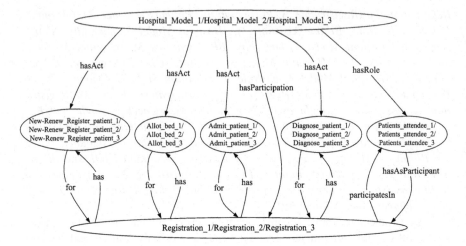

**Fig. 18.4** Matched graph legend: H_M: Hospital_Model, Reg: Registration, p: patient

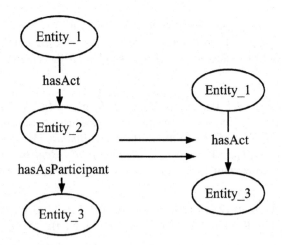

**Fig. 18.5** A sample transformation

represents the unique features of the Hospital_Model_2 and Hospital_Model_3. To provide more semantic matching similarity flooding technique is also used. For example, Nurse can play the Role of Registration_staff where as the Registration_staff cannot play the role of Nurse. Hence, the triples of $G_1$, $G_2$, and $G_3$ namely (Hospital_ Model_1, hasRole, Nurse_1), (Hospital_Model_2, hasRole, Registration_ staff_2), and (Hospital_Model_3, hasRole, Registration_staff_3) can be aligned based on its similarity. The aligned triple in this case will be (Hospital_Model_1, hasRole, Registration_staff_1), as Registration_staff can be Nurse .


## 3.1   Fitness of the Matched Graph

In this section, we present the fitness of the matched graph. $SG$ is returned along with a numeric score reflecting the fitness of the match between $G_1$, $G_2$, ..., $G_m$. This score is used in situations where one graph is matched with a set of graphs to select the best match. To calculate the fitness score, we apply (18.3) to $SG$ where

$$SG = \{(t_{1j1}, t_{2j1}, \ldots, t_{mj1}),$$

$(t_{1j2}, t_{2j2}, \ldots, t_{mj2}), \ldots, (t_{1jk}, t_{2jk}, \ldots, t_{mjk})\}$ and the $m$-tuples $(t_{1ji}, t_{2ji}, \ldots, t_{mji})$, for $j = 1, 2, \ldots, k$ are aligned triples of $G_1$, $G_2$, ..., $G_m$, respectively.

$$\text{Fitness (SG)} = \frac{\sum_{i=1}^{k} \text{score}(t_{ij}, t_{2j}, \ldots, t_{mj})}{\min\{(|G_1| + t_{r1}), (|G_2| + t_{r2}), \ldots, (|G_m| + t_{rm})\}} \qquad (18.3)$$

where $|G_s|$ is the number of triples in $G_s$ and trs is the number of additional triples added to $G_s$, for $s = 1 \ldots m$, by applying transformations. The score for each $m$-tuple of triple $(t_{1ji}, t_{2ji}, \ldots, t_{mji})$ is computed using a function score as given by (18.4)

$$\text{score}(t_{1ji}, t_{2ji}, \ldots, t_{mji}) = \frac{\frac{1}{x+1} + \frac{1}{y+1} + \frac{1}{z+1}}{3} \qquad (18.4)$$

where,

$x = taxdist (head_{1ji}, head_{2ji}, \ldots, head_{mji})$,
$y = taxdist (rela_{1ji}, rela_{2ji}, \ldots, rela_{mji})$,
$z = taxdist (tail_{1ji}, tail_{2ji}, \ldots, tail_{mji})$ and
$taxdist (C_1, C_2, \ldots, C_m) \Rightarrow taxdist (C_i, C_{j+1})$

The function $taxdist (C_i, C_{i+1})$ is the taxonomic (or semantics) distance [11] between two concepts $C_i$ and $C_{i+1}$. We calculate the taxonomic distance between two concepts as the minimum number of taxonomic edges that needs to be traversed to reach $C_i$ from $C_{i+1}$. After transformations have been applied, $SG$ is returned along with a numeric score reflecting the fitness of the match between $G_1, G_2, \ldots, G_m$. This score is computed based on the number of matched triples over the size of the graphs being matched [13]. The fitness score is also computed using a simple formula as given by (18.5). In our example, the fitness of SG is approximately 0.5.

$$\text{Fitness (SG)} = \frac{\text{No. of triples of SG}}{\text{Maximum of } n_1, n_2, \ldots, n_m} \qquad (18.5)$$

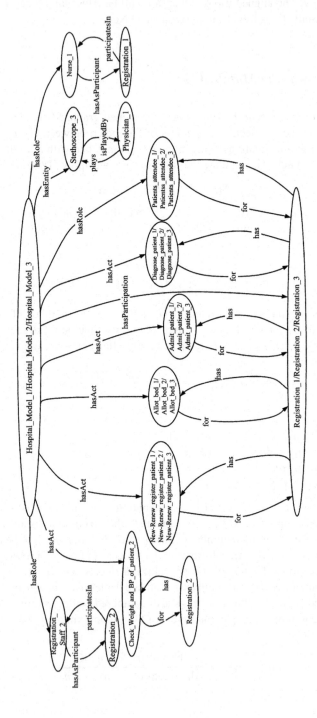

**Fig. 18.6** Merged graph Legent: H_M=Horizontal Model, Reg: Registration, P: Patient

## 4 Merging of $m$-Conceptual Graphs

In this section, we present an algorithm 18.2 to merge $m$-conceptual graphs after determining the matched graph. Let $p_1, p_2, \ldots, p_m$ be the number of unmatched triples of the graphs $G_1, G_2, \ldots, G_m$. Assume that the matched graph $SG$ as $G$ and consider each unmatched triple $t_{jij}$, of a graph $G_i$. Search $G$ to determine the node containing the head of $t_{jij}$. After determining such node, $t_{ji}$ is added as a subnode of the searched graph node. After adding all the unmatched triples in $G$, the resultant graph is the merged graph of $m$ graphs $G_1, G_2, \ldots, G_m$. Returning to our example, the merged graph of $G_1$, $G_2$, and $G_3$ is shown in Fig. 18.6.

## 5 Conclusion

Algorithms for matching and merging of ontologies using $m$-conceptual graphs have been developed. It is illustrated using examples from health care domain. In this work, similarity flooding and transformation techniques of conceptual graphs are incorporated in order to achieve semantic matching and merging. These algorithms can be implemented using any ontology management software to support its application to the health care domain.

## References

1. Yeh PZ, Porter B, Barker K (2003) Using transformations to improve semantic matching. In: K-CAP'03, Sanibel Island-florida, 23–25 Oct 2003
2. Bunke H, Jiang X, Kandel A (2000) On the minimum common super-graph of two graphs. Computing 65(1):13–25
3. Bunke H, Shearer K (1998) A graph distance metric based on the maximal common subgraph. Pattern Recog Lett 19:228–259
4. Messmer BT, Bunke H (1993) A network based approach to exact and inexact graph matching. In: Technical report IAM 93-021, Universitat Bern
5. Sanfeliu A, Fu K (1983) A distance measure between attributed relational graphs for pattern recognition. IEEE Trans SMC 13:353–362
6. Shapiro L, Haralick R (1981) Structural descriptions and inexact matching. IEEE Trans PAM 3:504–519
7. Tsai W, Fu K (1979) Error-correcting isomorphisms of attributed relational graphs for pattern analysis. IEEE Trans SMC 9:869–874
8. Genest D, Chein M (1997) An experiment in document retrieval using conceptual graphs. In: ICCS
9. Myacng S (1992) Conceptual graphs as a framework for text retrieval. In: Eklund P, Nagle T, Nagle J, Gerhortz L, Horwood E (eds) Current directions in conceptual structure research
10. Zhong J, Zhu H, Li J, Yu Y (2002) Conceptual graph for matching for semantic search. In: ICCS

11. Sowa JF (1984) Conceptual structures: information processing in mind and machine. Addison-Wesley, Reading
12. Yeh P, Porter B, Barker K (2003) Transformation rules for knowledge-based pattern matching. In: Technical report UT-AT-TR-03-299, University of Texas, Austin
13. Ganapathy G, Lourdusamy R (2011) Matching and merging of ontologies using conceptual graphs. In: Lecture notes in engineering and computer science: Proceedings of the world congress on engineering 2011. WCE 2011, London, 6–8 July 2011. pp 1829–1833

# Chapter 19
# Asymmetric Optimal Hedge Ratio with an Application

**Youssef El-Khatib and Abdulnasser Hatemi-J**

## 1 Introduction

Financial risk management and hedging against risk have become more important now due to the recent occurrence of the financial crisis worldwide and the consequent turmoil in financial markets. The optimal hedge ratio (OHR) has, therefore, important implication for investors in order to hedge against the price risk. Several different approaches have been suggested in the literature in order to estimate the OHR, among others, constant parameter and time-varying approaches have been applied.

The interested reader can refer to the following literature on the optimal hedge ratio: [3, 6–8, 10, 11, 15, 17].

But, one pertinent issue in this regard, which has not been investigated to our best knowledge, is whether the OHR has an asymmetric character or not. In another word, does a negative price change have the same impact as a positive price change of the same magnitude? This issue is addressed in the current paper by mathematically proving that the OHR is asymmetric.

In addition, we provide a method to deal with this asymmetry in the estimation of the underlying OHR. The asymmetric behaviour of returns and correlations among financial assets has been investigated by [2, 9, 13], as well as [1]. According to these publications, investors seem to respond more to negative shocks than the positive ones. Thus, the question is whether the issue of asymmetry matters in the

Y. El-Khatib (✉)
Department of Mathematical Sciences, UAE University, Al-Ain,
P.O. Box 17551, United Arab Emirates
e-mail: youssef_elkhatib@uaeu.ac.ae

A. Hatemi-J
Department of Economics and Finance, UAE University, Al-Ain,
P.O. Box 17555, United Arab Emirates
e-mail: Ahatemi@uaeu.ac.ae

S.-I. Ao and L. Gelman (eds.), *Electrical Engineering and Intelligent Systems*,
Lecture Notes in Electrical Engineering 130, DOI 10.1007/978-1-4614-2317-1_19,
© Springer Science+Business Media, LLC 2013

estimation of the OHR or not. The method suggested in this paper is applied to the
US equity market using weekly spot and future share prices during the period
January 5, 2006 to September 29, 2009. We find empirical support for an asymmet-
ric OHR. For a conference version of this paper see also [4].

The rest of the paper is structured as follows. Section 2 makes a brief discussion
of the optimal hedge ratio including a mathematical derivation of this ratio.
Section 3 describes the underlying methodology for estimating the asymmetric
OHR, and it also proves mathematically the asymmetric property of the OHR.
Section 4 provides an empirical application, and the last section concludes the
paper.

## 2   The Optimal Hedge Ratio

The main function of the OHR is to make sure that total value of the hedged
portfolio remains unaltered. The hedged portfolio includes the quantities of the spot
instrument as well as the hedging instrument and it can be expressed mathematically as follows:

$$V_h = Q_s S - Q_f F, \tag{19.1}$$

where $V_h$ represents the value of the hedged portfolio and $Q_s$ and $Q_f$ signify the
quantity of spot and futures instrument, respectively. $S$ and $F$ are the prices of
the underlying variables. Equation (19.1) can be transformed into changes because
the only source of uncertainty is the price. Thus, we can express (19.1) as follows:

$$\Delta V_h = Q_s \Delta S - Q_f \Delta F, \tag{19.2}$$

where $\Delta S = S_2 - S_1$ and $\Delta S = F_2 - F_1$. The ultimate goal of the hedging strategy
is to achieve $\Delta V_h = 0$, which results in having $\frac{Q_f}{Q_s} = \frac{\Delta S}{\Delta F}$. Now, let

$$h = \frac{Q_f}{Q_s}, \tag{19.3}$$

then we must have

$$h = \frac{\Delta S}{\Delta F}. \tag{19.4}$$

Thus, $h$ is the hedge ratio, which can be obtained as the slope parameter in a
regression of the price of the spot instrument on the price of the future (hedging)
instrument. This can be demonstrated mathematically also. Let us substitute (19.3)
into (19.2) that yields the following equation:

$$\Delta V_h = Q_s [\Delta S - h \Delta F]. \tag{19.5}$$

The OHR is the one that minimizes the risk of the change of the portfolio value that is hedged. The source of this risk is the variance of equation (19.5), which is given by the following equation:

$$\text{Var}[\Delta V_h] = Q_s{}^2\left[\sigma_s^2 + h^2\sigma_F^2 - 2h\rho\sigma_s\sigma_F\right].$$ (19.6)

In this case, $\sigma_s^2$ represents the variance of $\Delta S$, $\sigma_F^2$ signifies the variance of $\Delta F$, and the correlation coefficient between $\Delta S$ and $\Delta F$ is denoted by $\rho$. Thus, in order to obtain the OHR we need to minimize (19.6) with regard to $h$. That is,

$$\frac{\partial \text{Var}[\Delta V_h]}{\partial h} = Q_s{}^2\left[2h\sigma_F^2 - 2\rho\sigma_s\sigma_F\right] = 0.$$ (19.7)

This in turn gives the following:

$$h* = \rho\frac{\sigma_s}{\sigma_F}.$$ (19.8)

It should be mentioned that the OHR can also be obtained by estimating the following regression model:

$$\Delta S_t = \alpha + h\Delta F_t + U_t.$$ (19.9)

# 3 Methodology for Estimating Asymmetric Optimal Hedge Ratio

Assume that we are interested in investigating the relationship between the changes of the spot and future prices when each price index is random walk process. Thus, the changes $\Delta S_t$ and $\Delta F_t$ can be defined as follows:

$$\Delta S_t = \varepsilon_{1t},$$ (19.10)

and

$$\Delta F_t = \varepsilon_{2t},$$ (19.11)

where $t = 1, 2, \ldots T$, and the variables $\varepsilon_{1t}$ and $\varepsilon_{2t}$ signify white noise disturbance terms. We define the positive and the negative shocks as follows respectively:

$$\varepsilon_{1t}^+ = \max(\varepsilon_{1t}, 0), \ \varepsilon_{1t}^+ = \max(\varepsilon_{2t}, 0)$$
$$\varepsilon_{1t}^- = \min(\varepsilon_{1t}, 0), \text{ and } \varepsilon_{1t}^- = \min(\varepsilon_{2t}, 0).$$

It follows that the changes can be defined as

$$\Delta S_t^+ = \varepsilon_{1t}^+, \ \Delta S_t^- = \varepsilon_{1t}^-, \ \Delta F_t^+ = \varepsilon_{2t}^+ \text{ and } \Delta F_t^- = \varepsilon_{2t}^-.$$

Thus, by using these results we can estimate the following regression models:

$$\Delta S_t^+ = \alpha_1 + h_1 \Delta F_t^+ + u_t^+, \tag{19.12}$$

$$\Delta S_t^- = \alpha_2 + h_2 \Delta F_t^- + u_t^-. \tag{19.13}$$

Consequently, $h_1$ is the OHR for positive price changes and $h_2$ is the OHR for negative price changes. Per definition, we have the following results:

$$h_1 = \rho^+ \frac{\sigma_{S^+}}{\sigma_{F^+}} = \frac{\text{cov}\left(\Delta S_t^+, \Delta F_t^+\right)}{\sigma_{F^+}^2}$$

and

$$h_2 = \rho^- \frac{\sigma_{S^-}}{\sigma_{F^-}} = \frac{\text{cov}\left(\Delta S_t^-, \Delta F_t^-\right)}{\sigma_{F^-}^2}.$$

Given that there are positive as well as negative price changes then $h$ is different from $h_1$ as well as $h_2$. The following proposition shows the relationship between $h$, $h_1$, and $h_2$ and proves that $h$ is indeed different from $h_1$ as well as $h_2$.

**Proposition** *We have*

$$h = \left[ \begin{array}{c} h_1 \sigma_{F^+}^2 + h_2 \sigma_{F^-}^2 + \text{cov}\left(\Delta S_t^+, \Delta F_t^-\right) \\ + \text{cov}\left(\Delta S_t^-, \Delta F_t^+\right) \end{array} \right] \frac{1}{\sigma_F^2}. \tag{19.14}$$

*Thus, we deal with cases that are characterized by both price increases and price decreases during the underlying period.*

*Proof* The OHR is given by

$$h = \rho \frac{\sigma_S}{\sigma_F} = \frac{\text{cov}(\Delta S_t, \Delta F_t)}{\sigma_F^2}. \tag{19.15}$$

One can observe that

$$\Delta S_t = \Delta S_t^+ + \Delta S_t^- \\ \Delta S_t^+ \Delta S_t^- = 0 \tag{19.16}$$

and

$$\Delta F_t = \Delta F_t^+ + \Delta F_t^- \\ \Delta F_t^+ \Delta F_t^- = 0 \tag{19.17}$$

Using (19.16) and (19.17), the following is obtained:

$$\begin{aligned} \text{cov}(\Delta S_t, \Delta F_t) &= E\left[\left(\Delta S_t^+ + \Delta S_t^-\right)\left(\Delta F_t^+ + \Delta F_t^-\right)\right] - E\left[\Delta S_t^+ + \Delta S_t^-\right]E\left[\Delta F_t^+ + \Delta F_t^-\right] \\ &= \left(E\left[\Delta S_t^+ \Delta F_t^+\right] - E\left[\Delta S_t^+\right]E\left[\Delta F_t^+\right]\right) + \left(E\left[\Delta S_t^+ \Delta F_t^-\right] - E\left[\Delta S_t^+\right]E\left[\Delta F_t^-\right]\right) \\ &\quad + \left(E\left[\Delta S_t^- \Delta F_t^+\right] - E\left[\Delta S_t^-\right]E\left[\Delta F_t^+\right]\right) + \left(E\left[\Delta S_t^- \Delta F_t^-\right] - E\left[\Delta S_t^-\right]E\left[\Delta F_t^-\right]\right). \end{aligned}$$

This can be expressed as follows:

$$\begin{aligned} \text{cov}(\Delta S_t, \Delta F_t) &= \text{cov}\left(\Delta S_t^+, \Delta F_t^+\right) + \text{cov}\left(\Delta S_t^+, \Delta F_t^-\right) + \text{cov}\left(\Delta S_t^-, \Delta F_t^+\right) \\ &\quad + \text{cov}\left(\Delta S_t^-, \Delta F_t^-\right). \end{aligned} \tag{19.18}$$

By replacing (19.18) into (19.15) we obtain the following:

$$h = \left[\begin{array}{c} \text{cov}\left(\Delta S_t^+, \Delta F_t^+\right) + \text{cov}\left(\Delta S_t^+, \Delta F_t^-\right) \\ + \text{cov}\left(\Delta S_t^-, \Delta F_t^+\right) + \text{cov}\left(\Delta S_t^-, \Delta F_t^-\right) \end{array}\right] \frac{1}{\sigma_F^2}.$$

In order to derive (19.14) we make use of the fact that $\text{cov}\left(\Delta S_t^+, \Delta F_t^+\right) = h_1 \sigma_{F+}^2$ and $\text{cov}\left(\Delta S_t^-, \Delta F_t^-\right) = h_2 \sigma_{F-}^2$.

It should be pointed out that (19.14) clarifies the components that are normally used in estimating the OHR. However, if the specific investor has certain pertinent information that would indicate a price change in a given direction, then it would result in more precision if certain components of (19.14) are used and not all parts. As an example, assume that a price increase is expected at the maturity date. Based on this information we can conclude that the following conditions are fulfilled:

$$\Delta F_t^+ = \Delta F_t, \; \Delta F_t^- = 0 \text{ and } \sigma_F^2 = \sigma_{F+}^2$$

Utilizing this information and incorporating it into (19.14), the following OHR can be calculated:

$$\hat{h} = \left[h_1 \sigma_{F+}^2 + \text{cov}\left(\Delta S_t^-, \Delta F_t^+\right)\right] \frac{1}{\sigma_F^2} = h_1 + \frac{\text{cov}\left(\Delta S_t^-, \Delta F_t^+\right)}{\sigma_F^2}.$$

## 4   Empirical Findings [5, 12, 14, 16]

The dataset applied in this paper consists of weekly observations of spot and future prices for the USA during the period January 5, 2006 to September 29, 2009. The data source is DataStream. The positive and negative shocks of each variable were constructed by the approach outlined in the previous section and by using a program procedure written in Gauss that is available on requests from the authors.

**Table 19.1** The estimated
hedge ratios (the standard
errors are presented in the
parentheses)

| $h$ | $h_1$ | $h_2$ |
| --- | --- | --- |
| 0.9830 | 0.9745 | 0.9865 |
| (0.0003) | (0.0114) | (0.0088) |

**Table 19.2** The calculated
mean values

| $E[\Delta F_t^-]$ | $E[\Delta F_t^+]$ | $E[\Delta S_t^-]$ | $E[\Delta S_t^+]$ |
| --- | --- | --- | --- |
| −0.010639 | 0.009668 | −0.010451 | 0.009562 |

The estimation results for the optimal hedge ratios are presented in Table 19.1. Each value is statistically significant at any conventional significance level. It should be pointed out that the difference between the optimal hedge ratios is not huge in this particular case because the mean values of positive shocks and negative shocks are very close as is indicated in the Table 19.2.

## 5 Conclusion and Future Work

The optimal hedge ratio is widely used in financial markets to hedge against the underlying price risk. Different approaches have been suggested in the literature for estimating the OHR. This paper is however the first attempt, to our best knowledge, to take into proper account the potential asymmetric character of the underlying OHR that might prevail. It proves mathematically the asymmetric properties of the OHR. It also suggests a method to take into account this asymmetric property in the estimation. The approach is applied to estimating the OHR for the US equity market during the period January 5, 2006 to September 29, 2009. Weekly data are used. The OHR for positive shocks as well as the negative shocks are estimated separately. Our conjecture is that these separate hedge ratios could be useful to the investors in the financial markets in order to find optimal hedge strategies. This could be achieved by relying more on the OHR for positive cases, i.e. ($h_1$), if the investor expects a price increase at the maturity of the futures contract. On the other hand, it might be safer to rely on the OHR for negative chocks ($h_2$) in case the investor expects a price decrease at the maturity of the futures. However, if there are no expectations about the direction of any potential price change at the maturity the investor might just rely on the standard OHR ($h$). It should be mentioned that since we are making use of historical data to predict future, the general philosophical problem of the ex post and ex ante issue prevails, as for any other empirical calculation. Other potential applications than equity markets are among others energy (like crude oil or electricity), agriculture products (like beef, wheat or corn), crucial metals (like gold or silver), or currency exchange markets. Future applications will demonstrate whether the degree of asymmetry depends on the underlying market or not.

# References

1. Alvarez-Ramirez J, Rodriguez E, Echeverria JC (2009) A DFA approach for assessing asymmetric correlations. Phys A: Stat Mech Appl 388:2263–2270
2. Ang A, Chen J (2002) Asymmetric correlations of equity portfolios. J Financ Econom 63:443–494
3. Baillie RT, Myers RJ (1991) Bivariate GARCH estimation of the optimal commodity futures hedge. J Appl Econom 6(2):109–124
4. El-Khatib Y, Hatemi-J A (2011) An approach to deal with asymmetry in the optimal hedge ratio. In: Lecture notes in engineering and computer science: Proceedings of the world congress on engineering 2011, WCE 2011, London, 6–8 July 2011, pp 377–380
5. Ghosh A (1993) Hedging with stock index futures: Estimation and forecasting with error correction model. J Futures Markets 13(7):743–752
6. Ghosh A, Clayton R (1996) Hedging with international stock index futures: An intertemporal error correction model. J Financ Res 19(4):477–492
7. Hatemi-J A, Roca E (2006) Calculating the optimal hedge ratio: Constant, time varying and the Kalman filter approach. Appl Econom Lett 13(5):293–299
8. Hatemi-J A, El-Khatib Y (2012) Stochastic optimal hedge ratio: Theory and evidence. Appl Econom Lett 19(8):699–703
9. Hong Y, Zhou G (2008) Asymmetries in stock returns: Statistical test and economic evaluation. Rev Financ Stud 20:1547–1581
10. Kenourgios D, Samitas A, Drosos P (2008) Hedge ratio estimation and hedging effectiveness: The case of the S&P 500 stock index futures contract. Int J Risk Assess Manag 9(1/2):121–138
11. Kroner KF, Sultan J (1993) Time-varying distributions and dynamic hedging with foreign currency futures. J Financ Quant Anal 28(4):535–551
12. Lien D (1996) The effect of the cointegration relationship on futures hedging: a note. J Futures Markets 16(7):773–780
13. Longin F, Solnik B (2001) Extreme correlation of international equity markets. J Finance 56:649–676
14. Park TH, Switzer LN (1995) Time-varying distributions and the optimal hedge ratios for stock index futures. Appl Financ Econom 5(3):131–137
15. Sephton PS (1993) Hedging wheat and canola at the Winnipeg Commodity Exchange. Appl Financ Econom 3(1):67–72
16. Sim A-B, Zurbruegg R (2001) Optimal hedge ratios and alternative hedging strategies in the presence of cointegrated time-varying risks. Eur J Finance 7(3):269–283
17. Yang JW, Allen DE (2004) Multivariate GARCH hedge ratios and hedging effectiveness in Australian futures markets. Account Finance 45(2):301–321

# Chapter 20
# Computations of Price Sensitivities After a Financial Market Crash

**Youssef El-Khatib and Abdulnasser Hatemi-J**

## 1 Introduction

A financial derivative trader that sells an option to an investor in the over the counter encounters certain problems to manage the prevailing risk. This is due to the fact that in such cases the options are usually tailored to the needs of the investor and not the standardized ones that can easily be hedged by buying an option with the same properties that is sold. In such a customer tailored scenario, hedging the exposure is rather cumbersome. This problem can be dealt with by using the price sensitivities that are usually called "Greeks" in the financial literature. The price sensitivities can play a crucial role in financial risk management. The first price sensitivity is denoted by delta and it represents the rate of the value of the underlying derivative (in our case the price of the option) with regard to the price of the original asset, assuming the ceteris paribus condition. Delta is closely related to the Black and Scholes [1] formula for option pricing. In order to hedge against this price risk, it is desirable to create a delta-neutral or delta hedging position, which is a position with zero delta value. This can be achieved by taking a position of minus delta in the original asset for each long option because the delta for the original asset is equal to one. That is, buy $-\Delta$ of the original asset for each long position of the option. Therefore, calculating a correct value of the delta is of vital importance in terms of successful hedging. It should be mentioned that the

Y. El-Khatib (✉)
Department of Mathematical Sciences, UAE University, Al-Ain,
P.O. Box 17551, United Arab Emirates
e-mail: youssef_elkhatib@uaeu.ac.ae

A. Hatemi-J
Department of Economics and Finance, UAE University, Al-Ain,
P.O. Box 17555, United Arab Emirates
e-mail: Ahatemi@uaeu.ac.ae

S.-I. Ao and L. Gelman (eds.), *Electrical Engineering and Intelligent Systems*,
Lecture Notes in Electrical Engineering 130, DOI 10.1007/978-1-4614-2317-1_20,
© Springer Science+Business Media, LLC 2013

delta of an option changes across time and for this reason the position in the original asset needs to be adjusted regularly. Theta is another price sensitivity that represents the rate of the price of the option with respect to time. The third price sensitivity is gamma signifying the rate of change in delta with regard to the price of the original asset. Thus, if gamma is large in absolute terms then by consequence the delta is very sensitive to the price change of the asset, which implies that leaving a delta-neutral position unchanged during time is very risky. By implication, it means that there is a need for creating a gamma-neutral position in such a situation. The fourth price sensitivity is denoted as vega, which is the change of the option price with respect to the volatility of the original asset. If the value of vega is high in absolute terms it means that the option price is easily affected by even a small change in the volatility. Hence, it is important to create a vega neutral position in this case. Finally, the sensitivity of the option value with regard to the interest rate as a measure of risk free return is denoted by rho. To neutralize these price sensitivities is the ultimate goal of any optimal hedging strategy. For this reason the computation of these price sensitivities in a precise manner is an integral part of successful financial risk management in order to monitor and neutralize risk. The efficient estimation of the price sensitivities is especially important during the periods when the market is under stress like during the recent financial crisis. Economic agents, including investors and policymakers, are interested in finding out whether there are spill-over effects from one market to another during such a period, according to Forbes and Rigobon [8] and Hatemi-J and Hacker [10] among others. Because of globalization, with the consequent rise in integration between financial markets worldwide, this issue is becoming increasingly the focus of attention. It is during the crisis that the investors require to have access to precise calculations in order to deal with the increased level of risk and achieve immunization. Thus, to compute the price sensitivities correctly in such a scenario is crucial. The issue that this paper addresses is to suggest an approach to compute sensitivities during the crisis period based on the Malliavin calculus. This paper is the first attempt, to our best knowledge, to compute price sensitivities in a market that is suffering from a financial crash. For a conference version of this chapter the interested reader is referred to El-Khatib and Hatemi-J [5].

The rest of the paper is organized as follows. In Sect. 2 we present the model and we give an overview of the Malliavin derivative. Section 3 is devoted to the computation of the Greeks using Malliavin calculus. The last section concludes the paper.

## 2   The Model and the Malliavin Derivative

The first part of this section presents the model that we are using in order to tackle the issue of risk management during a financial crisis. In the second part of this section we give an overview of the Malliavin derivative in the Wiener space and of its adjoint, i.e. the Skorohod integral. We refer the reader to Nualart [13] and

Oksendal [14] for more details about the Malliavin calculus. See also El-Khatib and Hatemi-J [6] for the computation of the price sensitivities when the market is not under stress.

Our conjecture is that options pricing models coming from empirical studies on the dynamics of financial markets after the occurrence of a financial crash do not match with the stochastic models used in the literature. For instance, while the Black and Scholes [1] model assumes that the underlying asset price follows a geometric Brownian motion, the work of Sornette [16] shows empirically that the post-crash dynamics follow a converging oscillatory motion. On the other hand, the paper of Lillo and Mantenga [11] shows that financial markets follow power-law relaxation decay. Several ideas have been suggested to overcome this shortcoming of the Black–Scholes model. In fact, new option pricing models have been developed based on empirical observations (see for instance Savit [15], Deeba et al. [2], Tvedt [17], Baillie et al. [3] and McCauley [12]). Recently, in [3, 4], the authors suggest a newer model which extends the Black-Scholes model to take into account the post-crash dynamics as proposed by Sornette [16]. More specifically, Dibeh and Harmanani [4] derive the following stochastic differential equation that couples the post-crash market index to individual stock prices:

$$\frac{dS_t}{S_t} = \left(a + \frac{bg(t)}{S_t}\right)dt + \left(\sigma + \frac{\beta g(t)}{S_t}\right)dW_t,$$

where $t \in [0, T]$, $S_0 = x > 0$ and $g(t) = A + Be^{\alpha t}\sin(\omega t)$. The values $a, b, \beta, A$ and $B$ are real constants. The volatility of the original asset is denoted by $\sigma$. The authors obtain the following partial differential equation (PDE) for the option price:

$$\frac{\partial C}{\partial t} + rS\frac{\partial C}{\partial S} - rC + \frac{1}{2}(\sigma S + \beta g(t))^2\frac{\partial^2 C}{\partial S^2} = 0,$$

which has the terminal condition $C(S, T) = (S - K)^+$. It should be mentioned that $C$ is the call option price, $r$ the risk free rate and $K$ is the strike price. However, Dibeh and Harmanani [4] did not deal with the computation of price sensitivities. We consider a market with two assets—namely the risky asset, $S$, for which there is an underlying European call option and a riskless one that is defined by following expression:

$$dA_t = rA_t dt, \quad t \in [0, T], \quad A_0 = 1.$$

We work on a probability space $(\Omega, F, P)$, $(W_t)_{t \in [0, T]}$ denotes a Brownian motion and $(F_t)_{t \in [0, T]}$ is the natural filtration generated by $(W_t)_{t \in [0, T]}$. Recall that a stochastic process is a function of two variables, i.e. time $t \in [0, T]$ and the event $\omega \in \Omega$. However, in the literature it is common to write $S_t$ instead of $S_t(\omega)$. The same is true for $W_t$ or any other stochastic process in this paper. We assume that the probability $P$ is the risk-neutral probability and the stochastic differential equation for the underlying asset price under the risk-neutral probability $P$ is given as in [6] by

$$\frac{dS_t}{S_t} = rdt + \left(\sigma + \frac{\beta g(t)}{S_t}\right)dW_t,$$

where $t \in [0, T]$ and $S_0 = x$. Let $(D_t)_{t \in [0,T]}$ be the Malliavin derivative on the direction of $W = (W_t)_{t \in [0,T]}$. We denote by $V$ the set of random variables $F : \Omega \to R$, such that $F$ has the representation

$$F(\omega) = f\left(\int_0^T f_1(t)dW_t, \ldots, \int_0^T f_n(t)dW_t\right),$$

where $f(x_1, \ldots, x_n) = \sum c_m x^m$ is a polynomial in $n$ variables $x_1, \ldots, x_n$ and deterministic functions $f_i \in L^2([0,T])^m$. Let $\|\cdot\|_{1,2}$ be the norm

$$\|F\|_{1,2} := \|F\|_{L^2(\Omega)} + \|D.F\|_{L^2([0,T]\times\Omega)}, \quad F \in L^2(\Omega).$$

Thus, the domain of the operator $D$, $Dom(D)$, coincides with $V$ with respect to the norm $\|\cdot\|_{1,2}$. The next proposition will be useful.

**Proposition 1** *Given $F = f\left(\int_0^T f_1(t)\,dW_t, \ldots, \int_0^T f_n(t)\,dW_t\right) \in V$. We have*

$$D_t F = \sum_{k=0}^{k=n} \frac{\partial f}{\partial x_k}\left(\int_0^T f_1(t)\,dW_t, \ldots, \int_0^T f_n(t)\,dW_t\right)f_k(t).$$

*To calculate the Mallaivin derivative for integrals, we will utilize the following propositions.*

**Proposition 2** *Let $(u_t)_{t \in [0,T]}$ be a $\mathcal{F}_t$–adapted process, such that $u_t \in Dom(D)$, we have then*

$$D_s \int_0^T u_t\,dt = \int_s^T (D_s u_t)\,dt, \quad s < T.$$

*And*

**Proposition 3** *Let $(u_t)_{t \in [0,T]}$ be a $\mathcal{F}_t$-adapted process, such that $u_t \in Dom(D)$, we have*

$$D_s \int_0^T u_t dW_t = \int_s^T (D_s u_t)dW_t + u_s, \quad s < T.$$

*From now on, for any stochastic process $u$ and for $F \in Dom(D)$ such that $u.D.F \in L^2([0,T])$ we let*

$$D_u F := \langle DF, u \rangle_{L^2([0,T])} := \int_0^T u_t D_t F dt.$$

*Let $\delta$ represent the Skorohod integral in Wiener space. One can observe that $\delta$ is the adjoint of D as showing in the next proposition, which is an extension of the Itô integral.*

**Proposition 4**

(a) *Let $u \in Dom(\delta)$ and $F \in Dom(D)$, we have $E[D_u F] \le C(u) \parallel F \parallel_{1,2}$ and $E[F\delta(u)] = E[D_u F]$.*

(b) *Consider a $L^2(\Omega \times [0, T])$-adapted stochastic process $u = (u_t)_{t \in [0,T]}$. We have*
$$\delta(u) = \int_0^T u_t dW_t$$

(c) *Let $F \in Dom(D)$ and $u \in Dom(\delta)$ such that $uF \in Dom(\delta)$, thus*
$$\delta(uF) = F\delta(u) - D_u F.$$

# 3  Computations of Greeks

This section is dedicated to the computation of the price sensitivities. The computation of Greeks by Malliavin approach rests on the well-known integration by parts formula–cf. Fournié et al. [9] and El-Khatib and Privault [7]—given in the following proposition.

**Proposition 5** *Let I be an open interval of $\mathbf{R}$. Let also $(F^\zeta)_{\zeta \in I}$ and $(H^\zeta)_{\zeta \in I}$ be two families of random functionals continuously differentiable in Dom (D) in the parameter $\zeta \in I$. Assume that $(u_t)_{t \in [0,T]}$ is a process satisfying*

$$D_u F^\zeta \ne 0, \quad a.s. \ on \ \{\partial_\zeta F^\zeta \ne 0\}, \quad \zeta \in I,$$

*and such that $uH^\zeta \frac{\partial_\zeta F^\zeta}{D_u F^\zeta}$ is continuous in $\zeta$ in Dom ($\delta$). We have*

$$\frac{\partial}{\partial \zeta} E[H^\zeta f(F^\zeta)] = E\left[f(F^\zeta)\delta\left(uH^\zeta \frac{\partial_\zeta F^\zeta}{D_u F^\zeta}\right)\right] + E[f(F^\zeta)\partial_\zeta H^\zeta].$$

*for any function f such that $f(F^\zeta) \in L^2(\Omega)$, $\zeta \in I$.*

*Our aim is to compute the Greeks for options with payoff $f(S_T)$, where $(S_t)_{t \in [0,T]}$ denotes the underlying asset price given by*

$$S_T = x + r \int_0^T S_s \, ds + \int_0^T (\sigma S_s + \beta g(s)) \, dW_s.$$

*Let $\zeta$ be a parameter taking one of the following values: the initial asset price $x = S_0$, the volatility $\sigma$, the interest rate r or the maturity T. Let $C = e^{-rt}E[f(S_t^\zeta)]$ be the price of the option. We will compute the following Greeks:*

$$\text{delta} = \frac{\partial C}{\partial x}, \quad \text{gamma} = \frac{\partial^2 C}{\partial x^2}, \quad \text{rho} = \frac{\partial C}{\partial r}, \quad \text{vega} = \frac{\partial C}{\partial \sigma} \text{ and theta} = \frac{\partial C}{\partial T}.$$

## 3.1   Delta, Rho and Vega

By using Proposition 5 and (c) in Proposition 4 we obtain the following:

$$\frac{\partial}{\partial \zeta} E\left[H^{\zeta} f\left(S_T^{\zeta}\right)\right] = E\left[f(S_T)(L^{\zeta}\delta(u))\right] - E\left[f(S_T)(D_u L^{\zeta} - \partial_{\zeta} H^{\zeta})\right], \tag{20.1}$$

where

$$L^{\zeta} := \frac{H^{\zeta} \partial_{\zeta} S_T^{\zeta}}{D_u S_T^{\zeta}} \tag{20.2}$$

and

$$D_u L^{\zeta} = D_u \frac{H^{\zeta} \partial_{\zeta} S_T^{\zeta}}{D_u S_T^{\zeta}} = \frac{D_u\left(H^{\zeta} \partial_{\zeta} S_T^{\zeta}\right) - D_u D_u S_T^{\zeta}}{\left(D_u S_T^{\zeta}\right)^2}. \tag{20.3}$$

Note that delta and vega are the first-order derivatives of $E[H^{\zeta}f(S_T^{\zeta})]$ with respect to $\zeta = x$ and $\zeta = \sigma$, respectively, where $H^{\zeta} = e^{-rT}$ and so $\partial_{\zeta} H^{\zeta} = 0$. Hence, we have

$$\frac{\partial}{\partial \zeta} E\left[e^{-rT} f\left(S_T^{\zeta}\right)\right] = E\left[f\left(S_T^{\zeta}\right)(L^{\zeta}\delta(u) - D_u L^{\zeta})\right], \tag{20.4}$$

where $L^{\zeta}$ is given by (20.2). For instance, the delta can be computed as follows:

$$delta = e^{-rT} E\left[f(S_T)\left(\frac{\partial_x S_T}{D_u S_T}\delta(u) - D_u\left(\frac{\partial_x S_T}{D_u S_T}\right)\right)\right].$$

For rho and theta, we use equation (20.1) with $H^{\zeta} = e^{-rT}$, then $\partial_r e^{-rT} = -r e^{-rT}$ and $\partial_T e^{-rT} = -T e^{-rT}$. The rho is then given by

$$rho = e^{-rT} E\left[f(S_T)\left(\frac{\partial_r S_T}{D_u S_T}\delta(u) - D_u\left(\frac{\partial_r S_T}{D_u S_T}\right) - r\right)\right].$$

## 3.2   Gamma

The gamma is the second-order derivative of $C = E[e^{-rT}f(S_T)]$ with respect to $x$, and it is obtained by differentiating delta with respect to $x$. By using (20.1) twice the following result is obtained:

$$\begin{aligned}\frac{\partial^2}{\partial x^2} E[e^{-rT}f(S_T)] &= \frac{\partial}{\partial x} E[f(S_T^x)(L^x\delta(u) - D_u L^x)] \\ &= \frac{\partial}{\partial x} E[f(S_T)(G^x\delta(u) - D_u G^x + \partial_{\zeta} G^x)], \end{aligned} \tag{20.5}$$

where

$$G^x := \frac{(L^x\delta(u) - D_u L^x)\partial_x S_T^x}{D_u S_T^x}, \tag{20.6}$$

and $L^x$ is given by (20.2). In addition, one can obtain

$$D_u G^x = \frac{D_u\big((L^x\delta(u) - D_u L^x)\partial_x S_T^x\big) - D_u D_u S_T^x}{(D_u S_T^x)^2}. \tag{20.7}$$

In order to compute the Greeks based on (20.1–20.7), we need to find $D_u S_T$, $D_u D_u S_T$ and $D_u D_u D_u S_T$. This can be achieved by using mainly Proposition 2. The first derivative $D_u S_T$ can be computed as

$$D_u S_T = \int_0^T u_t D_t S_T \, dt.$$

The second derivative $D_u D_u S_T$ is obtained likewise as

$$\begin{aligned} D_u D_u S_T &= D_u\left(\int_0^T u_t D_t S_T \, dt\right) \\ &= \int_0^T u_s D_s\left(\int_0^T u_t D_t S_T \, dt\right) ds \\ &= \int_0^T u_s \int_s^T D_s(u_t D_t S_T) \, dt \, ds \\ &= \int_0^T u_s \int_s^T (u_t D_s D_t S_T + D_t S_T D_s u_t) \, dt \, ds. \end{aligned}$$

The third derivative can be obtained by differentiating one more time and using similarly Proposition 2. In order to make the final computation of the Greeks operational we need to obtain the first-order, second-order and third-order derivatives of $S_T$ with respect to $D$. Thus, we put forward the following proposition.

**Proposition 6** *For $0 \le t \le T$, we let*

$\xi_t = \exp\left[\left(r - \frac{\sigma^2}{2}\right)t + \sigma W_t\right]$ *We have $\partial_x S_T = \xi_T$ and*

$$\begin{aligned} D_t S_T &= (\sigma S_t + \beta g(t))\xi_{T-t} \\ D_s D_t S_T &= \sigma\{(\sigma S_s + \beta g(s))\xi_{T-s} 1_{s\le t} + (\sigma S_t + \beta g(t))\xi_{T-t} 1_{s\le T-t}\} \\ D_l D_s D_t S_T &= \sigma\{\xi_{T-s} 1_{s\le t}\sigma D_l S_s + (\sigma S_s + \beta g(s))1_{s\le t} \\ &\quad D_l \xi_{T-s} + \xi_{T-t} 1_{s\le T-t}\sigma D_l S_t + (\sigma S_t + \beta g(t))1_{s\le T-t}D_l \xi_{T-t}\}, \end{aligned}$$

*where*

$$D_s \xi_v = \sigma \xi_v 1_{s\le v},$$

*and $s, l, v$ are in $[0, T]$.*

*Proof* By the chain rule of $D_t$ and thanks to Propositions 2 and 3 we obtain

$$\partial_x S_t = 1 + r \int_0^t \partial_x S_\tau d\tau + \sigma \int_0^t \partial_x S_\tau dW_\tau.$$
$$D_t S_T = D_t x + D_t \int_0^T (aS_s + bg(s))ds + D_t \int_0^T (\sigma S_s$$
$$+\beta g(s))dW_s = \int_t^T D_t(aS_s + bg(s))ds$$
$$+ \int_t^T D_t(\sigma S_s + \beta g(s))dW_s$$
$$= a \int_t^T D_t S_s ds + \sigma \int_t^T D_t S_s dW_s + \sigma S_t + \beta g(t).$$

Using Itô Lemma, the processes $(\partial_x S_t)_{0 \leq t \leq T}$ and $(D_t S_T)_{0 \leq t \leq T}$ can be written as $\partial_x S_t = \xi_t$ and

$$D_t S_T = (\sigma S_t + \beta g(t))\xi_{T-t}.$$

For the second Malliavin derivative of $S_T$, we have for $0 \leq s, v \leq T$

$$D_s \xi_v = D_s \exp\left[\left(a - \frac{\sigma^2}{2}\right)v + \sigma W_v\right]$$
$$= \exp\left[\left(a - \frac{\sigma^2}{2}\right)v + \sigma W_v\right]\sigma D_s(W_v)$$
$$= \exp\left[\left(a - \frac{\sigma^2}{2}\right)v + \sigma W_v\right]\sigma D_s\left(\int_0^v dW_\alpha\right)$$
$$= \sigma \xi_v 1_{s \leq v}$$

Thus,

$$D_s D_t S_T = D_s((\sigma S_t + \beta g(t))\xi_{T-t})$$
$$= \sigma\xi_{T-t}D_s S_t + (\sigma S_t + \beta g(t))D_s\xi_{T-t}$$
$$= \sigma\xi_{T-t}(\sigma S_s + \beta g(s))\xi_{t-s}1_{s \leq t}$$
$$+(\sigma S_t + \beta g(t))\sigma\xi_{T-t}1_{s \leq T-t}$$
$$= \sigma\{(\sigma S_s + \beta g(s))\xi_{T-s}1_{s \leq t}$$
$$+(\sigma S_t + \beta g(t))\xi_{T-t}1_{s \leq T-t}\}$$

The third Malliavin derivative of $S_T$ can be computed as follows, for $0 \leq l \leq T$

$$D_l D_s D_t S_T = \sigma D_l\{(\sigma S_s + \beta g(s))\xi_{T-s}1_{s \leq t}+(\sigma S_t + \beta g(t))\xi_{T-t}1_{s \leq T-t}\}$$
$$= \sigma\{\xi_{T-s}1_{s \leq t}\sigma D_l S_s + (\sigma S_s + \beta g(s))1_{s \leq t}D_l\xi_{T-s} + \xi_{T-t}1_{s \leq T-t}\sigma D_l S_t$$
$$+(\sigma S_t + \beta g(t))1_{r \leq T-t}D_l\xi_{T-t}\}.$$

Proposition 6 provides also the derivative of $S_T$ with respect to $S_0 = x$, which is necessary for the computation of delta and gamma. For the rho, vega and theta, first derivatives can be computed in a similar way.

# 4  Conclusions

The calculation of the price sensitivities of a financial derivative (like an option or a portfolio of option contracts) is of paramount importance for implementing hedging strategies that are successful to neutralize the underlying risk. This is the case especially during a financial crisis in which the need for dealing with the increased level of risk is urgent. While different approaches have been utilized in the literature to calculate the price sensitivities during normal circumstance, none has focused on this issue during a financial crisis. This paper is the first attempt, to our best knowledge, to deal with this issue by suggesting a formula for computing each of the underlying price sensitivities in a more precise manner during a financial crisis based on the Malliavin calculus. Mathematical proof for each proposition that is necessary for the calculation of the underlying price sensitivities during a financial crisis is provided. Thus, the results obtained from this paper are expected to improve on the success of the hedging strategies that must be undertaken by the investor during a financial crisis that is a period in which the need for hedging is more imperative than normal circumstances.

# References

1. Black F, Scholes M (1973) The pricing of options and corporate liabilities. J Polit Econ 81:637–654
2. Deeba E, Dibeh G, Xie S (2002) An algorithm for solving bond pricing problem. Appl Math Comput 128(1):81–94
3. Baillie RT, Dibeh G, Chahda G (2005) Option pricing in markets with noisy cyclical and crash dynamics. Finance Lett 3(2):25–32
4. Dibeh G, Harmanani HM (2007) Option pricing during post-crash relaxation times. Phys A 380:357–365
5. El-Khatib Y, Hatemi-JA (2011a) On the price sensitivities during financial crisis. In: Proceedings of the world congress on engineering 2011. Lecture notes in engineering and computer science, WCE 2011, London, U.K., 6–8 July 2011, pp 401–404
6. El-Khatib Y, Hatemi-JA (2011b) On the calculation of price sensitivities with jump-diffusion structure. MPRA Paper 30596. University Library of Munich, Germany
7. El-Khatib Y, Privault N (2004) Computations of greeks in a market with jumps via the Malliavin calculus. Finance Stoch 8(2):161–179
8. Forbes KJ, Rigobon R (2002) No contagion, only interdependence: measuring stock market co-movements. J Finance 57:2223–2261
9. Fournié E, Lasry JM, Lebuchoux J, Lions PL, Touzi N (1999) Applications of Malliavin calculus to Monte Carlo methods in finance. Finance Stoch 3(4):391–412
10. Hatemi-J A, Hacker S (2005) An alternative method to test for contagion with an application to the Asian financial crisis. Appl Financ Econ Lett 1(6):343–347
11. Lillo F, Mantenga F (2003) Power-law relaxation in a complex system: Omori law after a financial market crash. Phys Rev E 016119
12. McCauley J (2004) The dynamics of markets: econophysics and finance. Cambridge University Press, Cambridge
13. Nualart D (1995) The Malliavin calculus and related topics. Springer, Berlin

14. Oksendal B (1996) An introduction to malliavin calculus with applications to economics. Working paper 3, Institute of finance and management science, Norwegian school of economics and business administration
15. Savit R (1989) Nonlinearities and chaotic effects in options prices. J Futures Mark 9 (6):507–518
16. Sornette D (2003) Why stock markets crash: critical events in complex financial markets. Princeton University Press, Princeton, NJ
17. Tvedt J (1998) Valuation of European futures options in the bifex market. J Futures Mark 18:167–175

# Chapter 21
# TRP Ratio and the Black–Litterman Portfolio Optimisation Method

**Gal Munda and Sebastjan Strasek**

## 1  Introduction

In their recent research Munda and Strasek [1, 2] observed characteristics of the "Target-to-Real-Price" ratio (TRP ratio) using 5-year timeline of 30 individual stocks across the Europe (developed and developing markets). They concluded that individual stocks have their unique mean-reverted values of the ratio between the stock's 6-months consensus target price from Bloomberg and the respective spot price.

Munda and Strasek [1, 2] have defined the TRP ratio as:

$$\text{TRP} = \frac{\text{TP}}{\text{PX}},\tag{21.1}$$

where TP represents consensus target price for a stock as published by Bloomberg professional terminal and PX as the equity's spot price at a given time $t_0$.

Upon proving the mean reversion of the sample, the authors are now suggesting the implementation of the TRP ratio for the portfolio optimization purposes.

In the portfolio, the TRP ratio is denoted as

$N$: items in the portfolio
$W_k^z$: weight of $k$th security during the investment period $z$

G. Munda (✉)
PricewaterhouseCoopers, Times Valley, Uxbridge UB8 1EX, UK
e-mail: gal.munda@uk.pwc.com

S. Strasek
The University of Maribor, 2000 Maribor, Slovenia
e-mail: sebastjan.strasek@uni-mb.si

S.-I. Ao and L. Gelman (eds.), *Electrical Engineering and Intelligent Systems*,
Lecture Notes in Electrical Engineering 130, DOI 10.1007/978-1-4614-2317-1_21,
© Springer Science+Business Media, LLC 2013

Under the assumption that all funds must be invested (cash cannot be held), the sum of weights has to equal one:

$$\sum_{k=1}^{N} W_k^z = 1 \quad W_k^z \geq 0 \ k = 1, ..., N. \tag{21.2}$$

where:

$P_k$: current price of security $k$
$\tilde{P}_k^\Delta$: average of expert predictions of price of security $k$ in time $\Delta$ from now
$\tilde{P}_k^\infty$: average of expert predictions of price of security $k$ in the distant future

The long-term TRP ratio is therefore

$$\text{TRP}_\infty = \frac{\tilde{P}_k^\infty}{P_k}. \tag{21.3}$$

The "potential," $\wedge_k$, for security $k$ is defined by

$$\wedge_k = \frac{\left(\frac{\tilde{P}_k^\Delta}{P_k}\right)}{\left(\frac{\tilde{P}_k^\infty}{P_k}\right)} = \frac{\tilde{P}_k^\Delta}{\tilde{P}_k^\infty}. \tag{21.4}$$

This is the ratio of the short-term and long-term average price predictions. The relative potential is defined as a normalization of each potential divided by the average potential

$$\frac{\wedge_k}{\left(\sum_{p=1}^{N} b_p = 1\right)}; \ b_k \geq 0 \quad k = 1, ..., N. \tag{21.5}$$

## 2 Implications of the TRP Ratio

This article focuses on the BL portfolio optimization. Others approaches will be used only as a benchmark which will enable us to assess the success of the augmented Black–Litterman's method.

### 2.1 Portfolio Optimization and Diversification

Every investor managing portfolios has a critical decision to make—what kind of optimization technique (if any) should be used? Several major methods are available and most of them have a quantitative ground. Some of them are extracted from the

**Table 21.1** Initial and final weights of the passive strategy

| Date | alv:gy | bay:gy | dbk:gy | ibe:sm | nok1v:fh |
|------|--------|--------|--------|--------|----------|
| 01/06/2004 | 10.00% | 10.00% | 10.00% | 10.00% | 10.00% |
| 05/07/2009 | 7.45% | 16.62% | 6.48% | 13.03% | 8.98% |

| Date | or:fp | rep:sm | sap:gy | sie:gy | tef:sm |
|------|-------|--------|--------|--------|--------|
| 01/06/2004 | 10.00% | 10.00% | 10.00% | 10.00% | 10.00% |
| 05/07/2009 | 8.22% | 9.31% | 8.23% | 8.01% | 13.68% |

modern portfolio theory (MPT) whereas the others are far more intuitive. According to Sharpe et al. [3], investment managers often avoid using complicated optimization procedures and rather implement qualitative approach which is based on experience.

Nevertheless, investment managers are aware of the phenomenon called diversification, and their goal is to find the proper combination of assets to create efficient portfolio for their clients. As Markowitz [4] emphasized: "A good portfolio is more than a long list of good stocks and bonds. It is a balanced whole, providing the investor with protections and opportunities with respect to a wide range of contingencies. The investor should build toward an integrated portfolio which best suits his needs."

## 2.2   Passive Portfolio Strategy

Every investor has to develop a portfolio strategy which will best suit his investment objectives. Portfolio strategies can be either active or passive. Passive portfolio strategy (often known as a buy-and-hold strategy) does not require additional inputs, such as return forecasting. Its main investment objective is to follow the performance of the benchmark index [5]. Passive strategy is the purest implementation of the efficient market hypothesis (EMH) as it assumes that markets will be able to reflect all available information in the stock prices.

In this article, the passive strategy was used to create a benchmark for other approaches to compare with. Starting point ($t_0$) was the equally weighted portfolio ($1/N$), and the weights were not adjusted throughout the investment period. It means that the number of shares was constant over time and weights were then changing daily according to stocks' price movements. Table 21.1 shows the initial and the final weights of the passive strategy based on the movement of the spot price.

## 2.3   Active (Dynamic) Portfolio Strategies

The common feature of all active portfolio optimization strategies is "expectations about the factors that influence the performance of the class of assets" ([5], p. 145).

Investor will therefore decide on using active or passive strategy by assessing the efficiency of the market. If investor believes that markets are totally efficient, there is no point in trying to outperform it by using an active strategy. In other cases, there are plenty of methods which are trying to systematically "beat the market."

### 2.3.1 Minimum-Variance Portfolio Optimization

The rationale behind the minimum-variance portfolio was introduced in the 1952 when Harry Markowitz presented the Modern Portfolio Theory. It suggests that rational investor should choose the optimal portfolio which will be developed from the trade-off between risk and expected return. The idea is to maximize the expected return for given level of risk (where risk is denoted as standard deviation of the portfolio).

This research uses minimum-variance instead of the mean-variance approach to create a benchmark. There are several reasons for implementing the minimum-variance instead of originally proposed mean-variance approach. One of the disadvantages of mean-variance is the requirement of choosing the expected return, which is hard to estimate. Errors in estimation of that parameter lead to inefficient portfolios. As a consequence, weights become highly unstable. The other pitfall of the mean-variance approach is the sensitivity to small changes in the mean returns of portfolio's assets. Michaud [6] concludes that mean-variance method is the "error-maximization method." In order to avoid the problems connected to mean-variance optimization, we concentrate on the minimum-variance portfolio.

If the distribution parameters of stocks are known, the weights ($W_{MV} = (w_{MV,1}, \ldots, w_{MV,N})'$) of the global minimum variance portfolio are given by

$$W_{MV} = \frac{\Sigma^{-1}[1]}{[1]'\Sigma^{-1}[1]}, \tag{21.6}$$

where [1] is a column vector of ones.

Optimization model was built in Excel spreadsheet as proposed. Inputs for the optimization were: 30-days average mean return for each share, variance–covariance matrix, and initial investment (at the beginning of each month). Excel add-in "Solver" was implemented into the macro and used to minimize portfolio's variance at the beginning of the each month. For each optimization, Solver was calibrated as the minimum of

$$\text{Portfolio Variance} = \left(\begin{bmatrix} \text{Benchmark} \\ \text{portfolio} \\ \text{proportions} \end{bmatrix}^T\right) \begin{bmatrix} \text{Variance–} \\ \text{covariance} \\ \text{matrix} \end{bmatrix} \begin{bmatrix} \text{Benchmark} \\ \text{portfolio} \\ \text{proportions} \end{bmatrix}. \tag{21.7}$$

Optimal proportions were calculated on a monthly basis. It is believed that more active approach could significantly increase transaction costs, although they were neglected in this research. This method clearly centralizes solutions and the optimum portfolio usually does not include all shares from the benchmark. In order to present centralized feature of the minimum variance approach, average number of different assets was calculated. Whereas the passive benchmark included all 10 assets over the whole investment period, minimum variance approach on average consisted only of 4.53 shares.

### 2.3.2  Equally Weighted Portfolios

As shown with the minimum-variance method, solution of implementing traditional portfolio optimization is often expressed in highly concentrated portfolio. One alternative to overcome such difficulties is to use the "equal weighting approach." This method is often considered as a "naive diversification strategy which attempts to capture some of the potential gains from international diversification" ([14], p. 229). Its major advantage is robustness as it does not require return or volatility forecasts, which is also one of the most important reasons for popularity of the EQW approach. Despite its simplicity and popularity, EQW certainly has some pitfalls. One of the most obvious is the fact that it does not account for volatilities and correlations between assets.

Equally weighted portfolios (EWP) are composed of selected securities where each of them represents the same portion of portfolio. This can be written as

$$Wi = \frac{1}{N} \tag{21.8}$$

where $N$ represents number of securities included into portfolio and $W_i$ is weight of $i$th security. As a result, all selected securities are included into portfolio.

This approach suggests that portfolio weights are rebalanced to original values at the end of each holding period. EQW method was implemented similarly to all other optimization methods, with the holding period of 1 month. Portfolio was therefore rebalanced at the beginning of each month. This approach is very similar to passive strategy, with one major difference—passive strategy keeps the number of shares (for respective assets) constant over time, whereas EQW method adjusts number of shares in order to keep weights at the same level.

### 2.3.3  The Black–Litterman Optimization Method

Although Modern Portfolio Theory (MPT) changed the way investors look at the investments, it is well documented that it has plenty of practical issues. One of the biggest pitfalls is the fact that optimization often proposes enormously large long and short positions, which are not achievable in practice. The main problem with the implementation of portfolio optimization is the fact that historical returns are not good predictors of future returns.

In contrast to MPT, researchers of Goldman Sachs—Fisher Black and Robert Litterman—proposed a technique that tackles most of the problems, commonly associated to the classical portfolio optimization methods. They start the process of optimization with the assumption that investor chooses his optimum portfolio within a *finite group of assets*. In essence, the BL model turns the MPT on its head—it does not compute the optimal portfolio from the historical data, but rather assumes that a given portfolio in fact is the optimal one. This idea is backed by several researches which show that it is very difficult for investor to systematically outperform well-diversified benchmark. BL then derive the expected returns for different positions in the portfolio. If investor agrees with the market assessment, benchmark becomes the optimal portfolio and the funds should be invested accordingly. On the other hand, if someone has different opinions about the expected returns of some of the stocks in the portfolio, the BL approach allows him to adjust the weights according to his projections. The result is the optimal portfolio, based on investor's individual assessment of market potential.

The last stage of the BL approach is the addition to standard BL procedure, as we will try to implement the use of TRP ratio. Simple rules for standardization of investor's views about the individual stocks in the benchmark portfolio will be developed. The optimization will be performed in Excel, as proposed.

The first part of the BL procedure is similar to MPT, where we start with the definition of an optimal portfolio. In contrast to MPT, BL is interested in expected portfolio returns. Therefore, an efficient portfolio has to solve

$$
\begin{bmatrix} \text{Expected} \\ \text{returns} \end{bmatrix} = \begin{bmatrix} \text{Variance--} \\ \text{covariance} \\ \text{matrix} \end{bmatrix} \begin{bmatrix} \text{Efficient} \\ \text{portfolio} \\ \text{proportions} \end{bmatrix} * \begin{pmatrix} \text{Normal Factor} \\ + \\ \text{Rf Rate} \end{pmatrix}, \quad (21.9)
$$

where

$$
\textit{Normal Factor} = \left( \frac{\textit{Expected Benchmark Return} - \textit{Rf Rate}}{\left( \begin{bmatrix} \textit{Benchmark} \\ \textit{portfolio} \\ \textit{proportions} \end{bmatrix}^T \right) \begin{bmatrix} \textit{Variance--} \\ \textit{covariance} \\ \textit{matrix} \end{bmatrix} \begin{bmatrix} \textit{Benchmark} \\ \textit{portfolio} \\ \textit{proportions} \end{bmatrix}} \right)
$$

$$(21.10)$$

therefore,

$$
\begin{bmatrix} \text{Benchmark} \\ \text{portfolio} \\ \text{returns} \end{bmatrix} = \begin{bmatrix} \text{Variance} - \\ \text{covariance} \\ \text{matrix} \end{bmatrix} \begin{bmatrix} \text{Benchmark} \\ \text{portfolio} \\ \text{proportions} \end{bmatrix}
$$

$$
* \left( \frac{\text{Expected Benchmark Return} - \text{Rf Rate}}{\left( \begin{bmatrix} \text{Benchmark} \\ \text{portfolio} \\ \text{proportions} \end{bmatrix}^{T} \right) \begin{bmatrix} \text{Variance} - \\ \text{covariance} \\ \text{matrix} \end{bmatrix} \begin{bmatrix} \text{Benchmark} \\ \text{portfolio} \\ \text{proportions} \end{bmatrix}} \right)
$$

$$
+ \text{Rf Rate.}
$$

(21.11)

In the absence of additional information about the market's expected returns, it is safe to assume that weights from the benchmark represent the efficient weights. Otherwise, we have to introduce our own opinions. Investor views can be expressed in absolute or relative terms (i.e., Allianz will outperform Telefonica by 0.2% in the next month is a relative view, and Allianz will earn 1.2% in the next month is an absolute view).

After assuming that the benchmark is efficient, it is possible to calculate expected returns for each stock in a portfolio. It has to be emphasized that due to the correlations between assets, changing one of the expected returns results in adjusted optimum weights of the whole portfolio. Having two or more opinions about the asset returns complicates the situation, as the problem cannot be easily implemented onto a spreadsheet. We use Excel's add-in "Solver" which we integrate in a macro in order to simulate the efficient portfolio weights.

### 2.3.4   Implementation of the Augmented BL Model

Implementation of the BL model was performed in five stages, as shown in Table 21.2 (Adopted from [11]).

*Stage 1*
Firstly, market weights for the benchmark have to be defined. This represents the efficient weights in cases when investors do not have specific views about the market. We use the so-called "*1/N*" strategy, which is also known as "*Equally weighted portfolio.*" This approach is set initial weights of our approach to 10% to reflect the 10 stock in the portfolio.

*Stage 2*
Secondly, it is required to estimate the equilibrium returns for the benchmark using (21.11).

**Table 21.2** Stages for implementation of the bl model ([11], p. 278)

| Stage | Description | Purpose |
|---|---|---|
| 1 | Define equilibrium market weights and variance–covariance matrix for all assets | Get inputs for calculating equilibrium expected returns |
| 2 | Back-solve equilibrium expected returns | Form the neutral starting point for formulating expected returns |
| 3 | Express own views about the expected returns in the next period | Reflect the investor's expectations for different assets |
| 4 | Calculate the view-adjusted market equilibrium returns | Form the expected return that reflects both market equilibrium and views |
| 5 | Run mean-variance optimization | Obtain efficient frontier and portfolio weights |

**Table 21.3** Long-term and one month trp ratios

| Stock | Long term Mean | One month average TRP ratio |
|---|---|---|
| alv:gy | 1.22 | 1.27 |
| bay:gy | 1.14 | 1.22 |
| dbk:gy | 1.19 | 1.24 |
| ibe:sm | 1.09 | 1.08 |
| nok1v:fh | 1.11 | 1.01 |
| or:fp | 1.08 | 1.18 |
| rep:sm | 1.08 | 1.01 |
| sap:gy | 1.14 | 1.04 |
| sie:gy | 1.18 | 1.35 |
| tef:sm | 1.15 | 1.17 |

*Stage 3*

The third stage of the adjusted BL method is *unique* and represents the highest added value of the article. We use publically available information about the analysts' target prices to calculate adjusted views about the market returns. This enables us to calculate the view-adjusted market returns in the next stage of the BL procedure.

We start the third stage by collecting daily target prices and the actual stock prices for individual stocks, included in the benchmark portfolio. We then calculate the TRP ratio by using (21.1).

After obtaining daily TRP ratios for all stocks in the portfolio, we calculate the 1 month's average TRP ratio for each asset. Then, we compare the current 1 month TRP ratio to the long-term TRP ratio mean, which was obtained by the autoregressive model (AR1).

Table 21.3 shows the long-term TRP values and current TRP values (at the end of August 2009) for stocks, included in this portfolio.

The actual Overweight (Underweight) is therefore

$$\frac{Ow}{Uw} = \frac{1m \ Avg \ TRP \ Ratio}{LTMean \ TRP \ Ratio}. \tag{21.12}$$

Applying (21.12) to the Deutsche Bank, on 4/8/2009:

$$Ow = 1.\frac{24}{1}.19 = 1.0420. \qquad (21.13)$$

The result obtained in (21.13) implies that we expect Deutsche Bank's stock to outperform the market's expected return by 4.2% over the next 6 months.

*Stage 4*
We are now able to calculate returns which include our opinions by using expected benchmark returns without opinions and adding them our opinions adjusted for covariances between the stocks.

Adjusted returns were calculated by using the following equation:

$$\begin{bmatrix} \text{Adjusted} \\ \text{portfolio} \\ \text{returns} \end{bmatrix} = \begin{bmatrix} \text{Benchmark} \\ \text{portfolio} \\ \text{returns} \end{bmatrix} + \begin{bmatrix} \text{Tracking} \\ \text{factors} \end{bmatrix} \begin{bmatrix} \text{Analyst} \\ \text{opinions} \\ \text{(delta)} \end{bmatrix}. \qquad (21.14)$$

The most important step here is to calculate deltas. The main feature is to minimize the sum of the squares of individual analyst opinions by adjusting them according to the following constraints:

– Individual adjusted return on stock has to equal our expectations
– Optimized benchmark proportions have to be positive (restriction of no short sales)

*Stage 5*
The last stage of the BL portfolio optimization is application of the mean-variance approach, where we obtain the efficient portfolio weights based on previously calculated "adjusted portfolio returns."

In order to calculate optimized benchmark proportions, the following formula was implemented:

$$\begin{bmatrix} \text{Optimized} \\ \text{portfolio} \\ \text{proportions} \end{bmatrix} = \frac{\begin{bmatrix} \text{Variance-} \\ \text{covariance} \\ \text{matrix} \end{bmatrix}^T \begin{bmatrix} \text{Adjusted} \\ \text{portfolio} \\ \text{returns} \end{bmatrix} - R_f}{\sum \begin{bmatrix} \text{Variance-} \\ \text{covariance} \\ \text{matrix} \end{bmatrix}^T \begin{bmatrix} \text{Adjusted} \\ \text{portfolio} \\ \text{returns} \end{bmatrix} - R_f}. \qquad (21.15)$$

All stages of the BL method were implemented onto spreadsheet by using VBA programming language.

# 3   Results

This research considered the problem of managing €10 million portfolio of stocks between 1 June 2004 and 4 August 2009. Portfolio optimization methods were subjects to various constraints, which accounted for different types of risks. The most important is nonnegativity. This restriction was introduced for different reasons, but the most important is qualitative—portfolio managers are usually not allowed to take significant short positions, especially when managing portfolios for noninstitutional clients.

Results of all optimization approaches will be presented in this section. Performance will be measured on three different factors:

- Return on Investment (ROI)
- Value at Risk (VaR) [12]
- Sharpe Ratio

## 3.1   Return on Investment

Return on Investment (ROI) was calculated as a plain one-period arithmetic return, as shown in (21.16):

$$\text{ROI}_y = \frac{V_{fy}V_{iy}}{V_{iy}}, \qquad (21.16)$$

where:

$V_i$: Initial value of the portfolio "y"
$V_f$: Final value of the portfolio "y"

## 3.2   Sharpe Ratio

In this research Sharpe ratio was calculated by dividing the return on a strategy by the standard deviation of return, as proposed by Damodaran [7]:

$$\text{Sharpe ratio} = \frac{R_p}{\sigma_p}, \qquad (21.17)$$

Where:

$R_p$: return on portfolio
$\sigma_p$: standard deviation of the portfolio

**Table 21.4** Summary of results

| Performance measure | MV | Passive | EQW | BL |
|---|---|---|---|---|
| ROI | −9.28% | 14.15% | 22.22% | *47.31%* |
| VaR99 as % of portfolio value | *2.39%* | 2.53% | 2.59% | 2.61% |
| Sharpe ratio | 0.0335 | 0.0388 | 0.0400 | *0.0431* |

Sharpe ratio was calculated each month and the final result represents the average of monthly ratios ([12], p. 352). Monthly Sharpe ratio was calculated each month by taking the average return of an optimization method and dividing it by the standard deviation from the same strategy that month. Standard deviation was calculated in the same manner as already presented in this section of the research.

## 3.3 Interpretation of Results

This part represents performance measures of all five portfolio optimization approaches. Results are summarized in Table 21.4.

Each financial crisis raises the question whether one can devise a strategy to obtain returns above the market whilst protecting the capital invested. Especially, the most recent developments have emphasized the importance of reliable risk management. It is therefore crucial to study the performance of risk factors, calculated from the optimization outputs.

According to Table 21.4, minimum-variance approach represents the safest optimization strategy. Standard deviation of the MV approach is considerable lower than the others. The MV approach aims to optimise portfolio weights so that the optimal solution will be the portfolio with the lowest variance. *The empirical results follow that fact.*

On the other hand, we developed a method that includes additional factor in the optimization—*analysts' opinions*. It is expected that the volatility of the opinions will increase the volatility of the whole portfolio. Higher "gross" risk is therefore expected.

Table 21.4 shows that the TRP strategy significantly outperformed the benchmark in relation to the Sharpe ratio and ROI. Although minimum variance was identified as the safest method, it achieved the lowest return per unit of risk between all five strategies. Benchmark passive investment and the EQW portfolio produced similar results.

We conclude that although *incorporating analysts' opinions increases the riskiness of our portfolio, it yields significantly higher return per amount of that risk.* Investor who trades off between the risk and return should therefore *choose TRP-based strategy*, as it will give him the highest reward for risk they take.

# 4 Conclusion

Equity analysts have become an influential factor on the capital markets. Some of the previous researches, such as Womack [8], Barber et al. [9], and Espahbodi et al. [10], even proved that analysts' coverage is associated with the positive abnormal returns on the stock. These studies focused on the "buy" ratings, issued by analysts.

In this research, different approach to "exploiting" the analysts' knowledge is proposed. The main focus is placed on the target-to-real-price ratio (TRP ratio). Based on the stationarity [13] of the TRP ratio, several approaches for its implementation can be developed. We tested whether it is possible to outperform the passive investment strategy and obtained extremely positive results.

In fact, we obtained results that are not consistent with the efficient market hypothesis (EMH). There are a few possible interpretations for the results; one could say that the investment period is too short and that results were obtained in the "non-normal" market conditions. Others might suggest that assumptions are not realistic. The fact is that our portfolio outperformed the best-performing benchmark by more than 25% and returned 47% ROI in the challenging market conditions. This suggests that there might be time in the future where portfolio managers and traders start considering TRP ratio as one of the factors when they place their buy/ sell orders.

In 1991, Schipper showed that the information analysts' produce improves the market efficiency by helping investors "to value companies' assets more accurately". In line with this statement, we presume that if everyone started using TRP ratio as the proper measure of stock's value, assets would be priced more efficiently and the opportunity of earning higher abnormal returns would disappear. We therefore conclude that there are clear indications the market currently operates inefficiently, but with more frequent use of this important information, it could become efficient.

# References

1. Munda G, Strasek S (2011) Standardised Black–Litterman optimisation using TRP ratio. In: Lecture notes in engineering and computer science: Proceedings of the world congress on engineering 2011, WCE 2011, 6–8 July, 2011, London, U.K., pp 391–396
2. Munda G, Strasek S (2011) Use of the TRP ratio in selected countries. Our Economy: review of current problems in economics ED-57-1:55–60
3. Sharpe WF, Alexander GJ, Bailey JV (2006) Investments, 6th edn. Prentice Hall, New York
4. Markowitz H (1991) Portfolio selection: efficient diversification of investments. Blackwell Publishing, New York
5. Merna T, Al-Thani FF (2008) Corporate risk management. Wiley Finance, Chichester
6. Michaud R (1989) The Markowitz optimization enigma: is optimized optimal? Financ Analyst J ED-45(1):31–42
7. Damodaran A (2003) Investment philosophies: successful strategies and investors who made them work. Wiley Finance, New Jersey

8. Womack KL (1996) Do brokerage analysts' recommendations have investment value? J Finance ED-51:137–167
9. Barber B, Lehavy R, McNichols M, Trueman B (2001) Can investors profit from the prophets? Consensus analyst recommendations and stock returns. J Finance ED-56:341–372
10. Espahbodi R, Dugar A, Tehranian H (2001) Further Evidence on Optimism and Underreaction in Analysts' Forecasts. Rev Financ Econ ED-10:1–21
11. Maginn JL (2007) Managing investment portfolios: a dynamic process. Wiley Finance, New Jersey
12. Dowd K (2008) Measuring market risk. Wiley, Chichester
13. Maddala GS, Kim IM (2002) Unit roots, cointegration, and structural change. Cambridge University Press, Cambridge
14. Stonehill AI, Moffett MH (1993) International financial management. Taylor & Francis, Chatham

# Chapter 22
# An Overview of Contemporary and Individualised Approaches to Production Maintenance

James Hogan, Frances Hardiman, and Michael Daragh Naughton

## 1 Introduction

A maintenance strategy can be described as a long-term plan that covers all attributes of maintenance management in addition to clear action plans and direction to attaining the desired maintenance function [1]. Maintenance is conducted in order to stop the deterioration of an asset and to hold the inherent value of the asset for the financial benefit of the enterprise. For manufacturing enterprises to increase their effectiveness in growing competitive markets, the importance of asset maintenance is continuously reinforced as the need for greater product quality and lower operational costs are becoming key economic areas of focus. Maintenance has no intrinsic value to an enterprise but is used to support the strategic objectives of the plant and the fundamental objectives of the organisation [2]. Due to the numerous different applications and mechanisms of production assets, contemporary strategies such as reliability centred maintenance (RCM) and total productive maintenance (TPM) may be suitable for one asset, but it may not be the most appropriate strategy for another asset. In order to apply the most suitable strategy, familiarity of the assets and the obtainable resources must be recognised along with a comprehensive awareness by top management of the affecting factors surrounding asset maintenance.

J. Hogan (✉) • F. Hardiman • M.D. Naughton
Department of Mechanical and Automobile Technology,
Limerick Institute of Technology, Limerick, Ireland
e-mail: james.hogan@lit.ie; frances.hardiman@lit.ie; daragh.naughton@lit.ie

S.-I. Ao and L. Gelman (eds.), *Electrical Engineering and Intelligent Systems*,
Lecture Notes in Electrical Engineering 130, DOI 10.1007/978-1-4614-2317-1_22,
© Springer Science+Business Media, LLC 2013

## 2  First Generation Maintenance

The demand for reliability and productivity has led to the creation and implementation of various maintenance management strategies. The evolution of maintenance strategies can be subdivided into three generations that have emerged since the 1940s and have developed to the present day as outlined by Moubray [3]. The first generation nucleated pre-World War II. During this time, manufacturing was highly un-mechanised, which lead to little downtime. The simplicity and over-design of assets made them reliable and easy to repair. Asset maintenance was not of high importance to most managers with only the need for simple lubrication, cleaning and servicing required. Run-to-failure was the maintenance strategy most prominently employed [3].

## 3  Second Generation Maintenance

The second generation (1950s–1970s) witnessed immense progression in the ability and complexity of industrial equipment. Awareness of maintenance as a value-added process grew in the 1950s as the cost of equipment failure escalated to a level that it was necessary for action to be taken in order to reduce production-related costs. The 1950s period was post World War II, and maintenance strategies being implemented at the time were not adequate particularly for equipment such as modernised commercial aircraft [4].

The development of preventive maintenance (PM) in 1951 introduced a periodic maintenance schedule based on time or asset utilisation where the asset requirements would be recognised and a time-frame would be planned for maintenance implementation [5]. Corresponding to the development of PM, predictive maintenance also evolved in the 1950s in response to asset deterioration under different circumstances.

This strategy allowed for asset conditions to be maintained and diagnosed by measuring physical characteristics such as temperature or vibration. The appropriate maintenance was planned and conducted only after a fault has been recognised [6].

In 1957, corrective maintenance (CM) was introduced and brought about changes in the design of assets allowing for improvements in reliability and ergonomics. For CM to be applied, a problem with the asset must first be established before any corrective action is taken. The knowledge gained during the CM could then be applied to the next generation of assets to further improve asset efficiency [5].

## 3.1   Development of Total Productive Maintenance

Additionally, the 1950s witnessed the commercial development and advance of TPM. TPM involves the cooperation of the entire organisation from top management to the staff on the production floor in an effort to reduce costs and improve workplace efficiency throughout the organisation. A Japanese enterprise "Nipondenso" was the first organisation to incorporate TPM plant wide in 1960 [7].

Core to the progression of TPM is the training and development of the personnel with the support of training programmes to help create an expert workforce within an organisation. To allow for a true TPM strategy to progress, funding in the required areas must be made available. The cost of implementing TPM can vary depending on different organisational factors and the pace at which TPM is being applied by the organisation. Other than the cost of training, the age and condition of assets will also determine the overall cost of implementation as older assets will occasionally require additional parts [8]. Although costs may be high, the results of implementation as described by Venkatesh [7] will encourage essential changes within an organisation that will generate valuable financial and personal growth such as:

(1) Reduction of maintenance cost
(2) Multi-skilled workforce
(3) Production of goods without the reduction in product quality

Although TPM can arguably to be one of the best ways of improving the total efficiency throughout an organisation, there are various areas within TPM that can cause complications during its implementation. Sinha [9] has extensively reviewed barriers to TPM which include:

(1) Resistance to change from the workforce
(2) Fear of job loss amongst employees
(3) Insufficient resources (money, time, skill level)

Due to the dramatic changes TPM can have on organisational behaviour, responsibilities, skill development and the additional use of information technology, the success rate for most organisations is less than 30% [10]. According to Hartmann [11], organisations that try to implement TPM a second time also typically result in failure.

## 3.2   Progression of Reliability Centred Maintenance

Later in the second generation of development, RCM originated and was utilised by the U.S. Department of Defence [12]. This strategy was adopted mainly by the aircraft industry and is used to determine the maintenance needed to ensure that assets fulfil their intended purpose while in operation [13]. The RCM strategy focuses on minimum safety standards and the development of maintenance plans and rules [3] and in 1978 when a report entitled "Reliability Centred Maintenance" published by Stanley Nowlan and Howard Heap became the report that all RCM approaches are now based on. It is a predictive methodology that is also used to improve asset performance as well as the reliability of the end product [14]. It is commonly used to remove inefficient PM tasks from existing maintenance plans [15]. The success of RCM leads to an increased understanding of cost effectiveness and risk levels [16].

Comparable to TPM, RCM similarly exhibits both advantages and disadvantages in its implementation. Moubray [3] outlined in his book RCM II, the seven basic questions to RCM. These questions are:

(1) What are the functions and associated performance standards of the asset in its present operating context?
(2) In what ways can it fail to fulfil its function?
(3) What causes each functional failure?
(4) What happens when each failure occurs?
(5) In what way does each failure matter?
(6) What can be done to predict or prevent the failure?
(7) What should be done if a suitable proactive task cannot be found?

To have a true RCM strategy in place, RCM maintenance should comply with the JA1011 (Evaluation criteria for RCM processes) standard [17]. Numerous benefits can be gained by using RCM including [18]: the lowering of maintenance costs by removing unnecessary maintenance and overhauls, the reduction in the frequency of maintenance implementation, increased reliability of components, increased emphasis on critical components and the use of root cause analysis to assess the cause of component failure. Difficulties generally associated with RCM include: the high start-up costs associated with the training of staff and the purchasing of equipment used for predictive maintenance.

The potential savings that can be achieved using RCM are sometimes not recognised by management, therefore preventing its initiation [19].

Due to the complexity and detail required to carry out RCM successfully, the success rate of implementation is in the range of 5–10% with about 90% of applications resulting in failure [20].

## 4  Third Generation Maintenance

By the end of the 1970s, maintenance moved into its third generation of development transition. The progression in technology and the emphasis on health/safety and the environment emerged as catalysts for new developments. The advances in technology lead to the development of smaller, faster computers which replaced their slower, bigger ancestors from the past [3] encouraging the growth of expert systems. The 1980s saw the advancement of atomisation and the growth of dependency on the reliability and availability of organisational assets. The focus was on having zero down time or no in-service breakdowns [21]. Advancements in maintenance support structures such as decision support systems (DSS) and failure mode effect analysis (FMEA) lead to greater strategy/policy selections along with developments in predictive maintenance technologies also aiding maintenance strategy selection. A reorientation in organisational thinking towards team work and participation further enhanced maintenance management strategies throughout the definable third generation period [3].

As a result of numerous strategies and technologies being available to today's managers', strategy selection is now more effective than ever before. It must be admitted, however, that there is no one standard solution and many of the concepts are only practical for specific industry or assets [22]. Contemporary maintenance concepts have many advantages such as a proven procedure to follow, as with RCM, however they also incur various disadvantages as follows.

First, a contemporary strategy, such as TPM, can be resource intensive. Its implementation requires long periods of time, intensive training programmes and total commitment from all staff members [23]. These drawbacks are also confirmed by a survey conducted in the UK of 36 small and medium sized manufacturing enterprises [24]. The survey identified some of the potential barriers that can be faced by maintenance managers. Of the enterprises surveyed, 80% of management said that the lack of finances constrained the adoption of new maintenance approaches and that 84% reported that detailed and continual training programs would be needed if a new maintenance approach was introduced. However, 100% of the respondents said that no finance would be made available for training programmes.

In addition to the available resources, implementation of a contemporary strategy typically requires asset data and information to be readily available in a compressed form that is accessible to all maintenance staff members [25]. However, many organisations do not have this information readily available.

## 4.1  Advancement of Individualised Maintenance Approaches

Due to the above limitations, organisations are moving from contemporary solutions that may not suit their resource and staff constraints and are now leaning towards the utilisation of their internal experience, knowledge and skill to design a maintenance strategy that suits their needs by using an individualised approach. An individualised approach to maintenance consists of "hand picking" from contemporary maintenance strategies and using their useful techniques and ideas to create a unique strategy for the enterprise [23]. It emphasises asset characteristics and allows for an appropriate maintenance strategy to be applied. It also allows for the decision making of key characteristics easier by allowing traditional concepts more available to choose from. Finally, an individualised approach allows for the utilisation of appropriate resources by carefully selecting assets that require maintenance and applying a suitable maintenance solution to each asset [26].

Many individualised maintenance strategies have been developed over the last number of decades and are applied to an array of industries. An example of individualised maintenance strategy development and implementation is shown by Waeyenbergh and Pintelon with the CIBOCOF framework [26, 27]. Centrum voor Industrieel Beleid Onderhouds Concept Ontwikkelings Framework (CIBOCOF) or in English (Centre for Industrial Management Maintenance Concept Development Framework) is an individualised maintenance strategy that is unique to the organisation/enterprise. The concept does not emphasise on one goal, but allows there to be emphasis on other areas such as: resource and asset examination, technical and functional inspection, maintenance policy selection, policy implementation and delivering feedback on the process, also providing procedures on how to share the information throughout the organisation. CIBOCOF uses a reiteration cycle of planning, doing, controlling, and adjusting (PDCA-approach), which when complete provides a firm maintenance plan.

It is outlined that the improvements of this strategy included: an increase in the profitability of the organisation, improvement in customer satisfaction and also workplace safety. The CIBOCOF maintenance framework was successfully applied to a light production enterprise where it was found that the modules could be used as a whole or used independently if required.

In addition to the development of CIBOCOF, an approach known as value driven maintenance policy (VDMP) was established in order to show the hidden values of maintenance and how organisations can benefit from these values [23]. In terms of VDMP, value is defined as "the delivery of maximum availability at minimum cost" [28]. The approach was developed using principles and concepts from TPM, RCM and risk-based inspection (RBI), and it requires an organisation to concentrate on the dynamic prospects for value creation using appropriate steps/ techniques instead of using a one method fits all approach [29]. The steps used to implement VDMP include [30]:

(1) Create a definition of the maintenance planning, tactical and central objectives of the production plant
(2) Categorise equipment locations into maintenance categories along with their functional necessities and requirements
(3) Select appropriate maintenance strategies to implement for asset maintenance

These steps encourage the continuous improvement of cost effectiveness associated with assets maintenance.

Central to VDMP is a method called experience-based reliability centred maintenance (EBRCM) that uses feedback data and decision logic to methodically select maintenance tasks for assets [31]. This approach utilises the internal expert and operating knowledge in collaboration with FMEA's and decision logic methods to create an updated maintenance plan that is unique to the requirements of an organisation. To complement the improvements made by a VDMP, a lifecycle analysis should be carried out before the purchase of any asset to estimate the total cost of an asset before installation.

## 4.2  Difficulties with Implementing Individualised Approach's

As shown, the structure of individualised maintenance strategies can be as diverse as contemporary strategies with numerous steps to be taken before implementation is successful. In general, organisations are leaning towards individualised approaches to avoid costly areas that are associated with contemporary strategies and use organisational strengths such as internal experience to make a strategy customised to their available resources; however, there are some drawbacks to individualised approaches.

Although the individualised approach requires organisations to utilise internal knowledge and experience, Naughton and Tiernan [25] indicate that there is a lack of self-belief in the practitioners own abilities and a lack of regard towards experience-based knowledge amongst personnel. With that, their research has highlighted further difficulties such as:

(1) Experience-based protectionism
(2) Fear of change amongst employees
(3) Lack of management support
(4) The overall cost of the change initiative
(5) The lack of plant/process knowledge
(6) Scepticism and/or low morale for the change initiative

A positive change in management support is necessary for maintenance strategies to deliver their economic benefits. A change in organisational culture is also essential to allow for a smooth transition of both individualised and contemporary strategies into an organisation. For individualised strategies to continue their effectiveness, it is suggested that key areas within the implemented strategy should be reviewed periodically to ensure consistency and quality of the concept and to account for changes within the surrounding environment. This aspect may cause drawbacks in the individualised strategy due to its time consumption [23].

## 5  Conclusion

Contemporary maintenance strategies such as RCM and TPM have been successfully implemented to an array of industries with numerous accounts of their benefits being acknowledged by diverse organisations. They typically have well-structured steps that allow for detailed implementation of maintenance to be carried out on industrial assets. However, contemporary maintenance strategies do have drawbacks that include: the length of time needed for implementation, the cost of implementation (training and purchasing of predictive equipment) and the amount of detail/information about the process required for the strategy to work at its full potential.

Due to the various requirements needed for successful implementation, a lot of organisations opt to go for an individualised strategy that will work with the resources available to their maintenance sector. Many organisations can find that contemporary strategies are not suitable to their requirements. It is known that the budget allocated to the maintenance sector of an organisation plays a huge role in strategy selection and may determine the strategy to be used [24]. Although the type of industry may also dictate the selection process, an individualised strategy could exploit the resources available to the maintenance sector. The workforce knowledge, skill and experience could also be utilised to create a strategy suited to the requirements of the organisation, especially if there is a lack of resources needed for particular aspects required for a true contemporary maintenance strategy.

Although individualised strategies have many benefits, they also incur some disadvantages including the lack of regard towards the experience-based knowledge amongst personnel. This can hinder the development of the strategy and will require full management support in order for the strategy to be effective. An organisation needs a flexible strategy that applies ideas and methods from contemporary concepts in order to tailor the requirements and resources of the organisation into the implemented strategy.

**Acknowledgements** The authors would like to thank Limerick Institute of Technology for the use of their library resources and also Kostal Ireland GmbH for their assistance throughout the research.

# References

1. Knackstedt T (2011) What is a maintenance strategy? Kwaliteg. http://www.kwaliteg.co.za/. Accessed 26 Jan 2011
2. Keeney R (1996) A path to creative decision making. Value-focused thinking. Harvard University Press, Cambridge
3. Moubray J (1997) Reliability-centered maintenance. Industrial Press Inc, New York, p 320
4. Kennedy R (2006) Examining the processes of RCM and TPM. Plant Maint. http://www.plantmaintenance.com/articles/RCMVsTPM.pdf. Accessed 19 Jan 2011
5. Ben-Daya MO, Duffuaa S, Raou A (2009) Handbook of maintenance management and engineering. Springer, New York, p 420
6. Girdhar P, Scheffer C (2004) Practical machinery vibration analysis and predictive maintenance. Elsevier, UK
7. Venkatesh J (2009) An introduction to total productive maintenance. Plant Maintenance Resource Center. http://www.plant-maintenance.com/articles/tpm_intro.shtml. Accessed 3 August 2009
8. Gupta S, Tewari PC, Sharma AK (2011) TPM concept an implementation approach. Maintenance World. http://www.maintenanceworld.com/articles/sorabh/research_paper.pdf. Accessed 2 Feb 2011
9. Sinha PK (2008) Manufacturing and operations management. Nirali Prakashan. Pune, India
10. Choy DS, Siam Y (2011) TPM implementation experiences. Maintenance Resources. http://www.maintenanceresources.com/referencelibrary/ezine/tpmimplementation.htm. Accessed 2 Feb 2011
11. Hartmann Ed (2000) Prescription for total TPM success. Maintenance Technology. http://www.mt-online.com/component/content/article/178-april2000/573-prescription-for-total-tpm-success.html?directory=90. Accessed 2 Feb 2011
12. August J (1999) Applied reliability-centered maintenance. PennWell, Tulsa

13. Dunn S (2010) Maintenance terminology—some key terms. Maintenance resources. http://www.maintenanceresources.com/referencelibrary/maintenancemanagement/keyterms.htm#RCM. Accessed 10 November 2010
14. Telang AD, Telang A (2010) Comprehensive maintenance management: Policies strategies and options. PHI, New Delhi
15. Rausand M (1998) Reliability Centered Maintenance. Reliab Eng Saf Syst 60(2):121–132
16. Cadick J, Capelli-Schellpfeffer M, Neitzel D (2006) Electrical safety handbook. McGraw-Hill, New York, pp 5–5
17. Moore R (2007) Selecting the right manufacturing improvement tools: What tool? When? Butterworth-Heinemann, St. Louis
18. Technology, reliability and PdM (2010) Reliability centered maintenance advantage and disadvantages. PM-PdM technology for manufacturing industries. http://preventive-predictive-maintenance.blogspot.com/2010/01/reliability-centered-maintenance.html. Accessed 16 Dec 2010
19. Lingham L (2011) Management consulting/maintainance management. AllExperts. http://en.allexperts.com/q/Management-Consulting-2802/2010/9/MAintainance-Management.htm. Accessed 3 Feb 2011
20. Bloom NB (2005) 3 Day in-house rcm training seminars. RCM training seminars. http://www.rcmtrainingseminars.com/index.html. Accessed 2 Feb 2011
21. Jabar HB, Bhd, S P Sdn (2003) Plant maintenance strategy: Key for enhancing profitability. Maintenance Resources. http://www.plant-maintenance.com/tzd.shtml. Accessed 10 Nov 2010
22. Waeyenbergh G, Pintelon L (2000) A framework for maintenance concept development. Int J Prod Econom 77:299–313
23. Murthy DNP, Kobbacy KAH (2008) Complex system maintenance handbook. Springer, New York, pp 34–35
24. Hogan J, Hardiman F, Naughton MD (2011) Asset management—a review of contemporary & individualised strategies. In: Lecture notes in engineering and computer science: Proceedings of the world congress on engineering 2011, WCE 2011, London, 6–8 July 2011, pp 545–549
25. Naughton MD, Tiernan P (2011) Individualised maintenance management: A proposed framework & case study. Int J Qual Maint (in press)
26. Waeyenbergh G, Pintelon L (2006) CIBOCOF: A framework for industrial maintenance concept development. Int J Prod Econom 121(2):633–640
27. Waeyenbergh G, Pintelon L (2002) A framework for maintenance concept development. Int J Prod Econom 77:299–313
28. Haarman M (2004) Value driven maintenance-creating shareholder value with maintenance. J Maint Asset Manag, 29–32
29. Gharachorlou K, Jonker R (2008) Getting lean through value-driven maintenance. The Engineer. http://www.theengineer.co.uk/channels/processengineering/getting-lean-through-value-driven-maintenance/306009.article. Accessed 18 November 2010
30. Rosqvist T, Laakso K, Reunanen M (2007) Value-driven maintenance planning for a production plant. Reliab Eng Syst Saf 94(1):97–110
31. Laakso K, Hänninen S, Simola K (1995) Experience based reliability centred maintenance—a case study of the improvement of the maintenance programme for valve drives. Int J Manag Phys Assets, 333–350
32. Baglee D (2008) Maintenance strategy development within SME's: the development of an integrated approach. International Maintenance Conference In: Proceedings of ASME 2007 International Design Engineering Technical Conferences and Computers and Information in Engineering Conference

# Chapter 23
# Weibull Prediction Limits for a Future Number of Failures Under Parametric Uncertainty

**Nicholas A. Nechval, Konstantin N. Nechval, and Maris Purgailis**

## 1 Introduction

Results of research about the prediction of random quantities have enormous potential for application in engineering and are, indeed, of fundamental importance, which is revealed when one addresses the question: if in a system with $m$ components, $k$ quantities failed in the time interval $[0, t_1]$, how many components will fail in a given future time belonging to the interval $[t_1, t_2]$? or, even more, what are the prediction limits (lower and upper) for the number of components that will fail in the time interval $[t_1, t_2]$? There are many systems in which the prediction of times of failure of parts is critical.

Considerable literature describes statistical prediction applications and methods. Nelson [1] provided simple prediction limits for the number of failures that will be observed in a future inspection of a sample of units. The past data consist of the cumulative number of failures in a previous inspection of the same sample of units. Life of such units is modeled with a Weibull distribution with a given shape parameter value. Nelson's prediction limits were motivated by the following application. Nuclear power plants contain large heat exchangers that transfer energy from the reactor to steam turbines. Such exchangers typically have 10,000–20,000

N.A. Nechval (✉)
Department of Statistics, EVF Research Institute, University of Latvia,
Raina Blvd 19, LV-1050 Riga, Latvia
e-mail: nechval@junik.lv

K.N. Nechval
Department of Applied Mathematics, Transport and Telecommunication Institute,
Lomonosov Street 1, LV-1019 Riga, Latvia
e-mail: konstan@tsi.lv

M. Purgailis
Department of Cybernetics, University of Latvia, Raina Blvd 19, LV-1050 Riga, Latvia
e-mail: marispur@lanet.lv

S.-I. Ao and L. Gelman (eds.), *Electrical Engineering and Intelligent Systems*,
Lecture Notes in Electrical Engineering 130, DOI 10.1007/978-1-4614-2317-1_23,
© Springer Science+Business Media, LLC 2013

stainless steel tubes that conduct the flow of steam. Due to stress and corrosion, the tubes develop cracks over time. Cracks are detected during planned inspections. The cracked tubes are subsequently plugged to remove them from service. To develop efficient inspection and plugging strategies, plant management can use a prediction of the added number of tubes that will need plugging by a specified future time. A prediction expressed as an interval indicates the magnitude of the possible prediction error and quantifies the confidence in the prediction. Nelson [1] has established three procedures for the prediction intervals of the number of tubes to fail in components of heat exchangers, namely (I) the procedure of ratio of probabilities, (II) the procedure of ratio of probabilities simplified, and (III) the likelihood ratio procedure. Nordman and Meeker [2] evaluated the coverage probability for each one of the procedures proposed by Nelson [1], concluding that the more appropriate is the likelihood ratio procedure.

Meeker and Escobar [3] developed a method to determine the prediction limits (upper and lower) for the future number of fails ($Y$) in the time interval $[t_c, t_w]$. Such procedure is based upon the conditional binomial distribution of $Y$ given that $X$ components have failed in the time interval $[0, t_c]$. Rostum [4] developed statistical models to predict the state of the pipelines in a network of water distribution. Starting from the historical of the past failures in a network of water supply, it was possible to predict the future number of failures in each network. These predictions were then used to make decisions about maintenance of the water network. In other words, with his work it may be possible to answer about the following question: shall we repair points of failure in the network (or sub-network) of water supply, or shall we change an entire pipeline in a given sub-network? Nagaraja [5] described prediction problems for the exponential distribution. He discussed various predictors proposed in the literature, and he studied their properties. Faulkenberry [6] suggested a method that can be applied when there is a sufficient statistic that can be used as a predictor. Nechval et al. [7] described a technique for using censored life data from extreme value distributions to construct prediction limits or intervals for future outcomes. Cox [8] presented a general approximate analytical approach to prediction based on the asymptotic distribution of maximum likelihood estimators. Atwood [9] used a similar approach. Efron and Tibshirani [10] described an approximate simulation/pivotal-based approach. Beran [11] gave theoretical results on the properties of prediction statements computed with simulated (bootstrap) samples. Kalbfleisch [12] described a likelihood-based method, Thatcher [13] described the relationship between Bayesian and frequentist prediction for the binomial distribution, and Geisser [14] presented a more general overview of the Bayesian approach.

Hahn and Nelson [15], Patel [16], and Hahn and Meeker [17], provided surveys of methods for statistical prediction for a variety of situations.

In this paper, we use a frequentist procedure, which is called "within-sample prediction of future order statistics," when the time-to-failure follows the two-parameter Weibull distribution indexed by scale and shape parameters $\beta$ and $\delta$. We consider the case when both parameters $\beta$ and $\delta$ are unknown. The technique proposed here for constructing prediction limits emphasizes pivotal quantities relevant for obtaining ancillary statistics and represent a special case of the method of invariant embedding of sample statistics into a performance index applicable whenever the statistical problem is invariant under a group of transformations, which acts transitively on the parameter space [7, 18–26].

## 2   Weibull Within-Sample Prediction Limits for Future Order Statistics

For within-sample prediction, the problem is to predict future events in a sample or process based on early data from that sample or process. For example, if $m$ units are followed until $t_k$ and there are $k$ observed failures, $t_1, \ldots, t_k$, one could be interested in predicting the time of the next failure $t_{k+1}$; time until $r$ additional failures, $t_{k+r}$; number of additional failures in a future interval.

**Theorem 1** (*Lower (upper) one-sided prediction limit h on the lth order statistic $Y_l$ in a sample of m observations from the two-parameter Weibull distribution on the basis of the early-failure data $Y_1 \leq \ldots \leq Y_k$ from the same sample*). Let $Y_1 \leq \ldots \leq Y_k$ be the first $k$ ordered early-failure observations from a sample of size $m$ from the two-parameter Weibull distribution

$$f(y|\beta,\delta) = \frac{\delta}{\beta}\left(\frac{y}{\beta}\right)^{\delta-1} \exp\left[-\left(\frac{y}{\beta}\right)^{\delta}\right] (y>0), \qquad (23.1)$$

where $\delta>0$ and $\beta>0$ are the shape and scale parameters, respectively. Then a lower one-sided conditional $(1-\alpha)$ prediction limit $h$ on the $l$th order statistic $Y_l$ $(l > k)$ in the same sample is given by

$$h = w_h^{1/\hat{\delta}} y_k, \qquad (23.2)$$

where $w_h$ satisfies the equation

$$\Pr\left\{Y_l>h|z^{(k)}\right\} = \Pr\left\{Y_l>w_h^{1/\hat{\delta}} y_k|z^{(k)}\right\}$$

$$= \left[ \int_0^\infty v_2^{k-2} \prod_{i=1}^{k} z_i^{v_2} \sum_{j=0}^{l-k-1} \binom{l-k-1}{j} \frac{(-1)^{l-k-1-j}}{m-k-j} \right.$$

$$\left. \times \left( (m-k-j)(w_h z_k)^{v_2} + j z_k^{v_2} + \sum_{i=1}^{k} z_i^{v_2} \right)^{-k} dv_2 \right]$$

$$\times \left[ \int_0^\infty v_2^{k-2} \prod_{i=1}^{k} z_i^{v_2} \sum_{j=0}^{l-k-1} \binom{l-k-1}{j} \frac{(-1)^{l-k-1-j}}{m-k-j} \right.^{-1}$$

$$\left. \times \left( (m-k)z_k^{v_2} + \sum_{i=1}^{k} z_i^{v_2} \right)^{-k} dv_2 \right] = 1-\alpha,$$

$$(23.3)$$

$$z^{(k)} = (z_1, \dots, z_k), \tag{23.4}$$

$$Z_i = \left(\frac{Y_i}{\widehat{\beta}}\right)^{\widehat{\delta}}, i = 1, \dots, k, \tag{23.5}$$

$$w_h = \left(\frac{h}{y_k}\right)^{\widehat{\delta}}, \tag{23.6}$$

where $\widehat{\beta}$ and $\widehat{\delta}$ are the maximum likelihood estimates of $\beta$ and $\delta$ based on the first $k$ ordered past observations $Y_1 \leq \dots \leq Y_k$ from a sample of size $m$ from the two-parameter Weibull distribution (23.1), which can be found from solution of

$$\widehat{\beta} = \left(\left[\sum_{i=1}^{k} y_i^{\widehat{\delta}} + (m-k)y_k^{\widehat{\delta}}\right] \Big/ k\right)^{1/\widehat{\delta}}, \tag{23.7}$$

and

$$\widehat{\delta} = \left[\left(\sum_{i=1}^{k} y_i^{\widehat{\delta}} \ln y_i + (m-k)y_k^{\widehat{\delta}} \ln y_k\right) \right.$$
$$\left. \times \left(\sum_{i=1}^{k} y_i^{\widehat{\delta}} + (m-k)y_k^{\widehat{\delta}}\right)^{-1} - \frac{1}{k}\sum_{i=1}^{k} \ln y_i\right]^{-1}, \tag{23.8}$$

(Observe that an upper one-sided conditional $\alpha$ prediction limit $h$ on the $l$th order statistic $Y_l$ based on the first $k$ ordered early-failure observations $Y_1 \leq \dots \leq Y_k$, where $l > k$, from the same sample may be obtained from a lower one-sided conditional $(1-\alpha)$ prediction limit by replacing $1-\alpha$ by $\alpha$ ($\alpha < 0.5$))

*Proof* The joint density of $Y_1 \leq \dots \leq Y_k$ and $Y_l$ is given by

$$f(y_1, \dots, y_k, y_l | \beta, \delta)$$
$$= \frac{m!}{(l-k-1)!(m-l)!} \prod_{i=1}^{k} \frac{\delta}{\beta}\left(\frac{y_i}{\beta}\right)^{\delta-1} \exp\left(-\left(\frac{y_i}{\beta}\right)^{\delta}\right)$$
$$\times \left[\exp\left(-\left(\frac{y_k}{\beta}\right)^{\delta}\right) - \exp\left(-\left(\frac{y_l}{\beta}\right)^{\delta}\right)\right]^{l-k-1}$$
$$\times \frac{\delta}{\beta}\left(\frac{y_l}{\beta}\right)^{\delta-1} \exp\left(-\left(\frac{y_l}{\beta}\right)^{\delta}\right)\exp\left(-(m-l)\left(\frac{y_l}{\beta}\right)^{\delta}\right). \tag{23.9}$$

Let $\widehat{\beta}$, $\widehat{\delta}$ be the maximum likelihood estimates of $\beta$, $\delta$, respectively, based on $Y_1 \leq \dots \leq Y_k$ from a complete sample of size $m$, and let

$$V_1 = \left(\frac{\widehat{\beta}}{\beta}\right)^{\delta}, \quad V_2 = \frac{\delta}{\widehat{\delta}}, \quad W = \left(\frac{Y_l}{y_k}\right)^{\widehat{\delta}}, \tag{23.10}$$

and $Z_i = \left(Y_i \big/ \hat{\beta}\right)^{\hat{\delta}}, i = 1(1)k$. Using the invariant embedding technique [7, 18–26], we then find in a straightforward manner, that the joint density of $V_1$, $V_2$, $W$, conditional on fixed $z^{(k)} = (z_1, \dots, z_k)$, is

$$
f\left(v_1, v_2, w | z^{(k)}\right) = \vartheta\left(z^{(k)}\right) v_2^{k-1} \prod_{i=1}^{k} z_i^{v_2} (w z_k)^{v_2} v_1^k \sum_{j=0}^{l-k-1} \binom{l-k-1}{j} (-1)^{l-k-1-j}
$$
$$
\times \exp\left[-v_1\left((m-k-j)(w z_k)^{v_2} + j z_k^{v_2} + \sum_{i=1}^{k} z_i^{v_2}\right)\right],
$$
$$
v_1 \in (0, \infty), v_2 \in (0, \infty), w \in (1, \infty),
$$

$$(23.11)$$

where

$$
\vartheta\left(z^{(k)}\right) = \left[\int_0^\infty v_2^{k-2} \prod_{i=1}^{k} z_i^{v_2} \sum_{j=0}^{l-k-1} \binom{l-k-1}{j} \frac{(-1)^{l-k-1-j} \Gamma(k)}{m-k-j} \right.
$$
$$
\left. \times \left((m-k) z_k^{v_2} + \sum_{i=1}^{k} z_i^{v_2}\right)^{-k} dv_2\right]^{-1}
$$

$$(23.12)$$

is the normalizing constant.

Using (23.11), we have that

$$
\Pr\{Y_l > h | z^{(k)}\} = \Pr\left\{\left(\frac{Y_l}{y_k}\right)^{\hat{\delta}} > \left(\frac{h}{y_k}\right)^{\hat{\delta}} \Big| z^{(k)}\right\}
$$
$$
= \Pr\{W > w_h | z^{(k)}\} = \int_0^\infty \int_{w_h}^\infty \int_0^\infty f\left(v_1, v_2, w | z^{(k)}\right) dv_1 dw dv_2
$$
$$
= \left[\int_0^\infty v_2^{k-2} \prod_{i=1}^{k} z_i^{v_2} \sum_{j=0}^{l-k-1} \binom{l-k-1}{j} \frac{(-1)^{l-k-1-j}}{m-k-j} \right.
$$
$$
\left. \times \left((m-k-j)(w_h z_k)^{v_2} + j z_k^{v_2} + \sum_{i=1}^{k} z_i^{v_2}\right)^{-k} dv_2\right]
$$
$$
\times \left[\int_0^\infty v_2^{k-2} \prod_{i=1}^{k} z_i^{v_2} \sum_{j=0}^{l-k-1} \binom{l-k-1}{j} \frac{(-1)^{l-k-1-j}}{m-k-j} \right.
$$
$$
\left. \times \left((m-k) z_k^{v_2} + \sum_{i=1}^{k} z_i^{v_2}\right)^{-k} dv_2\right]^{-1},
$$

$$(23.13)$$

and the proof is complete.

**Corollary 1.1** *(Lower (upper) one-sided prediction limit h on the lth order statistic $Y_l$ in a sample of m observations from the two-parameter Weibull distribution, with $\delta=1$, on the basis of the early-failure data $Y_1 \leq \dots \leq Y_k$ from the same sample)*

Let $Y_1 \leq \ldots \leq Y_k$ be the first $k$ ordered early-failure observations from a sample of size $m$ from the two-parameter Weibull distribution (23.1). Then a lower one-sided conditional $(1-\alpha)$ prediction limit $h$ on the $l$th order statistic $Y_l$ $(l > k)$ in the same sample is given by

$$h = w_h y_k, \tag{23.14}$$

where $w_h$ satisfies the equation

$$\Pr\{Y_l > h\} = \Pr\{Y_l > w_h y_k\}$$

$$= \frac{\displaystyle\sum_{j=0}^{l-k-1} \binom{l-k-1}{j} \frac{(-1)^{l-k-1-j}}{m-k-j} \left( (m-k-j)w_h z_k + j z_k + \sum_{i=1}^{k} z_i \right)^{-k}}{\displaystyle\sum_{j=0}^{l-k-1} \binom{l-k-1}{j} \frac{(-1)^{l-k-1-j}}{m-k-j} \left( (m-k)z_k + \sum_{i=1}^{k} z_i \right)^{-k}} = 1-\alpha, \tag{23.15}$$

$$z^{(k)} = (z_1, \ldots, z_k), \tag{23.16}$$

$$Z_i = \frac{Y_i}{\widehat{\beta}}, i = 1, \ldots, k, \tag{23.17}$$

$$w_h = \frac{h}{y_k}, \tag{23.18}$$

$\widehat{\beta}$ is the maximum likelihood estimates of $\beta$ based on the first $k$ ordered past observations $Y_1 \leq \ldots \leq Y_k$ from a sample of size $m$ from the two-parameter Weibull distribution (23.1), which can be found from solution of

$$\widehat{\beta} = \frac{\displaystyle\sum_{i=1}^{k} y_i + (m-k)y_k}{k}. \tag{23.19}$$

*Proof* The joint density of $Y_1 \leq \ldots \leq Y_k$ and $Y_l$ is given by

$$f(y_1, \ldots, y_k, y_l | \beta) = \frac{m!}{(l-k-1)!(m-l)!} \prod_{i=1}^{k} \frac{1}{\beta} \exp\left(-\frac{y_i}{\beta}\right)$$

$$\times \left[ \exp\left(-\frac{y_k}{\beta}\right) - \exp\left(-\frac{y_l}{\beta}\right) \right]^{l-k-1} \tag{23.20}$$

$$\times \frac{1}{\beta} \exp\left(-\frac{y_l}{\beta}\right) \exp\left(-(m-l)\frac{y_l}{\beta}\right)$$

Let $\widehat{\beta}$ be the maximum likelihood estimate of $\beta$, based on $Y_1 \leq \ldots \leq Y_k$ from a complete sample of size $m$, and let

$$V_1 = \frac{\widehat{\beta}}{\beta}, \quad W = \frac{y_l}{y_k}, \tag{23.21}$$

and $Z_i = Y_i \big/ \widehat{\beta}, i = 1(1)k$. Using the invariant embedding technique [7, 18–26], we then find in a straightforward manner that the joint density of $V_1, W$, conditional on fixed $z^{(k)} = (z_1, \ldots, z_k)$, is

$$f\left(v_1, w | z^{(k)}\right) = \vartheta\left(z^{(k)}\right) z_k v_1^k \sum_{j=0}^{l-k-1} \binom{l-k-1}{j} (-1)^{l-k-1-j}$$

$$\times \exp\left[-v_1\left((m-k-j)wz_k + jz_k + \sum_{i=1}^{k} z_i\right)\right], \tag{23.22}$$

$$v_1 \in (0, \infty), \quad w \in (1, \infty),$$

where

$$\vartheta\left(z^{(k)}\right) = \left[\sum_{j=0}^{l-k-1} \binom{l-k-1}{j} \frac{(-1)^{l-k-1-j}\Gamma(k)}{m-k-j} \left((m-k)z_k + \sum_{i=1}^{k} z_i\right)^{-k}\right]^{-1} \tag{23.23}$$

is the normalizing constant.

Using (23.22), we have that

$$\Pr\{Y_l > h | z^{(k)}\} = \Pr\left\{\frac{Y_l}{y_k} > \frac{h}{y_k} \Big| z^{(k)}\right\} = \Pr\{W > w_h | z^{(k)}\} = \int_{w_h}^{\infty}\int_{0}^{\infty} f\left(v_1, w | z^{(k)}\right) dv_1 dw$$

$$= \left[\sum_{j=0}^{l-k-1} \binom{l-k-1}{j} \frac{(-1)^{l-k-1-j}}{m-k-j}\left((m-k-j)w_h z_k + jz_k + \sum_{i=1}^{k} z_i\right)^{-k}\right]$$

$$\times \left[\sum_{j=0}^{l-k-1} \binom{l-k-1}{j} \frac{(-1)^{l-k-1-j}}{m-k-j}\left((m-k)z_k + \sum_{i=1}^{k} z_i\right)^{-k}\right]^{-1} \tag{23.24}$$

This ends the proof.

**Corollary 1.2** (*Lower (upper) one-sided prediction limit h on the lth order statistic $Y_l$ in a sample of m observations from the two-parameter Weibull distribution, with $\delta=1$, on the basis of the early-failure data $Y_1 \leq \ldots \leq Y_k$ from the same sample*) Let $Y_1 \leq \ldots \leq Y_k$ be the first $k$ ordered early-failure observations from a sample of size $m$ from the two-parameter Weibull distribution (23.1). Then a lower one-sided conditional $(1-\alpha)$ prediction limit $h$ on the $l$th order statistic $Y_l$ $(l > k)$ in the same sample is given by

$$h = y_k + w_h^\bullet \widehat{\beta}, \tag{23.25}$$

where $w_h^\bullet$ satisfies the equation

$$\Pr\{Y_l > h\} = \Pr\left\{Y_l > y_k + w_h^\bullet \widehat{\beta}\right\}$$

$$= \frac{1}{B(l-k, m-l+1)} \sum_{j=0}^{l-k-1} \frac{\binom{l-k-1}{j}(-1)^{l-k-1-j}}{(m-k-j)\left[1 + (m-k-j)w_h^\bullet/k\right]^k} \tag{23.26}$$

$$= 1 - \alpha,$$

$$w_h^\bullet = \frac{h - y_k}{\widehat{\beta}}, \tag{23.27}$$

$\widehat{\beta}$ is the maximum likelihood estimates of $\beta$ based on the first $k$ ordered past observations $Y_1 \leq \ldots \leq Y_k$ from a sample of size $m$ from the two-parameter Weibull distribution (23.1), which can be found from solution of (23.19).

*Proof* The joint density of $Y_1 \leq \ldots \leq Y_k$ and $Y_l$ is given by (23.20). Let $\widehat{\beta}$ be the maximum likelihood estimate of $\beta$, based on $Y_1 \leq \ldots \leq Y_k$ from a complete sample of size $m$, and let

$$V_1 = \frac{\widehat{\beta}}{\beta}, \quad W^\bullet = \frac{Y_l - y_k}{\widehat{\beta}}, \tag{23.28}$$

and $Z_i = Y_i / \widehat{\beta}$, $i = 1(1)k$. Using the invariant embedding technique [7, 18–26], we then find in a straightforward manner that the joint density of $V_1$, $W^\bullet$ is

$$f(v_1, w^\bullet) = \vartheta v_1^k \sum_{j=0}^{l-k-1} \binom{l-k-1}{j}(-1)^{l-k-1-j} \exp[-v_1((m-k-j)w^\bullet + k)],$$

$$v_1 \in (0, \infty), w^\bullet \in (0, \infty) \tag{23.29}$$

where

$$\vartheta = \left[\sum_{j=0}^{l-k-1} \binom{l-k-1}{j} \frac{(-1)^{l-k-1-j}\Gamma(k)}{(m-k-j)k^k}\right]^{-1} \tag{23.30}$$

$$= [B(l-k, m-l+1)\Gamma(k)/k^k]^{-1}$$

is the normalizing constant.

Using (23.29), we have that

$$\Pr\{Y_l > h|\} = \Pr\left\{\frac{Y_l - y_k}{\hat{\beta}} > \frac{h - y_k}{\hat{\beta}}\middle|\right\}$$

$$= \Pr\{W^\bullet > w_h^\bullet\} = \int\limits_{w_h}^{\infty}\int\limits_{0}^{\infty} f(v_1, w^\bullet)dv_1 dw^\bullet \qquad (23.31)$$

$$= \frac{1}{B(l-k, m-l+1)} \sum_{j=0}^{l-k-1} \frac{\binom{l-k-1}{j}(-1)^{l-k-1-j}}{(m-k-j)\left[1+(m-k-j)w_h^\bullet/k\right]^k},$$

and the proof is complete.

# 3 Weibull Within-Sample Prediction Limits for Future Failures

Consider the situation in which $m$ units start service at time 0 and are observed until a time $t_c$ when the available Weibull failure data are to be analyzed. Failure times are recorded for the $k$ units that fail in the interval $[0, t_c]$. Then the data consist of the $k$ smallest-order statistics $Y_1 < \ldots < Y_k \le t_c$ and the information that the other $m - k$ units will have failed after $t_c$. With time (or Type I) censored data, $t_c$ is prespecified and $k$ is random. With failure (or Type II) censored data, $k$ is prespecified and $t_c = Y_k$ is random.

The problem of interest is to use the information obtained up to $t_c$ to construct the Weibull within-sample prediction limits (lower and upper) for the number of units that will fail in the time interval $[t_c, t_w]$. For example, this $t_w$ could be the end of a warranty period.

Consider the situation when $t_c = Y_k$. Using the result (23.3) of Theorem 1, the lower prediction limit for the number of units that will fail in the time interval $[t_c, t_w]$ is given by

$$L_{\text{lower}} = l_{\text{max}} - k, \qquad (23.32)$$

where

$$l_{\text{max}} = \max_{k < l \le m} \arg\left(\Pr\{Y_l > t_w | z^{(k)}\} \le \alpha\right), \qquad (23.33)$$

$$
\Pr\left\{Y_l > t_w | z^{(k)}\right\} = \Pr\left\{Y_l > w_{t_w}^{1/\hat{\delta}} y_k | z^{(k)}\right\}
$$

$$
=
\left[
\begin{array}{c}
\displaystyle\int_0^\infty v_2^{k-2} \prod_{i=1}^{k} z_i^{v_2} \sum_{j=0}^{l-k-1} \binom{l-k-1}{j} \frac{(-1)^{l-k-1-j}}{m-k-j} \\
\times \left( (m-k-j)(w_{t_w} z_k)^{v_2} + j z_k^{v_2} + \sum_{i=1}^{k} z_i^{v_2} \right)^{-k} dv_2
\end{array}
\right]
$$

$$
\times
\left[
\begin{array}{c}
\displaystyle\int_0^\infty v_2^{k-2} \prod_{i=1}^{k} z_i^{v_2} \sum_{j=0}^{l-k-1} \binom{l-k-1}{j} \frac{(-1)^{l-k-1-j}}{m-k-j} \\
\times \left( (m-k) z_k^{v_2} + \sum_{i=1}^{k} z_i^{v_2} \right)^{-k} dv_2
\end{array}
\right]^{-1}
\le \alpha,
$$

(23.34)

$$
w_h = \left( \frac{t_w}{y_k} \right)^{\hat{\delta}}.
$$

(23.35)

The upper prediction limit for the number of units that will fail in the time interval $[t_c, t_w]$ is given by

$$
L_{\text{upper}} = l_{\min} - k - 1,
$$

(23.36)

where

$$
l_{\min} = \min_{k < l \le m} \arg\left( \Pr\{Y_l > t_w | z^{(k)}\} \ge 1 - \alpha \right).
$$

(23.37)

# 4  Conclusion and Future Work

Prepare Prediction of an unobserved random variable is a fundamental problem in statistics. The aim of this paper is to construct lower (upper) prediction limits under parametric uncertainty that are exceeded with probability $1-\alpha$ ($\alpha$) by future observations or functions of observations. The prediction limits depend on early-failure data of the same sample from the two-parameter Weibull distribution, the shape and scale parameters of which are not known.

The methodology described here can be extended in several different directions to handle various problems that arise in practice.

We have illustrated the prediction methods for log-location-scale distributions (such as the Weibull distribution). Application to other distributions could follow directly.

The results obtained in this work can be used to solve the service problems of the following important engineering structures: (1) transportation Systems and Vehicles – aircraft, space vehicles, trains, ships; (2) civil Structures – bridges, dams, tunnels; (3) power generation—nuclear, fossil fuel and hydroelectric plants; (4) high-value manufactured products—launch systems, satellites, semiconductor and electronic equipment; (5) industrial equipment—oil and gas exploration, production and processing equipment, chemical process facilities, pulp and paper.

**Acknowledgment** This research was supported in part by Grant No. 06.1936, Grant No. 07.2036, Grant No. 09.1014, and Grant No. 09.1544 from the Latvian Council of Science and the National Institute of Mathematics and Informatics of Latvia.

# References

1. Nelson W (2000) Weibull prediction of a future number of failures. Qual Reliab Eng Int 16:23–26
2. Nordman DJ, Meeker WQ (2002) Weibull prediction for a future number of failures. Technometrics 44:15–23
3. Meeker W, Escobar L (1998) Statistical methods for reliability data. Wiley, New York
4. Rostum J (1999) Decision support tools for sustainable water network management. In: A research project supported by the European commission under the fifth framework program. http://www.unife.it
5. Nagaraja HN (1995) Prediction problems. In: Balakrishnan N, Basu AP (eds) The exponential distribution: Theory, methods & applications. Gordon and Breach, London, pp 139–163
6. Faulkenberry GD (1973) A method of obtaining prediction intervals. J Am Stat Assoc 68:433–435
7. Nechval NA, Nechval KN, Purgailis M (2011) Statistical inferences for future outcomes with applications to maintenance and reliability. In: Lecture notes in engineering and computer science: Proceedings of the world congress on engineering 2011, WCE 2011, London, 6–8 July 2011, pp 865–871
8. Cox DR (1975) Prediction Intervals and Empirical Bayes Confidence Intervals. In: Gani J (ed) Perspectives in probability and statistics. Academic, London, pp 47–55
9. Atwood CL (1984) Approximate tolerance intervals based on maximum likelihood estimates. J Am Stat Assoc 79:459–465
10. Efron B, Tibshirani RJ (1993) An introduction to the bootstrap. Chapman and Hall, New York
11. Beran R (1990) Calibrating prediction regions. J Am Stat Assoc 85:715–723
12. Kalbfleisch JD (1971) Likelihood methods of prediction. In: Godambe VP, Sprott DA (eds) Proceedings of the symposium on the foundations of statistical inference. Holt, Rinehart, and Winston, Toronto, pp 378–390
13. Thatcher AR (1964) Relationships between Bayesian and confidence limits for prediction (with discussion). J R Stat Soc, Ser B 26:176–210
14. Geisser S (1993) Predictive inference: An introduction. Chapman and Hall, New York
15. Hahn GJ, Nelson W (1973) A survey of prediction intervals and their applications. J Qual Technol 5:178–188

16. Patel JK (1989) Prediction intervals—a review, communications in statistics. Theory Meth 18:2393–2465
17. Hahn GJ, Meeker WQ (1991) Statistical intervals: A guide for practitioners. Wiley, New York
18. Nechval NA, Nechval KN (1999) Invariant embedding technique and its statistical applications. In: Conference volume of contributed papers of the 52nd session of the international statistical institute, ISI—International Statistical Institute, Finland. http://www.stat.fi/isi99/procee-dings/arkisto/varasto/nech0902.pdf
19. Nechval NA, Vasermanis EK (2004) Improved decisions in statistics. SIA "Izglitibas soli", Riga
20. Nechval NA, Berzins G, Purgailis M, Nechval KN (2008) Improved estimation of state of stochastic systems via invariant embedding technique. WSEAS Trans Math 7:141–159
21. Nechval NA, Purgailis M, Berzins G, Cikste K, Krasts J, Nechval KN (2010) Invariant embedding technique and its applications for improvement or optimization of statistical decisions. In: Al-Begain K, Fiems D, Knottenbelt W (eds) Analytical and stochastic modeling techniques and applications, vol 6148, LNCS. Springer, Berlin, pp 306–320
22. Nechval NA, Purgailis M, Cikste K, Berzins G, Rozevskis U, Nechval KN (2010) Prediction model selection and spare parts ordering policy for efficient support of maintenance and repair of equipment. In: Al-Begain K, Fiems D, Knottenbelt W (eds) Analytical and stochastic modeling techniques and applications, vol 6148, LNCS. Springer, Berlin, pp 321–338
23. Nechval NA, Purgailis M, Cikste K, Berzins G, Nechval KN (2010) Optimization of statistical decisions via an invariant embedding technique. In: Lecture notes in engineering and computer science: Proceedings of the world congress on engineering 2010, WCE 2010, London, 30 June–2 July 2010, pp 1776–1782
24. Nechval NA, Purgailis M, Cikste K, Nechval KN (2010) Planning inspections of fatigued aircraft structures via damage tolerance approach. In: Lecture notes in engineering and computer science: Proceedings of the world congress on engineering 2010, WCE 2010, London, 30 June–2 July, 2010, pp 2470–2475
25. Nechval NA, Purgailis M (2010) Improved state estimation of stochastic systems via a new technique of invariant embedding. In: Myers C (ed) Stochastic control. Sciyo, Croatia, pp 167–193
26. Nechval NA, Purgailis M, Nechval KN, Rozevskis U (2011) Optimization of prediction intervals for order statistics based on censored data. In: Lecture notes in engineering and computer science: Proceedings of the world congress on engineering 2011, WCE 2011, London, 6–8 July 2011, pp 63–69

# Chapter 24
# Vendors' Design Capabilities Enabler Towards Proton Internationalization Strategy

**Ana Sakura Zainal Abidin, Rosnah Mohd. Yusuff, Nooh Abu Bakar, Mohd. Azni Awi, Norzima Zulkifli, and Rasli Muslimen**

## 1 Introduction

The Malaysian automotive market is about to reach saturation point at 700,000 to 800,000 output units [1]. The total vehicle sales in Malaysia for 2010 has already reached 605,156 units and is expected to grow about 1% to 1.3% in the following years [2]. All national automakers are urged to penetrate overseas markets. According to a survey conducted by the American Management Association has ranked quality control and research and development (R&D)/PD are in the top important area of focus in the attempt to compete in the global market [3]. To achieve internationalization, Proton has to concentrate on developing technological capabilities and be innovative to create exportable brands [4]; it is necessary to innovate and adopt new technologies [5].

A.S.Z. Abidin (✉) • R. Muslimen
Department of Mechanical and Manufacturing Engineering, Faculty of Engineering,
Universiti Malaysia Sarawak, 94300 Kota Samarahan, Sarawak, Malaysia
e-mail: anras_126@yahoo.com; rasna_126@hotmail.com

R.M. Yusuff • N. Zulkifli
Department of Mechanical and Manufacturing Engineering, Faculty of Engineering,
Universiti Putra Malaysia, Selangor, UPM Serdang 43400, Malaysia
e-mail: rosnah@eng.upm.edu.my; zimazul@yahoo.com

N.A. Bakar
School of Graduate Studies, UTM International Campus, Jalan Semarak, Kuala Lumpur, Malaysia
e-mail: noohab@gmail.com

M.A. Awi
Group Procurement, Perusahaan Otomobil Nasional Sdn. Bhd., Batu 3, 40918 Selangor, Malaysia
e-mail: AZNIA@proton.com

S.-I. Ao and L. Gelman (eds.), *Electrical Engineering and Intelligent Systems*,
Lecture Notes in Electrical Engineering 130, DOI 10.1007/978-1-4614-2317-1_24,
© Springer Science+Business Media, LLC 2013

Manufacturers have to find new ways of increasing their competitiveness. Currently, leading automakers have moved the development process upstream into the product development (PD) stage. PD offers more opportunities for greater competitive advantage, which enables automakers to determine customer-defined value, make strategic investment, and minimize cost. In comparison the ability of the manufacturing stage is limited to ensuring quality and productivity only [6]. Many researchers have found that PD is very important as decisions made during the PD stage have a great impact on the whole production process. Changes in the process during productions would cost more and be difficult to make [6–8].

A car requires more than 20,000 parts [9], which are impossible to be manufactured by the automaker alone. Suppliers account for a large amount of the total cost of production: over half [7]; more than 60% [9]; about 75% [10]. Consequently, products supplied by vendors have a direct impact on cost, quality, technology and time to market of new products [7], as well as the final price and quality of the product [9]. A large number of suppliers are involved and this has given the automotive industry a complex supply network structure [9].

Besides, innovation never happens in isolation; it is dependent on an entire network to improve or create a new product. Network innovation happens when there is participation from several suppliers with distinct knowledge bases converging for a specific PD project [11]; . Previously automakers had a dominant role in PD, design, testing and assembly. However, to improve the PD process, these important roles have now been outsourced to suppliers [9, 13]. Outsourcing levels are different among automakers; for example, the average of European automakers outsource 50–60% which Japanese automakers have a higher percentage of 70–75% of parts and assemblies [14]. Through outsourcing, fewer parts are manufactured in-house [8]; therefore automakers can focus on their core competencies.

However, a study on the experiences of directors and project managers has found mixed results of SIPD [15]. SIPD can lead to longer development time, worse product performance and increase a PD cost. Suppliers' incapacity has been identified as one of the reasons [7]. A prior study found that Malaysian suppliers were unable to perform adequately because they lack capability [10, 16]. Many automakers, especially the Japanese, pay serious attention to their suppliers development programs [17]. The Malaysian automotive industry is also following this trend of late [10]. Proton and its suppliers have to move together to develop design capability (DC). Collaboration between automakers and suppliers in PD, like pooling number of organizations together that would create bigger assets investment, combined resources and shared knowledge, leading to "supernormal profit" [18].

These have been few studies on the Malaysian automotive industry, especially DC. This study intends to study the current situation in Proton and its vendors' PD practice and performance and investigate how vendors' DC will enable the success of Proton internalization strategy. At the same time, it will examine the appropriateness of the critical success factors (CSFs) highlighted by prior research on Malaysian DC development. This study focuses on PD activities in Proton because

Proton is the first Malaysian automaker, has 12,000 employees, and covers the whole value chain of business from design to after-sales service [19]. Tier-1 suppliers were chosen for the study because DC is an expensive investment and only tier-1 suppliers can afford it.

## 2 Literature Review

Improved PD is the best business strategy [20] for the company's far-reaching success in the marketplace. Several studies strongly recommends the importance of improving PD in automobiles, as shown in Table 24.1. To remain in business, a company needs to have effective and efficient PD, which can be gained only through a sufficient level of DC that allows product differentiation, [24]; main feature to distinguish successful business [11]; and a key to pursue competitive advantage and sustainability [20].

SIPD has been identified as an effective approach to improve PD. Tremendous benefits can be gained by the automaker through the collaboration as reported in Table 24.2. Suppliers' DC becomes vital in SIPD, as it helps to vitalize collaboration [9]; it is a pre-requisite to become successful suppliers [27, 28]; it can be to the

**Table 24.1** The importance of improving PD

| Importance of PD | Author |
| --- | --- |
| Decisions made during PD have a greater impact on every part of the organization | [20] |
| PD offers maximum fluidity and allows for maximum possible options for changes | [8] |
| Any changes have minimum risk and cost, compared to the manufacturing stage | [8] |
| Minimizes risks as every department integrated in advance | [8, 21] |
| Eliminates waste through process improvement | [6, 7] |
| Able to fulfil customers' needs | [6, 18] |
| Reduces time to market | [22] |
| Optimizes resources used and cost | [6, 15, 22, 23] |
| Improves the manufacturing processes | [22] |
| Improves product quality | [22] |
| Improves business performance | [6] |

**Table 24.2** The benefits of SIPD

| Benefits of SIPD | Author |
| --- | --- |
| Improves product performance, cost, and quality | [7] |
| Reduces time to market | [21] |
| Influences automaker's competitiveness | [9, 25] |
| Identifies suppliers as one of the important resources of the automaker | [7, 12, 26] |
| Automakers can reduce the cost of technology investment | [15] |
| Automakers can have innovative technology with minimum technological risk when costs are shared with suppliers | [7, 15] |

**Table 24.3** Relevant CSFs for PD from prior studies [33]

|    | Critical Success Factors              | Frequency of authors reporting |
|----|---------------------------------------|--------------------------------|
| 1  | Manpower                              | 16                             |
| 2  | Technology and tools                  | 16                             |
| 3  | Established processes                 | 11                             |
| 4  | Business performance results          | 10                             |
| 5  | Organization structure                | 8                              |
| 6  | External support                      | 8                              |
| 7  | Financial capability                  | 8                              |
| 8  | Proximity                             | 6                              |
| 9  | Top management commitment and support | 4                              |
| 10 | Relationship                          | 4                              |
| 11 | Market size                           | 4                              |
| 12 | Culture                               | 2                              |

suppliers' competitive advantage [7, 29–31]; it offers uniqueness as it is difficult for competitors to imitate [29]. DC is also advantagous to suppliers as it empowers the companies to fulfil current customers' needs [29]; enables the suppliers to determine the price of the product; the ability to design allows them to design according to customers' target prices [22, 32]. Thus, it facilitates the suppliers to scale up business volume [9]; allows them to expand businesses, and improve their chances of becoming an Original Design Manufacturer [9].

However, it is more challenging to develop DC among the suppliers than the automakers. A great gap is noticed between automakers' and suppliers' capability, especially in Malaysia. Most of the CSFs mentioned have been taken from research work abroad, as there were few studies conducted in Malaysia. They needed careful selection and judgement, based on the importance and suitability to the Malaysian scenario. The factors highlighted can be grouped into 12 categories as shown in Table 24.3. Basically the CSFs can be divided into two main groups, the first, specific to PD activity and the second, to other related factors that have an indirect influence on the success of PD. For the first category, there are components of PD like process, manpower, technology and tools to run the PD. The remaining factors fall under influential factors. Some factors (such as process, manpower, technology, and tools and business performance results (BPR)) have been addressed frequently by authors. Some factors get only a little attention. However, other factors that have some significant value to this research because of the similar environment and nature of the industry, the author has decided to consider these factors.

# 3 Research Methodology

To fulfill the objectives of the study, primary and secondary data were collected. For primary data, case study via in-depth interviews was employed. This is an appropriate approach for this study as only a few studies have been conducted.

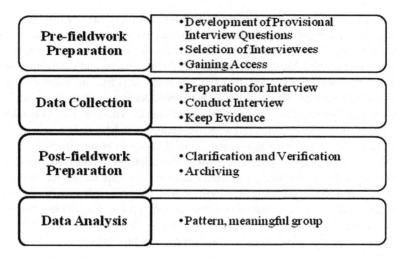

**Fig. 24.1** The interview process flow

**Table 24.4** Interviewees' profiles

| Interviewee | Position | Department |
|---|---|---|
| WA | General manager of engineering division | Engineering division |
| AN | Head of product service engineering | |
| MA | Section manager of strategic supplier management | Group procurement |
| BP | Section manager of vendor management development | |

In-depth interviews help to explore the actual scenario and the recent progress that has occured in the industry [26, 34]. The flow of the interview process adapted from [24] is shown in Fig. 24.1.

During the pre-fieldwork preparation, the design of this study was developed from literature review. Prior research provided the overview of design, significant CSFs towards successful PD activity, and collaboration between automaker and vendors, which has been used to structure the interview questions. All informants were given the same set of questions, to ensure consistency between interviews. At the same time, all the questions developed were open-ended, allowing for ample flexibility to explore new findings. Four officers from the managerial level were selected from Proton. Two were from the Engineering Division (ED), directly involved in PD and the other two interviewees were from group procurement (GP), responsible for suppliers. The background of the interviewees who were already involved in design activities, enabled the study to obtain fresh ideas and insights [34]. Proton as a buyers, enabled them to give a fair view of the performance of the entire suppliers' network. The interviewees' profile is shown in Table 24.4.

Arrangements for the interview session were set earlier, to ensure availability of the interviewees. The objective of the interview and questions for interview session were given in advance to the interviewees via e-mail. The interviews were conducted between February and March 2010 in three different sessions and the average time taken for each session was about two and half hours ($2^1/_2$ h). The interviews were conducted face-to-face. The sessions were recorded and transcribed, to ensure reliability and traceability of information [35]. In addition, each transcript once completed was sent to the respective interviewees for content validation. The interviewees corrected on any misunderstanding and clarified issues to ensure the validity and reliability of the information [35]. Interview transcripts were analyzed; direct and indirect answers were determined, and those answers that were similar were grouped together. The most appropriate terminology was used to represent the findings. Secondary data collected include a company annual reports, progress reports, press releases, electronics websites and research publications. Basically, the secondary data were used to enhance or sometimes to support primary findings.

# 4 Proton

Proton is an abbreviation for Perusahaan Otomobil Nasional, the first Malaysian automaker. Proton is a national company and was set up in 1983. Local suppliers (vendors in Malaysia) were also appointed to support the national automaker. The **Malaysian government** has implemented rules and policies to **help local companies** in the automotive industry, for instance, localization policies, mandatory deleted items (MDI), and local material content policy (LMCP) to ensure a certain percentage of local content [36]. Figure 24.2 shows the achievement of the localization program since it was launched in 1985. The trend has shown progressive increment especially since 1992 onwards, especially on the number of local parts produced. The data significantly showed that the localization program has been successfully implemented. Recently, Proton has introduced a new vendor tier system. The new system requires vendors to choose their own lower tier vendors according to the local quota as stated in the localization program, assisted by Proton. Even with the new vendor tier system, Proton still manages to pursue government localization programs.

Auto parts and component development began from the first Saga model launched. At first, localization activities concentrated on MDI. The growing number of localized parts and components has made the Malaysian automotive industry one of the key development indicators. Likely, the **National Automotive Policy (NAP)** and national car project have successfully industrialized the nation. As shown in Fig. 24.3, more parts/components have been produced since the *Iswara* model onwards.

The involvement of vendors in PD is still new to Proton. Unofficially, SIPD in Proton started since the *Wira* model, in 1993. As SIPD has shown **positive results**, the practice has been continued. In 2007, Proton managed to improve and develop a

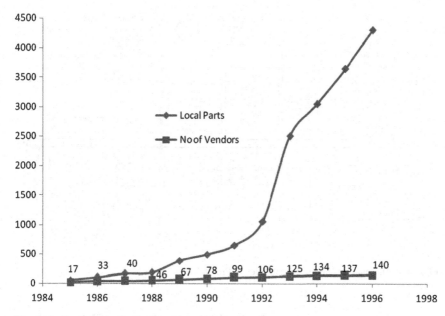

**Fig. 24.2** Proton's localization program achievement [10]

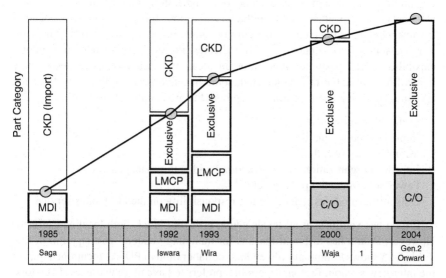

**Fig. 24.3** Evolution of auto parts and component development (source: Proton). CKD = complete knock down, MDI = mandatory deleted items, LMCP = local material content policy, c/o = corporate

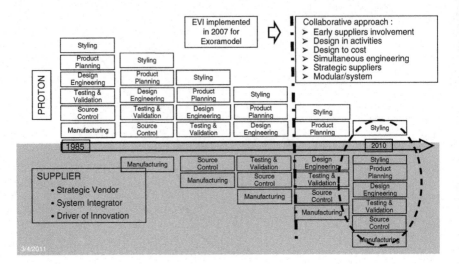

**Fig. 24.4** Sharing roles and responsibilities in Proton SIPD

proper SIPD program called as **Early Vendor Involvement (EVI)**. EVI is an expansion of the initial SIPD concept. EVI has a long lead time, which is normally 6–9 months for each project. The selected vendors worked with Proton from the beginning of the design process (concept design) until completion. The vendors' roles have now moved beyond manufacturing responsibility, as Proton has begun to involve vendors in PD (summarized in Fig. 24.4). More critical responsibilities are now outsourced to vendors, which indicates Protons' confidence in vendors' capabilities. EVI promotes innovation in ideas and technologies to design new items. From the interviews conducted, some benefits received by Proton from the EVI program have been identified. The **benefits** are summarized as follows:

- Deduced development time
- Improve product quality
- Minimal risks of failure
- Improved organization, more structured and better management
- Fewer staff allocated per project
- Outsourced modules reduce Proton responsibility on the after-sales service

Proton has **vendors**, at all **levels** white, gray and black, both local and foreign. However, Proton has different reasons for including vendors' participation in PD. Table 24.5 shows a summary of vendors' collaboration in PD. As mentioned during the interview session, Proton, if possible, prefers to have more white level vendors, so as not to relinquish control over design. However, due to limited capabilities, Proton has to outsource to vendors like power train system. Proton has outsourced many critical parts to vendors especially on the first Saga model (launched in 1985). At that time, Proton was still new and had no capability to develop high technical parts. Therefore, Proton had outsourced to foreign vendors abroad and relied on

**Table 24.5** Vendors' participation in Proton PD

| Level | Description | Specifications provided | Product complexity |
|---|---|---|---|
| White | Discussions are held with suppliers about specifications/requirements, but Proton makes all design and specifications decisions Proton provides details of part drawing and specification IPR belongs to automaker | Complete drawing | Simple parts and components e.g. BIW parts, floor and center console |
| Grey | Joint development effort between buyer and supplier May include information and technology sharing and joint decision making regarding design specifications Background IPR belongs to supplier, Foreground IPR belongs to automaker | Proton provides specification and supplier may enhance it if necessary | Simple assembly e.g. exhaust system, window regulator, control cables |
| Black | Proton has limited know-how, rely on supplier The supplier is informed of customer requirements and then is given almost complete responsibility for the purchased item Items are off shelf from supplier product range for other OEM with or without modification to suit Protons' specific needs IPR belongs to supplier | Supplier provide design and technical specifications | Complex assembly e.g. Brake system, air bag system, steering system, suspension system |

the joint venture partner, Mitsubishi Motor Corporation (MMC). Proton also encouraged local vendors to go for joint venture (JV) and technical assistance (TA) from established companies to speed up the development process.

Normally vendors from the *black box* level have a sufficient level of capability and capacity to develop a complex assembly. However, modular suppliers' roles are greater than that, and include managing lower tier vendors under them. A few capable modular suppliers are assigned to be in charge of whole modules or systems development. They are also responsible for service or warranty of the products developed. Proton **vendors' tier system** improved the structure and vendors' network management [19]. Tier-1 vendors are also known as direct vendors, working directly for Proton, responsible for part structure, especially big parts, assembly, complete subsystems or specific modules. All tier-1 vendors from *white* till *black* are involved in design, either **directly or indirectly** . Even vendors from the *white* box category, who do manufacturing work based on detailed drawing and specification, are consulted during the design stage to ensure manufacturability of the design. To support vendors' strategic development, several tier-1 vendors also act as one-stop production and develop **testing** and **R&D** facilities. Thus, it is possible to reduce Proton's responsibility, especially in managing the lower tier vendors and optimize Proton resources as well.

Since EVI is able to reduce the number Proton's of employees allocated for each project, Proton is able to optimize manpower and focus on their development. The collaboration also gives an opportunity for Proton to learn from experts, as expert vendors are already established and have **comprehensive capabilities**. At the same time, Proton also received technology transfer from **expert vendors** who are more familiar with **latest technology.** Sometimes, it is more economical for Proton to outsource rather than develop its own capability.

Modularity contributes significant advantages to the framework, as it is an efficient way to manage vendors' networks [37]. For modular practice, Proton has a different way of implementation from other automakers. As part of modular suppliers' responsibilities, proton expects vendors' manage lower tier suppliers. However, for lower tier supplier selection in Proton, the strategic decision is done by Proton. This is due to vendors' lack of **experience** in handling modular task, lack of networking to find capable lower tier vendors and in-competent **management capability** to handle bigger organization structure. Except for the well established vendors, who independently own their recognized **vendors' network** and well experienced in running lower tier vendors. Proton only needs to monitor and ensure the localization quota is followed. The accomplishment of SIPD requires close **supervision** from automakers. Proton through GP and ED are worked together to manage collaboration with vendors in PD. The success of collaborations has bolster up Proton capability to improve **lead time,** reduce **manpower** and also enhance local **capabilities.** *Waja* (launched in 2000) was the first model be developed in house took 36 months compared to *Exora* (launched in 2009)with only 18 months. *Exora* also developed by 250 local engineers, 75% manpower reduction compared to *Waja* that involved 1,000 manpower including some experts from Japan and Europe [23]. Through strategic development, Proton has gradually reduced dependency on abroad experts, replace by local ones.

# 5 Vendors

In Malaysia, vendors have different definitions based on their status. According to Companies Commission Malaysia (SSM), a local company is defined as a company that is registered and operated in Malaysia. It is divided into three categories; Bumiputra, non-Bumiputra and foreign. Bumiputra are the ethno-majority of prime ethnic Malays [16] and other indigenous ethnic groups in Malaysia [38]. The majority share (>50%) determines the status of the company. Foreign vendors are also considered local, if the majority of manpower hired are Malaysian citizens. Table 24.6 gives details of Proton vendors, according to their status, recorded in February 2010 (provided by Proton GP).

**Table 24.6** Proton vendors according to company status (source: Proton)

|  | Status* | Quantity | Percentage |  |
| --- | --- | --- | --- | --- |
| Malaysian | Bumiputra | 78 | 58% | Local |
|  | Non-Bumiputra | 75 |  |  |
| Non-Malaysian | Foreign | 57 | 22% |  |
|  | Oversea | 54 | 20% |  |
|  | Total | 264 | 100% |  |

In 2008, half of Malaysian vendors were sole suppliers to Proton [36], with 62.7% of them being SME [39]. In terms of market share, the majority are non-SME [36] catering to high technology base parts, as SME vendors have limited capabilities [40]. Through interviews, several barriers that prevent Malaysian vendors from developing DC (were identified) and can be summarized as follows:

- Limited financial capabilities that restrict technology development
- Difficulty in retaining experienced workers
- Lack of management commitment
- Limited allocation to /investment on R&D activities
- Lack of focus in the field, diversification to other businesses

However, competition and stringent demand from automakers have pushed vendors to move ahead. Since Proton has a implemented EVI, vendors also need to have a sufficient level of DC to ensure the success of the PD project. In addition, DC also brings more opportunities for Malaysian vendors to be:

- *Black box* vendors who own full capabilities to develop parts/components/ systems
- Able to design tools and equipment for manufacturing
- Proton also supports vendor DC development through other activities.
- *Guest Engineer* program is similar to what Toyota did. A number of engineers representing suppliers work together full time in Proton's R & D office on a specific project. Throughout the project, they get first-hand experience and "real base" problem training, guided by experts from Proton
- *Advance Product Quality Planning* (APQP) is a framework of procedures and techniques used to develop quality products. APQP must be audited and registered under ISO/TS 16949

Even though formal EVI was implemented only in 2007, some achievements have been recorded (shown in Tables 24.7 and 24.8). In general, Malaysian vendors have shown significant progress. Table 24.7 shows that Malaysian vendors have been awarded 40% to 50% of the total *black box* modules. However, the achievements in all aspects compared to foreign vendors showed that Malaysian vendors are still in the developing stage (based on interviews with Proton as shown in Table 24.8). When the new model *Exora*, a Proton first seven-seaters model was introduced, modules that used to be awarded to Malaysian vendors for sedan models

**Table 24.7** Summary of Proton's black box vendors (source: Proton)

| Model / Vendor \ Module | Persona (%) | | Saga (%) | | Exora (%) | |
|---|---|---|---|---|---|---|
| | $13^a$ | | $13^a$ | | 14 | |
| Overseas | 2 | 13 | 1 | 6 | 1 | 7 |
| Foreign | 7 | 47 | 7 | 44 | 8 | 57 |
| Malaysian | 6 | 40 | 8 | 50 | 5 | 36 |

[a]Total number of vendors greater than modules indicates that some modules are shared by more than one vendor

**Table 24.8** Proton vendors' performance based on interview summary

| | Foreign vendors | Malaysian vendors |
|---|---|---|
| Skills and knowledge | Well experienced experts | Successful in developing skills for low complexity parts and components |
| Facilities | Established current facilities and technology | Own basic DC, except for advance testing facilities, sometimes needs to be certified by third party |
| External support | Assisted by parent company | Need for TA to venture new technology and knowledge |
| SIPD | Able to work independently | Need close supervision |

were switched to foreign vendors. They were, for example, suspension system, braking system, ABS and instrument panel module. It seems that Malaysian vendors have not established the technology and are unable to bid with the foreign price.

# 6   Discussions

EVI has identified the Malaysian vendors with current PD activities and shared latest information, technology and facilities with the automotive industry. Proton and vendors have shown significant progress in developing DC. Gradually, Proton has managed to reduce the dependency on foreign vendors; at the same time Malaysian vendors are also improving their capabilities to replace foreign vendors [41]. Some Malaysian vendors have successfully expanded, even at the international level. Most of the tier-1 Malaysian vendors have the basic capability to create designs. However, without established DC, it is difficult to produce advanced and innovative products. For certain critical components related to the engine and transmission, Proton still depends on foreign experts [10]. The driver of the upgrading process can be countered by efficient suppliers [42]. Proton has chosen some reputable foreign vendors to join EVI projects. Their contributions undeniably have speeded up the development process [40]. The initiative of some of the Malaysian vendors to have TA or develop JV with other established multinational companies has expedited capabilities transfer and created a bigger export market.

In addition, the reviewed NAP policy emphasizes the success of the whole Malaysian automotive industry [16, 43]. NAP aims to attract more established automakers to invest in Malaysia. At the same time, NAP also encourages local participation via localization and offers more customers' proliferation, as many Malaysian vendors depended solely on Proton. This policy provides more opportunities for Malaysian vendors to receive technology transfer from foreign investors [5]. Vendors' valuable experience in working with other established automakers can improve vendors' technical proficiency and enable them to look into potential buyers other than Proton. There are a number of CSFs (highlighted in bold font) mentioned directly or indirectly in the findings. The CSFs from literature is specific to the Malaysian automotive industry, except for culture. The rank of the CSFs may not be the same as in Table 24.3 because of a different scenario. Therefore, authors have to further analyze the information and data provided, according to its importance, as *area of focus* for future DC development framework [44].

# 7  Conclusions and Future Work

Proton internationalization strategy needs supports from all vendors. International-ization has stringent standards that look for innovation and quality for each single part or product. Without a sufficient level of vendors' capabilities, the mission will be hard to accomplish. Consequently, Proton has to multiply its effort in developing its own DC and also that of other Malaysian vendors. Proton's national obligation to develop Malaysian vendors seems to delay the mission. Even though Proton has received ample privileges from the government in securing the local monopoly, Malaysia also has a commitment with other ASEAN countries towards developing a free trade area that appears to have a dateline for the development process. Data have showed that Proton and Malaysian vendors have developed significant prog-ress in DC. However, they have to move faster to catch up with other established automakers in the international arena.

In conclusion, an exploratory study of PD in the Malaysian automotive industry has been conducted successfully. The finding has contributed useful insight into the current achievement of Proton and vendors in PD. Proton and its vendors have understood the opportunity for internationalization and shared the common vision and act together on an aligned roadmap. Undeniably, vendors DC have significant potential to permit the success of Proton internationalization strategy. Therefore, a continuous, effective, and efficient development process is required to enhance vendors' DC development. The highlighted CSFs will be considered to develop a framework for DC enhancement. The CSFs have to be confirmed and ranked with other methods of study before applying them in the framework. The framework is expected to be able to enhance DC development and Proton can use the model to evaluate vendors' performance.

# References

1. Bernama (2010) Perodua hopes for less policy changes. The new straits times press (Malaysia), Berhad
2. Mahalingam E (2011) Record number of vehicles sold last year, The Star
3. Goetsch DL, Davis SB (2010) Quality management for organizational excellence: introduction to total quality, 6th edn. Pearson Education, New Jersey
4. Wad P (2009) The automobile industry of Southeast Asia: Malaysia and Thailand. J Asia Pac Econ 14(2):172–193
5. Bernama (2010) DRB-HICOM & Volkswagen pact will help make Malaysia an ASEAN auto hub. http://www.bernama.com/bernama/v5/newsbusiness.php?id=551576. Accessed 22 Dec 2010
6. Morgan JM, Liker J (2006) The Toyota product development system: integrating people, process, and technology. Productivity Press, New York
7. Handfield RB, Ragatz GL, Petersen KJ, Monczka RM (1999) Involving suppliers in new product development. Calif Manage Rev 42(1):59–82
8. Dieter GE (2000) Engineering design—a materials and processing approach, 3rd edn. McGraw-Hill International Editions, New York
9. Oh J, Rhee SK (2008) The influence of supplier capabilities and technology uncertainty on manufacturer–supplier collaboration: a study of the Korean automotive industry. Int J Oper Prod Manag 28(6):490–517
10. Abdullah R, Lall MK, Tatsuo K (2008) Supplier development framework in the Malaysian automotive industry: proton's experience. Int J Econom Manag 2(1):29–58
11. Schiele H (2006) How to distinguish innovative suppliers? Identifying innovative suppliers as new task for purchasing. Ind Market Manag 35(8):925–935
12. Cousins PD, Lawson B, Petersen KJ, Handfield RB (2011) Breakthrough scanning, supplier knowledge exchange, and new product development performance. J Prod Innov Manag 28(6):930–942
13. Doran D (2005) Supplying on a modular basis: an examination of strategic issues. Int J Phys Distrib Logist Manag 35(9):654–663
14. Lettice F, Wyatt C, Evans S (2010) Buyer–supplier partnerships during product design and development in the global automotive sector: who invests, in what and when? Int J Prod Econ 127(2):309–319
15. Wagner SM, Hoegl M (2006) Involving suppliers in product development: Insights from R&D directors and project managers. Ind Market Manag 35:936–943
16. Wad P, Govindaraju VGRC (2011) Automotive industry in Malaysia: an assessment of its development. Int J Automot Technol Manag 11(2):152–171
17. Sako M (2004) Supplier development at Honda, Nissan and Toyota: comparative case studies of organizational capability enhancement. Ind Corp Change 13(2):281–308
18. Handfield RB, Bechtel C (2002) The role of trust and relationship structure in improving supply chain responsiveness. Ind Market Manag 31:367–382
19. Hashim MA, Tahir SZASM (2008) Proton Annual Report 2008. Proton Holding Bhd. Shah Alam, Malaysia
20. Stevenson WJ (2009) Operations management, 10th edn. McGraw Hill, UK
21. Handfield RB, Lawson B (2007) Integrating suppliers into new product development. Ind Res Inst 50:44–51
22. Afonso P, Nenus M, Paisana A, Braga A (2008) The influence of time to market and target costing in the new product development success. Int J Prod Econ 115:559–568
23. Jilan AZ (2009) Tempoh 18 bulan bangunkan Exora. Utusan, Kuala Lumpur
24. Yahaya SY (2008) New product development decision making process at selected technology based organizations in Malaysia. PhD Thesis, Universiti Teknologi Malaysia

25. Bennett D, Klug F (2009) Automotive supplier integration from automotive supplier community to modular consortium. In: Proceedings of the 14th annual logistics research network conference, Cardiff, 9–11 Sept 2009
26. Enkel E, Gassmann O (2010) Creative imitation: Exploring the case of cross-industry innovation. R&D management 40(3):256–270
27. Sturgeon TJ, Biesebroeck JV (2010) Effects of the crisis on the automotive industry in developing countries: A global value chain perspective. The World Bank, pp 1–31
28. Wagner S (2008) Supplier traits for better customer firm innovation performance. Institute for the Study of Business Markets (ISBM), Pennsylvania, p 38
29. Teece DJ (2007) Explicating dynamic capabilities: the nature and microfoundations of (sustainable) enterprise performance. Strateg Manag J 28(13):1319–1350
30. Trappey AJC, Hsiao DW (2008) Applying collaborative design and modularized assembly for automotive ODM supply chain integration. Comp Ind 59(2–3):277–287
31. Bonjour E, Micaelli J-P (2009) Design core competence diagnosis: a case from the automotive industry. IEEE Trans Eng Manag 57(2):323–337
32. De Toni A, Nassimbeni G (2001) A method for the evaluation of suppliers' co-design effort. Int J Prod Econ 72(2):169–180
33. Abidin ASZ, Yusuf RM, Ismail MY, Muslimen R (2010) Identification of critical success factors in developing design capabilities for Malaysian vendors in automotive industry. King Mongkut's University of Technology North Bangkok Press, Bangkok, pp 151–156
34. Kotabe M, Parente R, Murray JY (2007) Antecedents and outcomes of modular production in the Brazilian automobile industry: a grounded theory approach. J Int Bus Stud 38:84–106
35. Binder M (2008) The importance of collaborative frontloading in automotive supply networks. J Manufact Technol Manag 19(3):315–331
36. Mohamad N (2008) Parts suppliers involvement in customer's product development activities. PhD Thesis, Universiti Teknologi Malaysia
37. Frigant V (2007) Between internationalisation and proximity: the internationalisation process of automotive first tier suppliers. Research group on theoretical and applied economics: GREThA UMR CNRS 5113. Pessac, France
38. SRM (2005) Doing business in Malaysia: maritime defence and security; marine related industries, Sea Resources Management. pp 1–58
39. MITI (2004) Signing ceremony of MOU between SMIDEC, AFM, JAMA & JAPIA on the technical experts programme for the automotive industry. pp 1–5
40. Rosli M, Kari F (2008) Malaysia's national automotive policy and the performance of proton's foreign and local vendors. Asia Pac Bus Rev 14(1):103–118
41. Abidin ASZ, Yusuff RM, Awi MA, Zulkifli N, Muslimen R, Assessing proton and vendors design capabilities towards internationalization. In: Lecture notes in engineering and computer science: proceedings of the World congress on engineering 2011, WCE 2011, London, pp 728–733
42. Wad P (2008) The development of automotive parts suppliers in Korea and Malaysia: A global value chain perspective. Asia Pac Bus Rev 14(1):47–64
43. MITI (2009) Review of national automotive policy. In: Book review of national automotive policy, MITI, pp 1–13, Kuala Lumpur
44. Oh Y, Suh E-h, Hong J, Hwang H (2009) A feasibility test model for new telecom service development using MCDM method: a case study of video telephone service in Korea. Exp Syst Appl 2009(36):6375–6388

# Chapter 25
# Effects of Leaked Exhaust System on Fuel Consumption Rate of an Automobile

Peter Kayode Oke and Buliaminu Kareem

## 1 Introduction

A car exhaust system consists of a series of pipes that links the burnt exhaust gasses in the engine cylinder through an exhaust manifold, catalytic converter, silencer, and muffler to the atmosphere [1–4]. The exhaust systems consist of tubing, which are used for discharging or expelling burnt gasses or steam through the help of a controlled combustion taking place inside the engine cylinder [5, 6]. The major components used in a typical automobile exhaust system are: exhaust manifold, resonator, catalytic converter, exhaust pipe, muffler, tail pipe, "Y" pipe and ball flanges [7, 8]. The products of combustion from internal combustion engines contain several constituents that are considered hazardous to human health, including CO, UHCs, $NO_x$, and particulates (from diesel engines). In a bid to reduce the effects caused by these gasses, exhaust system's components are designed to provide suitable and effective exhaust flow, reduction of noise and emission levels, and conversion of the gasses to water vapor and carbon (iv) oxide at the exhaust [9, 10]. The exhaust system is well designed to sustain engine performance. In an attempt to reduce these emissions, several devices have been developed to arrest the dangerous emissions [9]. A thermal reactor is seldom used to oxidize UHC and CO [2]. Catalytic converters utilize a catalyst, typically a noble metal such as platinum, rhodium, or palladium, to promote reactions at lower temperatures. In all cases, an arrangement which requires that the engine be operated with a rich mixture which decreases fuel economy is emphasized.

P.K. Oke (✉) • B. Kareem
Department of Mechanical Engineering, The Federal University of Technology,
P.M.B. 704, Akure, Ondo State, Nigeria
e-mail: okekayode@yahoo.com; karbil2002@yahoo.com

S.-I. Ao and L. Gelman (eds.), *Electrical Engineering and Intelligent Systems*,
Lecture Notes in Electrical Engineering 130, DOI 10.1007/978-1-4614-2317-1_25,
© Springer Science+Business Media, LLC 2013

It is worthy of note that, first, the exhaust gasses or moisture must be at or above a certain temperature [1]. This is why the converter is placed close to the engine. Second, there must be a certain minimum surface area of catalyst for the gasses to come in contact with. This is the reason for the honeycomb design. It provides a large surface area in a small space. Third, the ratio of exhaust gas to air must be maintained within very rigid limits [2, 3]. These limits are maintained by placing a special sensor in the exhaust just before the converter. This sensor detects variations in the ratio and signals the fuel supply system to increase or decrease the amount of fuel being supplied to the engine.

The dominant factor in automobile activity operational efficiency and profitability is maintenance philosophy [11, 12]. Engine designers have so far done great works in ensuring that exhaust gasses are reduced to the minimal, and as much, harmless, bearing in mind fuel efficiency, engine's and engine components' life [13]. Fuel economy, exhaust emission, and engine noise have become important parameters not only for engine competitiveness, but also are subjected to legislation, and it is becoming more severe every few years. Over the years, many have argued about the actual effect(s) of a leak in the exhaust system on the fuel consumption rate of an automobile system. This research work seeks to provide reliable and technical reasons on the point of discourse by developing a model to evaluate the fuel consumption rate with respect to leakage diameter and length on the exhaust system.

However, many research efforts have been carried out to measure emission and fuel consumption rates in the exhaust system and many models have been developed in this direction. These include:

## 1.1  Average-Speed Models

Average-speed emission functions for road vehicles are widely applied in regional and national vehicular operational inventories, but are currently used in a large proportion of local air pollution prediction models [4, 7]. Average-speed models are based upon the principle that the average emission factor for a certain pollutant and a given type of vehicle varies according to the average speed during a trip [7]. The emission factor is measured in grams per vehicle-kilometer (g/km). The continuous average-speed emission function is fitted to the emission factors measured for several vehicles over a range of driving cycles (with each cycle representing a specific type of driving) including stops, starts, accelerations, and decelerations [7]. The measured data for the identified exhaust emission components are plotted, modeled and then compared to assess their variability.

A number of factors have contributed to the widespread use of average-speed approach. It is one of the oldest approaches, the models are comparatively easy to use, and there is a reasonably close correspondence between the required model inputs and the output data made available to the end users. However, there are a number of limitations associated with average-speed models. These are: (1) trips having different vehicle operational characteristics and emission levels, may have the same average speed; (2) all the types of operation associated with a given average speed cannot be accounted for by the use of a single emission factor; (3) at higher average speeds, the possible variations in vehicle operation are limited, while it is greater at low average speed. Besides, the shape of an average-speed function is not fundamental, but depends on, among other factors, the types of cycle used in development of the functions [4]. Each cycle used in the development of the functions represents a given real-world driving condition, while actual distribution of these driving conditions is not taken into consideration. Average-speed models do not allow for detailed spatial resolution in emission predictions, and this is an important drawback in dispersion modeling.

One of the limitations of average-speed models mentioned earlier was the inability to account for the ranges of vehicular operation and emission behaviors, which can be observed for a given average speed; this has made the concept of "cycle dynamics" useful for emission model developers [11]. The term "vehicular operation" refers to a wide range of parameters, which describe the way in which a driver controls a vehicle (average speed, maximum speed, acceleration pattern, gear-change pattern), as well as the way in which the vehicle responds (engine speed, engine load).

## *1.2  Multiple Linear Regression Models (VERSIT + Model)*

The VERSIT+ model [2] employs a weighted-least-square multiple regression approach to model emissions, based on tests on a large number of vehicles over more than 50 different driving cycles. Within the model, each driving cycle used is characterized by a large number of descriptive parameters (average speed, number of stops) and their derivatives. For each pollutant and vehicle category a regression model is fitted to the average emission values over the various driving cycles, resulting in the determination of the descriptive variables, which are the best predictors of emissions. A weighting scale is also applied to each emission value, based on the number of vehicles tested over each cycle and the inter-dependence of cycle variables. The VERSIT+ model requires a driving pattern as the input, from which it calculates the same range of descriptive variables and estimates emissions based on the regression results. As with the other models requiring a driving pattern as the input, the use of the model is currently restricted to a comparatively small number of users because of complexity in determining the actual physical variables involved in it.

## 1.3 Instantaneous Models

The aim of instantaneous emission modeling is to map emission measurements from tests on a chassis dynamometer or an engine test bed in a neutral way [7]. In theory, the advantages of instantaneous models include the following: emissions can be calculated for any vehicle operation profile specified by the model user, and new emission factors can be generated without the need for further testing; and some instantaneous models, especially the older ones, relate fuel consumption and/ or emissions to vehicle speed and acceleration during a driving cycle. Other models use some description of the engine power requirement. However, it must be noted that there are a number of fundamental problems associated with the older genera- tion of instantaneous models. It is extremely difficult to measure emissions on a continuous basis with a high degree of precision, and then it is not straightforward to allocate those emission values to the correct operating conditions [11]. It is stated in [10] that, during measurement in the laboratory, an emission signal is dynamically delayed and smoothed, and this makes it difficult to align the emissions signal with the vehicle operating conditions. Such distortions have not been fully taken into account in instantaneous models until recently. In order to apply instantaneous models, detailed and precise measurements of vehicular oper- ation and location are required, which may be difficult to attain by many public model users. Consequently, the use of instantaneous models has largely been restricted to the research community.

## 1.4 Passenger Car and Heavy-Duty Emission Model

Is a model capable of accurately simulating emission factors for all types of vehicles over any driving pattern, vehicle load, and gradient. The resulting tool, passenger car and heavy-duty emission model (PHEM), estimates fuel consumption and emissions based on the instantaneous engine power demand and engine speed during a driving pattern specified by the user [7]. The PHEM model thus has the capability of providing suitable emission resolution for use with micro-simulation traffic models. The main inputs are a user-defined driving pattern and a file describing vehicle characteristics. For every second of the driving pattern, PHEM calculates the actual engine power demand based upon vehicle driving resistances and transmission losses, and calculates the actual engine speed based upon trans- mission ratios and a gear-shift model. The engine power and speed are then used to reference the appropriate emission (and fuel consumption) values from steady-state engine maps. The emission behavior over transient driving patterns is then taken into consideration by "transient correction functions," which adjust the second-by-second steady-state emission values according to parameters describing the dynamics of the driving pattern. The outputs from the model are engine power, engine speed, fuel consumption, and emissions of CO, $CO_2$, HC, $NO_x$, and PM every second, as well as average values for the entire driving pattern.

## 1.5   Comprehensive Modal Emission Model

Barth et al. [9] describe the development of a "comprehensive modal emissions model" (CMEM). The model is capable of predicting second-by-second exhaust (and engine-out) emissions and fuel consumption, and is comprehensive in the sense that it is able to predict emissions for a wide range of vehicle and technology categories, and in various states of condition (properly functioning, deteriorated, malfunctioning). The main purpose of CMEM is to predict vehicle exhaust emissions associated with different modes of vehicle operation such as idle, cruise, acceleration, and deceleration. The model is more detailed than others; it takes into account engine power, vehicle operation including variable starting conditions (cold-start, warm start), and off-cycle driving.

CMEM uses a "physical power-demand" modal modeling approach based on a "parameterized analytical representation of emissions production" [9]. That is the production of emissions is broken down into components, which correspond to different physical processes, and each component is then modeled separately using various parameters which are characteristics of the process. These parameters vary according to the vehicle type, engine, and emission technology. The majority of the parameters are known (vehicle mass, engine size, aerodynamic drag coefficient), but other key parameters are deduced from a test program. Using this type of modeling approach entails the establishment of models for the different engine and emission-control technologies in the fleet of vehicles. Once these models have been established, it is necessary to identify the key parameters in each component of the models for characterizing vehicle operation and emissions production. A critical component of the approach is that emission-control malfunction and deterioration are explicitly modeled. The correct modeling of high-emitting vehicles is also an important part of the approach. In order to predict emission rates, the next step is to combine the models with vehicle operating parameters that are characteristics of real-world driving, including environmental factors (ambient temperature and air density) as well as dynamic factors (commanded acceleration, road gradient and the use of auxiliaries (air conditioning, electric loads)) [9].

The complete model is composed of two groups of input: (1) input operating variables; and (2) model parameters. There are also four operating conditions in the model: (1) variable soak time start; (2) stoichiometric operation; (3) enrichment; and (4) "enleanment." The model determines in which condition the vehicle is operating at a given moment by comparing the vehicle power demand with threshold values [9].

From the literature, it is clear that the identified emission and fuel consumption models take into account the various factors affecting emissions and fuel consumption rate in automobile. It is further established that these factors have affected vehicular emissions and fuel consumption in varying degree. However, efforts on fuel consumption have since continued but still the effects of a leakage on any part of the exhaust system in relation with fuel consumption rate have not been investigated. Sequel to this, effects of leaks on the motor vehicle, specifically on fuel consumption rate shall be fully investigated, and a mathematical model shall be formulated. The rest of the paper is presented as follows: methodology adopted for the research is present in Sect. 2; results and discussion is in Sect. 3; while Sect. 4 hosts conclusion and future work.

## 2 Methodology

A vehicle with a new exhaust system (comprising of resonator, catalytic converter, muffler and its pipes) was used for the research. Before the commencement of the research, the vehicle was serviced; also, the complete exhaust system of the vehicle was replaced with the new one to ensure a good condition and a reliable result during the experiment. During the period of the experiment, the vehicle was stationary and the engine allowed operating in slow running mode in order to eliminate the effects of speed, acceleration, drive pattern, and cruise control on the fuel consumption rate. The following items were used during the course of the experiment; two hoses (with diameter 1.5 cm), two tightening rings, a measuring can, veneer caliper, a stop watch, and 50 l of fuel. The hose from the fuel tank was connected to the external measuring can with the 1.5 cm diameter hose with two tightening ring clip. The external fuel tank was filled with 1 l of fuel before the commencement of the experiment. Four tests were conducted with two trials per test on the exhaust pipe at two different times: first, the effect of varying the diameter of the leak on fuel consumption rate; second, the effect of varying the length of exhaust system at which the leak occurs on fuel consumption rate.

The following describe the various steps taken while carrying out the experiment (the vehicle used is a Honda CRV jeep).

*Step 1*: The vehicle was made to stay in a fixed position. The default exhaust pipe of the vehicle was removed and replaced with a new one.

*Step 2*: The pipe leading to the fuel tank was disconnected after the fuel pump. The pipe from the fuel pump was then extended with the first hose, using one of the tightening rings.

*Step 3*: The fuel return pipe was disconnected from the injector. The second hose was then connected to the injector at the fuel return outlet, using the other tightening ring.

*Step 4*: The measuring can was used as the external tank. A little quantity of fuel was poured into the external tank and with the intake and return hoses put into the external tank; the engine was allowed to operate under slow running for 45 s. This was done so that the intake and return hoses will retain some amount fuel in order that our results may not be affected.

*Step 5*: The overall length of the complete exhaust system was measured, and its value was recorded as $L_0$.

**Fig. 25.1** A complete
exhaust system

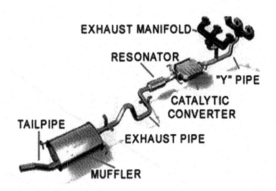

*Step 6*: The measuring can was filled with 1 l of fuel and the engine was made to start under slow running without any puncture on the exhaust pipe; the time taken to use up the 1 l of fuel was taken (using a stop watch) and its value recorded as $t_0$.

*Step 7*: Step 6 was repeated for leak diameters 5, 10, 15, and 20 mm on the following locations of the exhaust system: between the exhaust manifold and catalytic converter (43.7 cm from the exhaust manifold outlet); between the catalytic converter and silencer (138.40 cm from exhaust manifold); very close to the silencer outlet (233.70 cm from exhaust manifold), and muffler mouth (355.60 cm from exhaust manifold)

*Step 8*: The leaks were repaired after every puncture for each location on the exhaust pipe using oxy-acetylene gas welding process.

*Step 9*: Steps 6–8 were carried out again and their average values were used.

During the course of this experiment, it was ensured that: the vehicle used was serviced shortly before the commencement of the experiment; the correct quantity of fuel was used and the accurate time taken; the engine was made to run at a constant speed; and the leaks were properly repaired after each exercise. The configuration of the exhaust system used for the experiment is shown in Fig. 25.1.

## 3   Results and Discussion

The results obtained from four trials experimentation on exhaust system of overall length, $L_0$ of 419.10 cm are shown by Fig. 25.2. A leak/hole in an automobile exhaust system affects not only the health of its driver and passengers but also the fuel consumption rate and the engine performance. From the experiment and the results obtained from the experiment, the following deductions can be made: (1) the rate at which fuel is consumed increases with the hole diameter, regardless of its location on the exhaust system. That is, the rate of fuel consumption varies directly (although this might not be a proportionate variance) with the hole diameter; (2) considering the entire exhaust assembly, the region of the catalytic converter

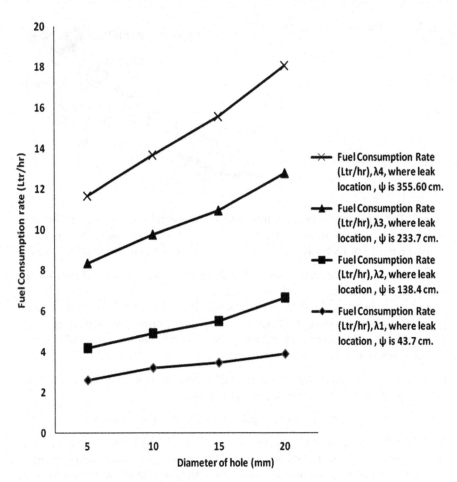

**Fig. 25.2** Fuel consumption rate with varying leak location on the manifold

consumes fuel the most. The catalytic converter is responsible for the increased fuel consumption rate at point "2." This is because it uses oxidation catalyst made up of ceramic beads coated with platinum to reduce HC and CO emissions. Due to catalytic action, the converter takes more fuel (for burning to achieve the desired essence); to convert HC, CO, and other pollutants to water vapor and $CO_2$; (3) the rate of fuel consumption decreases with the length of the leak from the exhaust manifold; and (4) as the length and hole approaches the muffler, noise level reduces.

Since this research considers two major parameters-diameter and length of leak-hole with respect to a leaking exhaust system, the general effects of hole (diameter) and length variations are as follows: (1) an increase in back pressure-design factor designers has been battling with for years—trying to reduce it to its barest minimum; (2) for vehicles using fuel injectors, the leak alters or interferes with the oxygen sensor reading. Thereby sending a wrong reading to the ACS

**Table 25.1** Experimental results for modeling

| Diameter of leak, $\alpha$, mm | 5.00 | 10.00 | 15.00 | 20.00 |
|---|---|---|---|---|
| Fuel consumption rate, $\lambda_1$, where leak location , $\psi$ is 43.7 cm | 2.60 | 3.20 | 3.46 | 3.89 |
| Fuel consumption rate, $\lambda_2$, where leak location , $\psi$ is 138.4 cm | 1.59 | 1.69 | 2.04 | 2.78 |
| Fuel consumption rate, $\lambda_3$, where leak location , $\psi$ is 233.7 cm | 4.15 | 4.86 | 5.43 | 6.10 |
| Fuel consumption rate, $\lambda_4$, where leak location , $\psi$ is 355.60 cm | 3.31 | 3.92 | 4.64 | 5.29 |

(automatic control system)—this results in inefficient combustion, poor fuelling and poor power; (3) enhances catalytic converter damage—which is quite expensive to replace; (4) in a more severe case, may cause backfire; and (5) more fuel is consumed. The issue of length or location is also paramount. For instance, if the hole is before the sensor, it will affect the reading. Also if it is before the catalytic converter, it could damage it, else no effect; if it is on the muffler-noisier exhausts results.

The summary of the results of the four trials is shown in Table 25.1. The outcomes are modeled using multiple linear regression analysis (25.1) with the following parameters [12–14]:

$\alpha$ = the diameter of the hole/leak on the exhaust system
$\psi$ = the location (length from the exhaust manifold outlet) where the hole/leak occurs
$\lambda$ = the fuel consumption rate
$\beta$ = coefficient of entity
$i$ = the $i$th terms in each trial
$n$ = number of terms being considered

$$n\beta_0 + \beta_1 \sum_{i=1}^{n} \alpha_i + \beta_2 \sum_{i=1}^{n} \psi_i = \sum_{i=1}^{n} \lambda_i \tag{25.1}$$

The linear multiple regression analysis results for the first, second, third, and fourth trials are, respectively modeled as,

$$\lambda_1 = 2.190000 + 0.084000\alpha - 0.00000282\psi \tag{25.2}$$

$$\lambda_2 = 0.920000 + 0.0894000\alpha + 0.00000578\psi \tag{25.3}$$

$$\lambda_3 = 3.510000 + 0.128400\alpha - 0.00000414\psi \tag{25.4}$$

$$\lambda_4 = 2.630000 + 0.135000\alpha + 0.00000811\psi \tag{25.5}$$

From the models, fuel consumption rates can normally be predicted ($R^2 \geq 0.9$) with a given leakage diameter and location on the exhaust system.

# 4   Conclusion and Future Work

From the research work carried out, it has been established that the rate of fuel consumption increases, approximately linearly, with increase in diameter of hole on the exhaust system; also that the location of leakage has a somewhat negligible effect on fuel consumption rate. Therefore, leakage location can only be considered for design purposes to maintain design accuracy. The knowledge of the effect of a leaking exhaust system on the rate of fuel consumption will help car owners and drivers to make better choices of maintenance policies for their exhaust system. The modeling results would help the vehicle owners in monitoring and controlling their fuel expenses. Vehicle owners could say, the hole is here or there, or on this or that, so it can still be tolerated. The models would also help vehicle designers to look into some design considerations at such points where the effect of leak is disastrous. This will promote better fuel economy and cleaner exhaust output. Having known the size of leaks and the locations, the rate at which fuel is consumed can be quantified. The study may be extended to other exhaust systems of automobile to verify the efficacy of the outcomes of this research.

**Acknowledgment** The authors thank the management of the Federal University of Technology, Akure, Nigeria for providing an enabling environment for carrying out this study.

# References

1. Hung WT, Tong HY, Cheung CS (2005) A modal approach to vehicular emissions and fuel consumption model development. J Air Waste Manag Assoc 55:1431–1440
2. Taylor MA, Young TM (1996). Developing a set of fuel consumption and emissions models for use in traffic network modelling. In: Proceedings of the 13th international symposium on transportation and traffic theory, Lyon, France, July 1996, pp 24–26
3. Challen BJ (2004) Vehicle noise and vibration: recent engineering developments (Pt series). SAE Society of Automotive Engineers – Surface Vehicle Emissions Standards Manual, vol 93. SAE, Warrendale, 207 pp
4. Roumégoux JP, André M, Vidon R, Perret P, Tassel P (2008) Fuel consumption and CO2 emission from the auxiliary equipment: air conditioning and alternator. In French, Bron, France: INRETS, report LTE0428, 28 pp
5. Obert EF (1973) Internal combustion engines and air pollution, 3rd edn. Harper & Row, New York, pp 97–106, 314–317
6. Owen K, Coley T (1995) Automotive fuels reference book, 2nd edn. Society of Automotive Engineers, Inc., Warrendale
7. Huai T, Shah SD, Miller JW, Younglove T, Chernich D, Ayala A (2006) Analysis of heavy-duty diesel truck activity and emissions data. Atmos Environm 40:2333–2344
8. Information on http://www.wikianswers.com/how_will_a_hole_in_the_exhaust_affect_fuel_-consumption. Accessed 12 Jan 2011
9. Barth M, Younglove T, Scora G, Levine C, Ross M, Wenzel T (2007) Comprehensive Modal Emissions Model (CMEM). version 2.02: User's Guide. University of California. Riverside Center for Environmental Research and Technology

10. Henein NA (1972) Emissions from combustion engines and their control. Ann Arbor Science Publishers, Ann Arbor
11. Highways Agency, Scottish Executive Development Department, Welsh Assembly Government and the Department for Regional Development Northern Ireland, Design Manual for Roads and Bridges, Volume 11, Section 3, Part 1, Air Quality, Highways Agency, Scottish Industry Department, The Welsh office and the Department of Environment for Northern Ireland, Stationary Office London, 2007
12. Information on http://peugeot.mainspot.net/fault_find/index.shtml. Accessed 12 Jan 2011
13. Aderoba AA (1995) Tools of engineering management, Engineering project management, vol 1. Besade Publishing Press, Nigeria
14. Oke PK, Kareem B, Alayande OE (2011) Fuel consumption modeling of an automobile with leaked exhaust system. Lecture notes in engineering and computer science. In: Proceedings of the world congress on engineering, WCE 2011, London, UK, 6–8 July 2011, pp. 909–912

# Chapter 26
# Designing a Robust Post-Sales Reverse Logistics Network

Ehsan Nikbakhsh, Majid Eskandarpour, and Seyed Hessameddin Zegordi

## 1 Introduction

During the past two decades, green supply chain management (GrSCM) has attracted attention of academics and professionals in logistics and supply chain management. This is highly due to governmental regulations [1], economical benefits of green supply chains [2], and customers' awareness and nongovernmental organizations [3]. From an operations management perspective, GrSCM models consider flows from final customers back to different supply chain members such as retailers, collection centers, manufacturers, and disposal centers. One of the first important steps in greening a supply chain is to consider environmental/ecological impacts during logistics network design stage. This problem, known as the reverse logistics network design problem, is comprised of three main decisions: number and location of facilities specific to a reverse logistics system (e.g., collection centers, recovery facilities, repair facilities, and disposal centers), capacities of facilities, and flows of material. A well-designed reverse logistics network can provide cost savings in reverse logistics operations, help retaining current customers, and attract potential customers [4].

In recent years, various researchers have studied the reverse logistics network design problem. Recent examples of these studies are:

- A multiproduct 4-tier reverse logistics network design model [5, 6]
- A two-stage stochastic programming model for a 3-tier reverse logistics network design problem with stochastic demand and supply [7]
- An MINLP model for the two-echelon reverse logistics network design [8]

E. Nikbakhsh • M. Eskandarpour • S.H. Zegordi (✉)
Department of Industrial Engineering, Faculty of Engineering, Tarbiat Modares University,
P.O. Box: 14117-13116, Tehran, Iran
e-mail: nikbakhsh@modares.ac.ir; eskandarpour@modares.ac.ir; zegordi@modares.ac.ir

S.-I. Ao and L. Gelman (eds.), *Electrical Engineering and Intelligent Systems*,
Lecture Notes in Electrical Engineering 130, DOI 10.1007/978-1-4614-2317-1_26,
© Springer Science+Business Media, LLC 2013

- 3-tier and 4-tier bi-objective post-sales reverse logistics network design models [9, 10]
- An integrated remanufacturing reverse logistic network design model [11]
- A 3-tier reverse logistics network design model while determining prices of financial incentives [12]
- A single echelon reverse logistics network design model to maximize profits obtained from returned products by customers [13]
- A single echelon reverse logistics network design model for locating hybrid distribution-collection facilities and material flows [2]
- A dynamic stochastic reverse logistics network design model [14]
- A dynamic 7-tier reverse logistics network design model for maximizing profits of a truck tire-remanufacturing network [15]

To the best of authors' knowledge, despite various applications of post-sales networks in real-world situations, the design of post-sales reverse logistics network has received little attention from the researchers [9, 10]. Therefore, in this study, the robust closed-loop reverse logistics network design problem is considered for a third party logistics service provider offering post-sales services for multiple products belonging to different manufacturers. Multiple decisions including location of repair facilities, allocation of repair equipments, and material flows between different tiers of the reverse logistics system are considered. In this study, amounts of defective product returns are assumed uncertain that is compatible with characteristics of reverse logistics systems in which amounts of product returns are usually unknown beforehand. In addition, various assumptions such as flows of spare parts and new products in the network, limited capacities of plants and repair facilities, and the limit on the maximum number of repair equipments assignable to each repair facility are considered. These assumptions are well matched with characteristics of post-sales service providers for electronics products in which various products designs are similar while the used components are different. For dealing with this problem, a robust bi-objective mixed integer linear programming model is proposed. In addition, the ε-constraint method is used for obtaining a list of Pareto-optimal solutions for the proposed model.

This paper is organized as follows. In Sect. 2, the research problem is defined. The deterministic mathematical model of the research problem, the robust counterpart model, and ε-constraint method are presented in Sect. 3. In Sect. 4, a numerical example is solved for demonstrating the application of the proposed model. Finally, the conclusion and future research directions of this study are given in Sect. 5.

## 2  Problem Definition

The 4-tier post-sale reverse logistics network presented in this study consists of production plants, repair facilities, collection centers, and disposal centers (Fig. 26.1). A third party logistics service provider (3PL) is responsible for providing

**Fig. 26.1** The research
problem post-sale reverse
logistics network

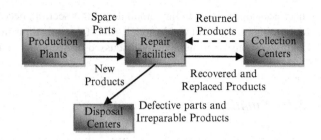

the post-sales logistical operations. The 3PL uses its distribution centers and local warehouses as repair facilities and collection centers, respectively.

In this post-sale network, defective products are returned to the collection centers by the customers. It is assumed that amounts of returned products by customers are uncertain parameters with known nominal and shift values. Then the returned products are shipped to repair facilities for initial inspection and repair. The inspection unit in each repair facility is responsible for determining whether a returned product is irreparable or repairable. Irreparable returned products are sent to disposal centers for disposing and the respective customers will be provided with new products as replacements. Repairable returned products are sent to repair facilities for repairing in which the defective parts of repairable products are replaced with necessary spare parts. Then the repaired returned products are shipped back to the collection centers for delivering to the customers. In addition, the defective parts are shipped to disposal centers for disposal. Finally, the production plants are responsible for providing spare parts and new products for repairing returned products and replacing irreparable returned products, respectively. The material flows in the aforementioned network are demonstrated in Fig. 26.1.

In this network, a limited number of equipments with limited repairing capacity are available for repairing the returned products, namely $NT_{max}$. In other words, at most, $NT_{max}$ equipments can be allocated to the repair facilities and each one of equipments can repair repairable returned products for a limited time, namely $b$. Finally, each candidate repair facility can accommodate a limited number of equipments, namely $N_j$.

## 3   Proposed Mathematical Model

Considering the problem description given in Sect. 2, the purpose of the proposed model is the determination of repairing equipment assignment to candidate repair facilities and material flows between 1) collection centers and repair facilities, 2) repair facilities and production plants, and 3) repair facilities and disposal centers in order to minimize total fixed costs and transportation costs as well as total tardiness of shipping returned products back to collection centers after the necessary operations. In the remainder of this section, necessary notation, proposed deterministic mathematical model, robust optimization model used in this study, and the ε-constraint method are discussed in Sects. 3.1–3.4, respectively.

## 3.1 Notation

- $j$: Index of candidate repair facilities, $j = 1, \ldots, J$
- $i$: Index of collection centers, $i = 1, \ldots, I$
- $h$: Index of plants, $h = 1, \ldots, H$
- $l$: Index of disposal centers, $l = 1, \ldots, L$
- $k$: Index of products, $k = 1, \ldots, K$
- $b$: An equipment repairing capacity expressed in time units
- $NT_{max}$: Maximum number of available equipments
- $N_j$: Maximum number of equipments that can be allocated to $j$th repair facility
- $\theta_{kj}$: Maximum number of $k$th product type that can be assigned to $j$th repair facility
- $a_{ik}$: Number of $k$th product type returned from $i$th collection center to repair facilities
- $rt_k$: Amount of time required for repairing a $k$th product type at any repair facility
- $\gamma_k$: Percentage of irreparable products for $k$th product type
- $t_{ij}$: Time required for a round trip between $i$th collection center and $j$th repair facility
- $\tau$: Maximum allowed time for returning repaired/new products to the collection sites
- $\sigma_{hk}$: Spare parts capacity of $h$th plant for $k$th product type
- $cr_{ij}$: Cost of shipping a returned product from $i$th collection center to $j$th repair facility and shipping it back to $i$th collection center after repairing/replacing it
- $cs_{hj}$: Cost of shipping a spare part from $h$th plant to $j$th repair facility for replacing defective parts of repairable products
- $cn_{hj}$: Cost of shipping a new product from $h$th plant to $j$th repair facility for replacing irreparable products
- $cd_{jl}$: Cost of shipping an irreparable returned product from $j$th repair facility to disposal center $l$
- $cp_{jl}$: Cost of shipping a defective part of repairable product from $j$th repair facility to disposal center $l$
- $\rho$: Percentage of returned products requiring spare parts
- $f_{jn}$: Fixed cost of installing $n$ repairing equipments at $j$th repair facility
- $\alpha$: The exponent measuring ratio of incremental to the costs of a unit of equipment representing economies of scale in repair facilities ($0 < \alpha < 1$)
- $X_{in}$: Binary variable representing the assignment of $n$ repairing equipments to $j$th repair facility
- $Y_{ijk}$: Percentage of returned $k$th product type from the $i$th collection center assigned to the $j$th repair facility

- $W_{hjk}$: Amount of spare parts shipped from $h$th plant to $j$th repair facility for repairing repairable $k$th product type
- $U_{hjk}$: Amount of new $k$th product type shipped from $h$th plant to $j$th repair facility for replacing irreparable products
- $P_{jlk}$: Amount of defective parts of repairable $k$th product type shipped from $j$th repair facility to disposal center $l$
- $V_{jlk}$: Amount of irreparable returned $k$th product type shipped from $j$th repair facility to disposal center $l$

Similar to Du and Evan [9], the equipment installing fixed cost scheme originally proposed by Manne [16] is considered. In this scheme, the fixed cost of installing single equipment in the $j$th facility is assumed to be $f_{j1} = \kappa b^{\alpha}$, where $\kappa$ is a constant coefficient, then the fixed cost of installing $n$ equipments in that facility is $f_{jn} = \kappa(nb)^{\alpha} = n^{\alpha}\kappa b^{\alpha} = n^{\alpha}f_{j1}$. Therefore, the fixed cost of locating a repair facility at the $j$th candidate location, $FC_j$, can be defined as follows provided that $\sum_{j=1}^{N_j} X_{jn}$ is at most 1.

$$FC_j = f_{j1}X_{j1} + f_{j1}2^{\alpha}X_{j2} + \ldots + f_{j1}N_j^{\alpha}X_{jN_j} = \sum_{n=1}^{N_j} n^{\alpha}f_{j1}X_{jn}$$

## 3.2   The Deterministic Mathematical Model

Objective function (26.1) tries to minimize repair equipments fixed installation costs and transportation costs. The transportation costs include costs of shipping (1) products from collection centers to repair facilities and vice versa, (2) spare parts from plants to repair facilities, (3) new products from plants to repair facilities, (4) irreparable returned products from repair facilities to disposal centers, and (5) defective parts of repairable products from repair facilities to disposal centers.

$$
\begin{aligned}
Z_1 = & \sum_{j=1}^{J}\sum_{n=1}^{N_j} n^{\alpha}f_{j1}X_{jn} + \sum_{i=1}^{I}\sum_{j=1}^{J}\sum_{k=1}^{K} cr_{ij}a_{ik}Y_{ijk} \\
= & + \sum_{h=1}^{H}\sum_{j=1}^{J}\sum_{k=1}^{K} cs_{jh}W_{hjk} + \sum_{h=1}^{H}\sum_{j=1}^{J}\sum_{k=1}^{K} cn_{jh}U_{hjk} \\
= & + \sum_{j=1}^{J}\sum_{l=1}^{L}\sum_{k=1}^{K} cd_{jl}V_{jlk} + \sum_{j=1}^{J}\sum_{l=1}^{L}\sum_{k=1}^{K} cp_{jl}P_{jlk}
\end{aligned} \tag{26.1}
$$

Objective function (26.2) minimizes the total weighted tardiness of returning repairable products and new products to the collection centers.

$$Z_2 = \sum_{i=1}^{I} \sum_{j=1}^{J} \sum_{k=1}^{K} \left( \begin{array}{l} \text{Max}\left[(rt_k + t_{ij} - \tau), 0\right](1 - \gamma_k) \\ + \text{Max}\left[(t_{ij} - \tau), 0\right]\gamma_k \end{array} \right) a_{ik} Y_{ijk} \qquad (26.2)$$

It is noteworthy that due to the nature of equipment allocation scheme, the first objective function tends to allocate the necessary equipments in a centralized manner and the second objective function tends to decentralize the required equipments allocation different repair facilities.

The objective function of the proposed bi-objective mathematical model and its corresponding constraints would be as follows [10]:

$$\text{Min}.\{Z_1, Z_2\} \qquad (26.3)$$

Subject to:

$$\sum_{n=1}^{N_j} X_{jn} \leq 1 \quad \forall j \qquad (26.4)$$

$$\sum_{i=1}^{I} \sum_{k=1}^{K} (1 - \gamma_k) a_{ik} rt_k Y_{ijk} \leq \sum_{n=1}^{N_j} nb X_{jn} \quad \forall j \qquad (26.5)$$

$$\sum_{i=1}^{I} a_{ik} Y_{ijk} \leq \theta_{jk} \quad \forall j, \forall k \qquad (26.6)$$

$$\sum_{j=1}^{J} \sum_{n=1}^{N_j} n X_{jn} \leq NT_{\text{max}} \qquad (26.7)$$

$$\sum_{j=1}^{J} Y_{ijk} = 1 \quad \forall i, \forall k \qquad (26.8)$$

$$\sum_{h=1}^{H} W_{hjk} = \rho(1 - \gamma_k) \sum_{i=1}^{I} a_{ik} Y_{ijk} \quad \forall j, \forall k \qquad (26.9)$$

$$\sum_{h=1}^{H} U_{hjk} = \gamma_k \sum_{i=1}^{I} a_{ik} Y_{ijk} \quad \forall j, \forall k \qquad (26.10)$$

$$\sum_{j=1}^{J} W_{hjk} \leq \sigma_{hk} \quad \forall h, \forall k \qquad (26.11)$$

$$\sum_{l=1}^{L} P_{jlk} = \rho(1 - \gamma_k) \sum_{i=1}^{I} a_{ik} Y_{ijk} \quad \forall j, \forall k \qquad (26.12)$$

$$\sum_{l=1}^{L} V_{jlk} = \gamma_k \sum_{i=1}^{I} a_{ik} Y_{ijk} \quad \forall j, \forall k \qquad (26.13)$$

$$X_{jn} \in \{0, 1\} \quad \forall j, n = 1, ..., N_j \qquad (26.14)$$

$$Y_{ijk} \geq 0 \quad \forall i, \forall j, \forall k \qquad (26.15)$$

$$W_{hjk} \geq 0, \quad U_{hjk} \geq 0 \quad \forall h, \forall j, \forall k \qquad (26.16)$$

$$V_{jlk} \geq 0, \quad P_{jlk} \geq 0 \quad \forall j, \forall l, \forall k \qquad (26.17)$$

Constraint (26.4) ensures that at most only one of the allowed numbers of equipments is assigned to each repair facility. Constraint (26.5) limits the time required for repairing products in each repair facility to its corresponding equipments capacity. Constraint (26.6) limits the maximum number of products assignable to each repair facility for inspection. Constraint (26.7) limits the sum of number of equipments assigned to repair facilities to the maximum number of equipments available to assign to the repair facilities. Constraint (26.8) ensures the complete assignment of each collection centers demand to the repair facilities. Constraints (26.9) and (26.10) ensure the flow conservation between plants and repair facilities for the required spare parts and new products, respectively. Constraint (26.11) limits the number of spare parts shipping from each plant to the repair facilities considering the plant spare part capacity. Constraints (26.12) and (26.13) ensure the flow conservation between repair facilities and disposal centers for the defective parts of repairable products and irreparable returned products, respectively. Finally, constraints (26.14)–(26.17) define the decision variables types.

## 3.3 The Robust Mathematical Model

Contrary to typically deterministic mathematical models available in the literature, real-world applications are usually surrounded with uncertainty. The two main approaches for dealing with uncertainty are stochastic programming and robust optimization. For developing stochastic programming models, probability distributions of uncertain parameters should be known in advance. However, in many practical situations, there is no information or enough information for obtaining probability distribution of uncertain parameters. Robust optimization models are viable answers to these situations via providing solutions that are always feasible as well as performing well for all possible realizations of uncertain parameters. Various researchers [17–21] have tried to develop mathematical models to cope with such situation. In this study, the concept of uncertainty budget proposed by [21] is used to develop a robust mathematical model for the research problem. Next, the robust optimization model of [21] is discussed.

First, without loss of generality, consider the following linear programming model with uncertain technology matrix:

$$\text{Max. } c^T x \tag{26.18}$$

Subject to:

$$\tilde{A}x \geq b \tag{26.19}$$

$$x \in \mathbf{R}_+^n \tag{26.20}$$

Each uncertain parameter, $\tilde{a}_{ij}$, is denoted by a nominal value, $\bar{a}_{ij}$, and a shift value, $\hat{a}_{ij}$. Therefore, uncertain parameter $\tilde{a}_{ij}$ could be demonstrated via the closed interval $[\bar{a}_{ij}-\hat{a}_{ij}, \bar{a}_{ij} + \hat{a}_{ij}]$. Contrary to common robust optimization models where it is assumed that all uncertain parameters can deviate from their nominal values, [21] assumed that probability of all uncertain parameters deviating from their nominal values is very small. Therefore, the decision-maker should decide a number $\Gamma$, known as uncertainty budget, which limits maximum number of uncertain parameters deviating from their nominal values. The selection of this number is highly dependent on the level of decision-maker's conservativeness. As the value of $\Gamma$ increases, the level of decision-maker's conservativeness increases, and the robust model becomes more similar to minimax model of [17]. Considering $J_i$ as the set of uncertain parameters belonging to the $j^{\text{th}}$ row of $\tilde{A}$, the robust counterpart of problem (26.18)–(26.20) could be written as follows [21]:

$$\text{Max. } c^T x \tag{26.21}$$

Subject to:

$$\sum_j \bar{a}_{ij}x_j + \max_{\left\{S_i \cup \{t_i\} \mid S_i \subseteq J_i, |S_i| = \lfloor \Gamma_i \rfloor, t_i \in J_i \backslash S_i\right\}}$$

$$\times \left\{\sum_{j \in S_i} \hat{a}_{ij}y_j + (\Gamma_i - \lfloor \Gamma_i \rfloor)\hat{a}_{it_i}y_t\right\} \leq b_i \tag{26.22}$$

$$x_j \leq y_j \quad \forall j \tag{26.23}$$

$$x \in \mathbf{R}_+^n, \ y \in \mathbf{R}_+^n \tag{26.24}$$

Nonlinear constraint (26.2) could be easily linearized using the proposition in [21]. In addition, this model could be easily used for modeling robust integer and mixed integer programming problems.

## 3.4   ε-Constraint Method

Solving model (26.1)–(26.17) directly via conventional single-objective optimization exact methods is not possible due to the bi-objective nature of the model. Therefore, a multiobjective optimization technique such as weighted sum, weighted minimax, weighted product, global criterion, ε-constraint, or lexicographic method should be used. In this study, the ε-constraint method [22] is used for transforming the bi-objective optimization problem into a single-objective optimization problem.

For illustrating the application of ε-constraint method, consider the following generic bi-objective optimization problem.

$$\text{Min. } f_1(X) \tag{26.25}$$

$$\text{Min. } f_2(X) \tag{26.26}$$

Subject to:

$$g_i(X) \leq 0 \quad \forall i \tag{26.27}$$

$$X \geq 0 \tag{26.28}$$

Supposing that $f_1(X)$ is the most important objective function, then one can simply rewrite problem (26.25)–(26.28) as follows to obtain a Pareto-optimal solution [23]:

$$\min f_1(X) \tag{26.29}$$

Subject to:

$$f_2(X) \leq \varepsilon_2 \tag{26.30}$$

$$g_i(X) \leq 0 \quad \forall i \tag{26.31}$$

$$X \geq 0 \tag{26.32}$$

Constraint (26.30) ensures that the second objective function would not be more than a pre-defined value, $\varepsilon_2$. In order to create a list of Pareto-optimal solutions, the above procedure has to be repeated for different values of ε.

# 4 Numerical Example

In this study, a numerical example was created to evaluate the behavior of the proposed robust model. For this matter, a random instance of the research problem was generated similar to a generation scheme available in the literature [9, 10]. The nodes of the considered post-sale reverse logistics network were generated on a $100 \times 100$ Euclidean space with 5 production plants, 10 repair facilities, 20 collection centers, and 5 disposal centers. In addition, the number of products is considered 3. In the randomly generated instance, the nominal value of products returned from each collection center to repair facilities is randomly generated from $U[10, 100]$ and its corresponding shift value is assumed 20% of the nominal value. The fixed costs of installing single equipment in each repair facility were generated from the uniform interval of $[5b, 10b]$. In addition, percentage of returned products requiring spare parts, the economies of scale in each facility, and maximum allowed time for returning repaired/new products to the collection sites are assumed 0.2, 0.8, and 30, respectively. Other parameters such as facility capacities, transportation costs, and travel times of the considered network are generated according to Table 26.1. Finally, necessary mathematical models were coded and solved via commercial optimization software, LINGO9.0.

Before solving the bi-objective research problem via the $\varepsilon$-constraint method, the ideal points of the deterministic and robust network design problems are obtained via single-objective optimization of the research problem (see Table. 26.2). Corresponding robust network design problems are solved for different values of $\Gamma$, namely 1, 5, 10, 20, and 60. Table 26.2 demonstrates the increase in the amount of objective function value for the robust model along the increase of $\Gamma$. After setting the first objective function to be the most important objective function, the values of $\varepsilon_2$ are increased sequentially 500 units to be further from best value of second objective function. The results, as shown in Fig. 26.2 for the deterministic model and the robust model after setting $\Gamma$ to be 5, demonstrate the success of $\varepsilon$-constraint method in creating a list of Pareto-optimal solutions as well as the conflict between the two considered objective functions.

**Table 26.1** Parameter generation scheme for the numerical example

| Parameter | Generation method | Parameter | Generation method |
|---|---|---|---|
| $rt_k$ | $\{8, 10, 12\}$ | $\gamma_k$ | $\{0.15, 0.2, 0.25\}$ |
| $\sigma_{hk}$ | $U[50, 80]$ | $\theta_{jk}$ | $U[400, 600]$ |
| $t_{ij}$ | $0.6 \times \text{Distance}(i,j)$ | $cn_{hj}$ | $0.075 \times \text{Distance}(h,j)$ |
| $cd_{jl}$ | $0.05 \times \text{Distance}(j,l)$ | $cr_{ij}$ | $0.1 \times \text{Distance}(i,j)$ |
| $cs_{hj}$ | $0.05 \times \text{Distance}(h,j)$ | $cp_{jl}$ | $0.085 \times cd_{jl}$ |
| $b$ | $\frac{0.3}{J} \sum_{i=1}^{I} \sum_{kl=1}^{K} a_{ik}$ | $N_j$ | $\left\lceil \frac{1}{b} \sum_{k=1}^{K} (1 - \gamma_k) rt_k \sum_{j=1}^{J} \theta_{jk} \right\rceil$ |
| $NT_{\max}$ | $1.1 \times \left\lceil \frac{1}{b} \sum_{i=1}^{I} \sum_{k=1}^{K} a_{ik} \right\rceil$ | | |

**Table 26.2** Ideal points of the deterministic and robust network design models

|  | $\Gamma$ |  | Best obj. function value | Corresponding other obj. function |
|---|---|---|---|---|
| *Deterministic* | – | $Z_{d1}$ | 62947.59 | 6903.97 |
|  | – | $Z_{d2}$ | 1618.21 | 356783.4 |
| *Robust* | 1 | $Z_{r1}$ | 65295.31 | 6831.35 |
|  |  | $Z_{r2}$ | 1695.408 | 356751.5 |
|  | 5 | $Z_{r1}$ | 66485.29 | 8205.00 |
|  |  | $Z_{r2}$ | 1902.968 | 357191.2 |
|  | 10 | $Z_{r1}$ | 66758.40 | 9527.93 |
|  |  | $Z_{r2}$ | 1941.855 | 356831.7 |
|  | 20 | $Z_{r1}$ | 66888.53 | 11538.89 |
|  |  | $Z_{r2}$ | 1941.855 | 356963.0 |
|  | 60 | $Z_{r1}$ | 67166.91 | 19491.13 |
|  |  | $Z_{r2}$ | 1941.855 | 357462.5 |

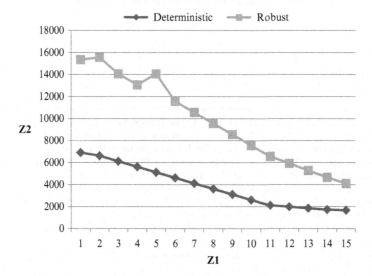

**Fig. 26.2** The graph of obtained Pareto-optimal solutions for the deterministic model and the robust model with $\Gamma = 5$

## 5  Conclusion

An important step in greening supply chains is to consider their environmental and ecological impacts during their own initial design stages. In this study, a comprehensive robust model for a 3PL multiproduct post-sales reverse logistics network consisting of collection centers, repair facilities, production plants, and disposal centers was considered. Various supply chain network design decisions such as location of repair facilities, allocation of repair equipments, and the material flows in the network were considered. The considered assumptions are well matched with characteristics of post-sales service providers in the electronics industry for various products such as cell phones and televisions. For this problem, a bi-objective robust mixed integer linear programming model was presented. In addition, a numerical example was designed to illustrate successful application of $\varepsilon$-constraint method in obtaining a list of Pareto-optimal solutions for the proposed model. While obtained robust solutions of the proposed model have greater objective function values compared to the ones of deterministic version, they are both always feasible and performing well for all possible realization of uncertain parameters.

At the end, future research opportunities of this study are integration of various tactical decisions of the reverse logistics networks with the research problem, extending the proposed model for incorporating risk via stochastic programming, application of other types of robust optimization models, and application of metaheuristics specifically evolutionary algorithms with variations of the NSGA, which could be of interest due to their effectiveness in creating a Pareto list.

# References

1. Sheu JB (2011) Bargaining framework for competitive green supply chains under governmental financial intervention. Transport Res E Logist Transport Rev 47:573–592
2. Lee D-H, Dong M (2008) A heuristic approach to logistics network design for end of lease computer products recovery. Transport Res E Logist Transport Rev 44:455–474
3. Kong N, Salzmann O, Steger U, Ionescu-Somers A (2002) Moving business/industry towards sustainable consumption: The role of NGOs. Eur Manag J 20:109–127
4. Srivastava SK (2008) Network design for reverse logistics. Omega 36:535–548
5. Jayaraman V, Patterson RA, Rolland E (2003) The design of reverse distribution networks: Models and solution procedures. Eur J Oper Res 150:128–149
6. Pishvaee MS, Kianfar K, Karimi B (2010) Reverse logistics network design using simulated annealing. Int J Adv Manuf Technol 47:269–281
7. Listeş O, Dekker R (2005) A stochastic approach to a case study for product recovery network design. Eur J Oper Res 160:268–287
8. Min H, Ko HJ, Ko CS (2006) A genetic algorithm approach to developing the multi-echelon reverse logistics network for product returns. Omega 34:56–69
9. Du F, Evans GW (2008) A bi-objective reverse logistics network analysis for post-sale service. Comput Oper Res 35:2617–2634
10. Zegordi SH, Eskandarpour M, Nikbakhsh E (2011) A novel bi-objective multi-product post-sales reverse logistics network design. Lecture notes in engineering and computer science. In: Proceedings of the world congress on engineering, WCE 2011, London, UK, 6–8 July 2011, pp 716–720
11. Demirel NO, Gökçen H (2008) A mixed integer programming model for remanufacturing in reverse logistics environment. Int J Adv Manuf Technol 39:1197–1206
12. de Figueiredo JN, Mayerle SF (2008) Designing minimum-cost recycling collection networks with required throughput. Transport Res E Logist Transport Rev 44:731–752
13. Aras N, Aksen D (2008) Locating collection centers for distance- and incentive-dependent returns. Int J Prod Econ 111:316–333

14. Lee D-H, Dong M (2009) Dynamic network design for reverse logistics operations under uncertainty. Transport Res E Logist Transport Rev 45:61–71
15. Sasikumar P, Kannan G, Haq AN (2010) A multi-echelon reverse logistics network design for product recovery: a case of truck tire remanufacturing. Int J Adv Manuf Technol 49:1223–1234
16. Manne A (1967) Investments for capacity expansion. MIT Press Cambridge, MA
17. Soyster AL (1973) Convex programming with set-inclusive constraints and applications to inexact linear programming. Oper Res 21:1154–1157
18. Mulvey JM, Vanderbei RJ, Zenios SA (1995) Robust optimization of large-scale systems. Oper Res 43:264–281
19. Ben-Tal A, Nemirovski A (1999) Robust solutions to uncertain programs. Oper Res Lett 25:1–13
20. Ben-Tal A, Nemirovski A (2000) Robust solutions of linear programming problems contaminated with uncertain data. Math Program 88:411–424
21. Bertsimas D, Sim M (2004) The price of robustness. Oper Res 52:35–53
22. Haimes YY, Lasdon LS, Wismer DA (1971) On a bicriterion formulation of the problems of integrated system identification and system optimization. IEEE Trans Syst Man Cybern 1:296–297
23. Marler RT, JS Arora (2004) Survey of multi-objective optimization methods for engineering. Struct Multidiscip O 26:369–395

# Chapter 27
# A Case Study of Lean Manufacturing Implementation Approach in Malaysian Automotive Components Manufacturer

Rasli Muslimen, Sha'ri Mohd. Yusof, and Ana Sakura Zainal Abidin

## 1 Introduction

The lean manufacturing (LM) or Toyota production system (TPS) is one of the proven productivity and quality improvement initiatives in the automotive industry. TPS was pioneered by a Japanese automotive company, Toyota motor corporation (TMC), during 1950s. TPS was born through various efforts of TMC to catch up with the automotives industries of western advanced country after the end of World War II. TPS has been initiated, created, and implemented from actual practices in the factories of TMC, and it has a strong feature of emphasizing practical effects, and actual practices over theoretical analysis. Due to its global superiority in cost, quality, flexibility and quick respond, LM was transferred across countries and industries [1]. LM has become a widely acceptable and adoptable best manufacturing practice across countries and industries [2]. The primary goals of LM were to reduce the cost of product and increase productivity by eliminating all kinds of wastes or nonvalue added activities [3]. Hence, LM or TPS is a productivity and quality improvement initiative that hailed as a cost of reduction mechanism [3–8].

In order to success in LM implementation, there are several factors and approaches. Prior study has identified four critical success factors: leadership and management, financial, skills and expertise, and supportive organizational culture of the organization [9]. Other researchers also suggested that applying the full set of

R. Muslimen (✉) • A.S.Z. Abidin
Department of Mechanical and Manufacturing Engineering, Faculty of Engineering, Universiti Malaysia Sarawak, 94300 Kota Samarahan, Sarawak, Malaysia
e-mail: anras_126@yahoo.com; rasna_126@hotmail.com

S.M. Yusof
Department of Manufacturing and Industrial Engineering, Faculty of Mechanical Engineering, Universiti Teknologi Malaysia, UTM Skudai 81310, Johor, Malaysia
e-mail: shari@fkm.utm.my

S.-I. Ao and L. Gelman (eds.), *Electrical Engineering and Intelligent Systems*, Lecture Notes in Electrical Engineering 130, DOI 10.1007/978-1-4614-2317-1_27, © Springer Science+Business Media, LLC 2013

lean principles and tools also contribute to the successful LM transformation [10, 11]. However, in reality not many companies in the world are successful to implement this system [12, 13]. Furthermore, previous researchers insist that there is no "cookbook" to explain step by step of the LM process and how exactly to apply the tools and techniques [14–16]. Many manufacturing companies have implemented LM in many different ways and names in order to suit with their environment and needs. Therefore, it is important to conduct the research in order to identify the approaches and processes in LM implementation.

In Malaysia, some studies have been done in manufacturing industries regarding LM implementation. Wong et al. [17] focused to examine the adoption of LM in the Malaysian electrical and electronics industries. Nordin et al. [18] focused on exploring the extent of LM implementation in Malaysian automotive manufacturing industries. Both studies found that most of the Malaysian manufacturing industries have implemented LM up to a certain extent and in-transition toward LM. However, the findings based on Malaysian manufacturing industries do not provide on how to implement and what approach to be used to successfully implement LM. Hence, this research is very important to give more detail sequences and steps in implementing LM.

Therefore, the purpose of this study is to investigate on how to implement and what suitable approach to be used in order to successfully implement LM in Malaysian manufacturing industries. The investigation focuses on the LM implementation approach in Malaysian automotive components manufacturer. From this study, it will give one of the several approaches in implementing LM that has been practiced in Malaysian automotive components manufacturer. This study will present and highlight the early stage of the LM implementation approach by the case study company. The next stage of the LM implementation approach will be presented in future publication that will highlight the continuous improvement of LM implementation approach in order to sustain the efforts and success.

## 2  Research Methodology

The research methodology used in this research is a case study methodology. Through this case study, it enabled several sources of evidences and practices to be highlighted. The case study also provides better understanding of the problems faced by the Malaysian automotive components manufacturer. The case study method allows researchers to retain the holistic and meaningful characteristics of the real-life events. Furthermore, the use of case study as a research method based on three conditions as follows [19]:

1. The type of research question posed: typically to answer questions such as"how" or "why"
2. The extent of control an investigator has over actual behavioral events: when investigator has no possibility to control the events
3. General circumstances of the phenomenon to be studied: contemporary phenomenon in a real-life context

A case study was performed in one of the automotive components manufacturers in Malaysia. This company selected was based on its achievement as a TPS model company awarded by Malaysia Japan automotive industries cooperation (MAJAICO) in year 2007. MAJAICO is a 5-year project from 2006 until 2011 initiated under the Malaysia Japan economic partnership agreement (MJEPA) to develop and improve the Malaysian automotive industry to become more competitive as global automotive players. The main function of MAJAICO is to introduce continuous improvement activities in manufacturing companies mainly through total implementation of lean manufacturing. Under MAJAICO project, TPS has known as lean production system (LPS) where the activities have been conducted by the Japanese experts and local experts from perusahaan otomobil nasional sendirian berhad (PROTON) and perusahaan otomobil kedua sendirian berhad (PERODUA).

Interview was conducted at the case study company with two interviewees at managerial level; manager of safety environment & quality management, and assistant manager of TPS & skill development. Both of them are from total quality management department and very wide experience in conducting LM implementation projects. Interview was conducted through prepared semistructured and open-ended questionnaires. The semistructured interview and open-ended questions were used where interviewees were encouraged to explain why the line operated in a certain way [20].

The semistructured and open-ended questionnaires were utilized to gain insights into the status of LM implementation approach in this case study company. For this case study company, the semistructured and open-ended questionnaires have been divided into three sections as follows:

(a) The company's background information—Year of establishment, start of production, ownership, number of employees, products, customers, and achievements.
(b) The understanding of lean manufacturing.
(c) The implementation of lean manufacturing.

In order to find out the approach of LM implementation from this company, a number of questions were tailored to enable the extraction of ideas that give a true reflection on the interviewee's practices. The questions attempt to investigate the company's understanding of LM and LM implementation. For example, the key questions in section (b) and (c) of the semistructured and open-ended questionnaires were as follows:

Section b: The understanding of lean manufacturing

- When did your company started to implement LM?
- What is your understanding about LM?
- Who has motivated your company to implement LM?
- How long it takes to complete the first implementation project of LM in your company?
- Do you think it is necessary to hire consultant to assist the implementation of LM? How about your company's practice?

Section c: The implementation of lean manufacturing

- Who is the person responsible to lead the implementation of LM in your company?
- Where has LM been implemented in your company?
- What were the criterions for choosing that specific area?
- How many people involved in the project?
- What kind of waste does LM eliminated in the project?

During the interview session, it was tape recorded with the permission from the interviewees to avoid any missing points of information given by them. Finally, the overall information obtained from the interview session was summarized and verified with the interviewees. Findings from the interview were analyzed and discussed in the findings and discussion section.

## 3 Background of Company

From the section (a) of the semistructured and open-ended questionnaires, the company's background information was gained and illustrated in Table 27.1. The name of this company is changed to MJ Sdn. Bhd. in terms of confidential issues. The company was established on 3rd April 1980 and starts their production on 1st July 1983. They have two manufacturing plants; thermal systems plant and electronics plant. In thermal systems plant, they have three product divisions: air-conditioning, cooling systems, and wiper & motor division where they produce nine products namely condenser, compressor, hose, piping, heater, ventilator, blower, radiator, and washer. And in electronics plant, they have four product divisions: industrial systems, electronics, body electronics, and engine control division where they produce four products namely programmable controller, engine electronic control unit, air-con amplifier, and CDI amplifier.

Currently, the number of employees of this company is 1,200 persons. This company is an industry specialist in high quality and technologically advanced automotive components with original equipments manufacturer status. This company has manufactured a total of 13 products from these 2 plants. This company is a major automotive components supplier to national car in Malaysia. Their major customers are Toyota, their own group companies, Perodua, Honda, Proton, and others.

## 4 Findings and Discussions

In 1996, the first lean manufacturing implementation initiative in this company was started. At the beginning, the concept of lean manufacturing is still new and the knowledge in this company is still at a very low level. In 2002, the president of the company from headquarter in Japan came and asked to continue lean

**Table 27.1** Company's profile

| Company name | MJ SDN.BHD. | |
|---|---|---|
| Establishment | 3rd April 1980 | |
| Start of prod. | 1st July 1983 | |
| Employees | 1,200 | |
| Land area | 70,100 M | |
| Build up area | 17,410 M (office + thermal systems plant) | |
| | 14,060 M (electronics plant) | |
| Manufacturing product | Product division | Products |
| Thermal systems plant | Air-conditioning | Condenser, compressor, hose, piping, heater, ventilator, blower |
| | Cooling systems | Radiator |
| | Wiper & motor | Washer |
| Electronics plant | Industrial systems | Programmable controller |
| | Electronics | Engine electronic control unit |
| | Body electronics | Air-con amplifier |
| | Engine control | CDI amplifier |
| Customers | Toyota, MJ Group Companies, Perodua, Honda, Proton, Others | |
| Achievements | 1994 – ISO 9002 certification from SIRIM | |
| | 2000 – ISO 14001 certification from SIRIM | |
| | 2003 – ISO/TS 16949 certification from SIRIM | |
| | 2006 – Company group president award | |
| | 2006 – Achieved zero emission | |
| | 2007 – TPS model company by MAJAICO | |
| | 2007 – Environment award from selangor government | |
| | 2007 – Achieved quality management excellent Award from MITI | |
| | 2007 – Proton best overall performance award | |
| | 2008 – ISO 9001:2000 certification from SIRIM | |
| | 2008 – OHSAS 18001/MS 1722 certification from SIRIM | |
| | 2009 – The winner of ministry of international Trade and industry Malaysia (MITI) | |
| | 2009 – The winner of prime minister hibiscus award | |

manufacturing activities in proper way where one team was formed with five full-time members. At the early stage of lean manufacturing implementation in this company, the project-based approach was used. The project is based on a small-scale project where the focus of LM implementation in this company is to solve the problem at the small area. From the interview, the authors have formulated the lean manufacturing implementation approach by this company as shown in Fig. 27.1.

From Fig. 27.1, this company forms a small team with five full-time members to run the lean manufacturing implementation project. A few Japanese experts from headquarter in Japan came to teach and shared their knowledge of lean manufacturing implementation with the team members.

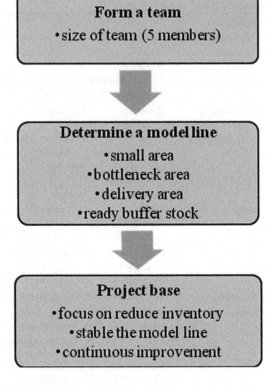

**Fig. 27.1** Lean manufacturing implementation approach

The next stage, a small team determines one model line in order to run the lean manufacturing implementation project. There are a few criterions to determine the selection of a model line. The selection of the model line was based on the following characteristics: small area, bottleneck area, and delivery area. Before running the lean manufacturing implementation, the buffer stock was ready and prepared at the model line for any shortages of the product during lean manufacturing implementation.

Finally, at the project-based approach by this company, the focus of lean manufacturing implementation is reducing the level of inventory. For this company, inventory is the mother of other wastes. There are several wastes that have been identified by the prior research. The seven main types of wastes identified by the father of TPS are as follows [21]:

- Waste of over production
- Waste of waiting inventory
- Waste of unnecessary transportation
- Waste of waiting times
- Waste of unnecessary processing
- Waste of unnecessary motion
- Waste of defected products

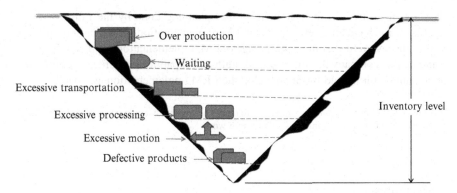

**Fig. 27.2** The level of inventory

In lean manufacturing implementation approach by this company, the level of inventory is visualized similar to the level of water in a river. When they reduced the level of inventory, this means that they will be able to lower down the level of water in the river. Consequently, this action will highlight other wastes hiding at the lower level. The other wastes at the lower level are namely over production, waiting times, excessive transportation, excessive processing, excessive motion, and defective products. This scenario of reducing inventory level can be best illustrated by the authors as shown in Fig. 27.2.

In the project-based approach by this company, they did the continuous improvement effort at the selected model line. This continuous improvement effort is continued until a saturated level of major improvement is made, and they reached the stable condition of the model line. In certain cases stabilizing the model line, the interviewee highlighted they did the major improvements for up to ten times. The duration to complete the LM implementation project by this case study company is within 3–6 months. The same approach of implementing LM as shown in Fig. 27.1 will be used continuously in the next LM implementation project to another area [22].

This direction and approach in LM implementation is similar with a traditional Toyota approach where they begin with a model line. In Toyota, they helped their external suppliers to implement TPS through their operation management consulting group lead by Taiichi Ohno [23]. However, findings from the interview session regarding the assistant from the consultant show different approach. In this case study company, they did not hire any external consultant. They solely depend on the internal consultant from their own group companies and their skill workers that have been trained in Japan. They also used their own facilities and their companies' facilities in order to implement lean manufacturing tools and techniques. For this case study company, they did the basic LM implementation largely common sense and suit with their environment and needs.

The analysis done by [24] found that the major difficulties companies encounter in attempting to apply lean are a lack of direction, a lack of planning, and a lack of adequate project sequencing. Consequently, this case study company has clear direction from the top management, proper planning done by the full-time team members, and has a long-term project in LM implementation. This long-term project will be discussed further in the next stage of LM implementation approach. It can be said that this company has their own strength and capabilities in order to implement lean manufacturing and further develop their LM implementation approach.

**Acknowledgement** The authors would like to thank the case study company in Malaysia for their willingness to participate in this study and future work.

# References

1. Schonberger RJ (2007) Japanese production management: An evolution - with mixed success. J Oper Manag 25:403–419
2. Holweg M (2007) The genealogy of lean production. J Oper Manag 25:420–437
3. Womack JP, Jones DT, Roos D (1990) The machine that change the world. Rawson Associates Scribner, New York
4. Womack JP, Jones DT (1996) Lean thinking banish waste and create wealth in your corporation. Simon & Schuster, New York
5. Creese RC (2000) Cost management in lean manufacturing enterprises. In: Proceedings of AACE international transactions. West Virginia University, Morgantown
6. Achanga P, Taratoukhine V, Nelder G (2004) The application of lean manufacturing within small and medium sized enterprise: What are the impediments? In: Proceedings of the second international conference on manufacturing research (ICMR 2004). Sheffield
7. Bicheno J (2004) The new lean toolbox towards fast and flexible flow. PICSIE Books, Buckingham
8. Achanga P, Shehab E, Roy R, Nelder G (2005a) Lean manufacturing to improve cost effectiveness of SMEs. Seventh international conference on stimulating manufacturing excellence in small and medium enterprises. University of Strathclyde, Glasgow
9. Achanga P, Shehab E, Roy R, Nelder G (2006) Critical success factors for lean implementation within SMEs. J Manuf Tech Manag 17(4):460–471
10. Herron C and Braiden PM (2007) Defining the foundation of lean manufacturing in the context of its origins. IET international conference on agile manufacturing. pp 148–157
11. James T (2006) Wholeness as well leanness. IET Manuf Eng 85:14–17
12. Balle M (2005) Lean attitude - Lean application often fail to deliver the expected benefits but could the missing link for successful implementations be attitude? Manuf Eng 84:14–19
13. Papadopoulou TC, Ozbayrak M (2005) Leanness: experiences from the journey to date. J Manuf Tech Manag 16:784–807
14. Allen JH (2000) Making lean manufacturing work for you. J Manuf Eng 2000:1–6
15. Oliver N (1996) Lean production practices. British Journal of Management: 1–10
16. Nanni A, Gregory M, Platt K (1995) Performance measurement system design. Int J Oper Prod Manag 15:80–116
17. Wong YC, Wong KY, Ali A (2009) A study on lean manufacturing implementation in the Malaysian electrical and electronics industry. Eur J Sci Res 38(4):521–535
18. Nordin N, Deros BM, Wahab DA (2010) A survey on lean manufacturing implementation in Malaysian automotive industry. Int J Innovat Tech Manag 1(4):374–380
19. Yin RK (2009) Case study research: design and methods. In: Applied social research methods series, 4th edn., SAGE Publications Inc., USA
20. Simons D, Zokaei K (2005) Application of lean paradigm in red meat processing. Br Food J 107(4):192–211

21. Ohno T (1988) Toyota production system beyond large - Scale production. Productivity Press, Oregon
22. Muslimen R, Yusof SM, and Abidin ASZ (2011) Lean manufacturing implementation in malaysian automotive components manufacturer: A case study. In: In: Proceedings of the world congress on engineering. Lecture notes in engineering and computer science, WCE 2011, London, UK, 6–8 July 2011, pp 772–776
23. Liker JK (1998) Becoming lean: Inside stories of US manufacturers. Productivity Press, New York
24. Bhasin S, Burcher P (2006) Lean viewed as a philosophy. J Manuf Tech Manag 17(1):56–72

# Chapter 28
# Investing in the Sheep Farming Industry: A Study Case Based on Genetic Algorithms

**Iracema del Pilar Angulo-Fernandez, Alberto Alfonso Aguilar-Lasserre, Magno Angel Gonzalez-Huerta, and Constantino Gerardo Moras-Sanchez**

## 1 Introduction

Mexico is a country with gastronomic richness, and meat is an important ingredient in its cuisine [1]. About 84% of Mexican households are fond of some type of meat[1] [2], but meat demand cannot be supplied by the production of the country, having to import mainly chicken meat, pig meat, and sheep meat [3]. The main livestock-producing states in Mexico are Jalisco and Veracruz, in which sheep production is below cattle, chicken, and pig production [4, 5]. Taking this premise as a starting point, we know that Mexico needs investment projects to increase meat production, especially of sheep, and Veracruz plays an important role for this purpose due to its agro-ecological features and productive potential. Economic evaluation studies done about sheep farming in Mexico have established the profitability of this activity [6, 7], but this research is trying a different approach: to optimize the profitability by means of a genetic algorithm (GA).

GAs [8] are used as a sophisticated tool of optimization, which is based on the principles of Darwin's theory of evolution by natural selection, and nowadays have been used to approach problems in the agricultural sector, mainly in disease control areas; however, its application can be extended to business, simulation, and identification areas [9].

---

[1] Fish and seafood consumption is excluded for this estimate.

I. del Pilar Angulo-Fernandez (✉) • A.A. Aguilar-Lasserre • M.A. Gonzalez-Huerta • C.G. Moras-Sanchez
Division de Estudios de Posgrado e Investigacion, Instituto Tecnologico de Orizaba,
Av. Oriente 9 No. 852, Col. Emiliano Zapata, 94320 Orizaba, Veracruz, Mexico
e-mail: irapily@yahoo.com.mx; aaguilar@itorizaba.edu.mx; mgonzalezh@itorizaba.edu.mx; cmoras@itorizaba.edu.mx

S.-I. Ao and L. Gelman (eds.), *Electrical Engineering and Intelligent Systems*,
Lecture Notes in Electrical Engineering 130, DOI 10.1007/978-1-4614-2317-1_28,
© Springer Science+Business Media, LLC 2013

To make an investment decision based on a GA about optimizing profitability, it is necessary that it must be supported in detailed economic studies, in this case, the complex system of sheep farming has been approached from a dynamic point of view, and simulation is a powerful tool which allows dealing with these kinds of problems.

This work begins representing the fattening-sheep semiextensive farming process, from the initial acquisition of the animals to their commercialization, using a model of simulation which supplies information for a second stage of the analysis, in which income statements for a period under consideration are presented and the net present value (NPV) is estimated. In a third stage of this study, a GA is run to optimize the NPV and to support the investment decision-making regarding the setting of variables analyzed.

This analysis proposal allows studying influential factors such as: proliferation, loss of life, animal weight, and market price, to name just a few. The factors have been represented in the simulation model through probability distributions, and the annual initial and final stocks were previously estimated with Monte Carlo simulation. The aim of optimizing the project profitability is according to four key variables: initial number of ewes, type of breed, planning horizon, and type of sale.

## 2 Methods and Tools

The economic evaluation studies are based on normative criteria, independent of subjective opinions, which indicate preferences in any course of action. Every alternative is independent of the other ones and the measure of the economic preferences is made separately. Renkema and Berghout [10] distinguish four basic approaches to evaluate investment alternatives: the financial approach, the multicriteria approach, the ratio approach, and the portfolio approach. Methods from the financial approach are usually recommended for the evaluation and selection of investment alternatives. Often used financial approach methods are: the payback period (PP) method, the internal rate of return (IRR) method, and the NPV method. The latter is used to evaluate this project.

The NPV method compares the alternatives using an $n$ years period of time, which not necessarily considers the total duration of the alternatives. This period is called planning horizon. The NPV function is defined as:

$$NPV = -P + \sum\nolimits_1^n \frac{CF}{(1 + MARR)^n} + \frac{SV}{(1 + MARR)^n}, \qquad (28.1)$$

where $P$ is the principal (present sum), "$CF$" is the cash flow, "$MARR$" is the minimum acceptable rate of return, "$SV$" is the salvage value, and $n$ is the planning horizon (number of periods in which the project is expected to exists).

As the aim of this research is to maximize the profitability of fattening-sheep farming, the objective function is established as:

$$Max\,NPV = -P + \sum_{1}^{n} \frac{CF}{(1+MARR)^n} + \frac{SV}{(1+MARR)^n}. \qquad (28.2)$$

Before putting the GA into practice to solve the function, the initial investment and cash flows must be estimated. This was carried out as shown in Fig. 28.1a, b.

The data included in the cash flow were estimated using a Monte Carlo simulation, considering initial investment and uncertain parameters related to the animals such as: mortality rate, production (birth rate) and animal replacements, among others. These were represented according to established probability distributions to deal with decision-making under risk. The data used to perform the simulation were obtained from Mexico's Secretariat of Agriculture, Livestock, Rural Development, Fisheries, and Food (SAGARPA stands in Spanish) [11].

The approach of simulation applied for the economic evaluation is summarized in the following stages [12]:

- Stage 1: To create alternatives. For this research, the alternatives studied depended on the combination of six types of sheep breed (Dorper, Damara, Pelibuey, Katahdín, Blackbelly, and Suffolk), an initial number of ewes (between 100 and 300), the planning horizon (between 2 and 10 years), and two types of sales ("on-the-hoof" and carcass). We have considered these four variables as "key variables," due to the rest of the variables behavior depends on them.
- Stage 2: To identify parameters with variation. This is one of the fundamental parts of the work, because the number of uncertain variables involved in complex systems must be limited (according to resources to quantify them) for the purpose of being studied, but this not means that the analysis of the variables is not detailed, in other words, every variable able to be quantified must be included in the evaluation.

  The parameters which present variation were identified. These are listed in Table 28.1, and some useful terms to understand them are defined in the appendix of this work.
- Stage 3: To determine the probability distributions. Depending on the type of variables, discrete or continuous, the corresponding best fit probability distribution was chosen, as indicated in Table 28.1. In this stage, the selection of the probability distribution was based on information from different sources. The available information in government databases was not enough to apply goodness-to-fit tests; however, the information used was supported by manuals for farming education [13].
- Stage 4: To run a pseudorandom sampling. The procedure to generate pseudorandom samples for the Monte Carlo simulation was performed. The results obtained until this step fed the economic evaluation database, which included the

**a**

**Initial investment:**

  Ewes

  Rams

  Land area (with body of water)

  Hammer mill

  Forage grinder

  Shed

  Lambing shed

  Sheep keeper's house

  Medium duty truck

  Barbed wire

  Barbed wire staple

**Total Initial investment**

**OPERATING EXPENSES STATEMENT**


Feedstock:

  Food

  Minerals and salts

  **Total feedstock costs**

Labor costs:

  Sheep keeper salary

  Veterinary fees

  Machinery technical expert fees

  Butcher fees

  **Total labor costs**

General expenses:

  Variables costs

    Vaccines

    Antiparasitic remedies

    Vitamins and supplements

  Fixed costs

    Hammer mill depreciation

    Forage grinder depreciation

    Supplies and maintenance of machinery  (replacement parts, grease, etc.)

  **Total general expenses**

  **Total operating expenses**

**Fig. 28.1** (a) Outline to estimate initial investment and cash flows; (b) outline to estimate initial investment and cash flows

**b**

> **COSTS OF GOODS SOLD STATEMENT**
>
> Petrol
>
> Vehicle depreciation
>
> Mechanic labor fees
>
> Maintenance supplies of the vehicle (engine oil, spark plugs, etc.)
>
> "Tenencia" tax
>
> "Verificación" tax
>
> - Total operating expenses
>
> **Total costs of goods sold**
>
> **INCOME STATEMENT**
>
> Sales revenue
>
>     Ewes
>
>     Rams
>
>     Lambs
>
> Total sales revenue
>
> - Total costs of goods sold
>
> Gross profit
>
> Net income (loss)

**Fig. 28.1** (continued)

initial investment, operating expenses and costs of goods sold with the planning horizon chosen, obtaining this way the NPV for the alternative considered. The next step was the application of the GA.

A GA allows an initial population composed of individuals with particular features to evolve under setting rules to get the best result of the objective function. A GA starts with the pseudorandom generation of individuals or chromosomes represented by bits, and the group of individuals is known as initial population. In the second part of the GA process, the initial population is subjected to an evaluation according to the function to be optimized, usually referred to as cost function. In this case, the cost function corresponds to maximization, so two individuals which obtain maximum values in the cost function, called parents, are selected in a third step to transfer by this way genetic information to a new generation known as offspring, with the purpose of having an individual with best fit characteristics for the cost function. In the last part of the process, crossover and mutation operators are performed. Cross-over consists of the creation of one or more offspring, using genetic information

**Table 28.1** Parameters which have uncertainty in the fattening-sheep farming and the distributions that better represented them

| Parameters | Units | Probability distribution used |
|---|---|---|
| Animals to leave mortality | Percentage | Normal probability distribution |
| Average weight of animals when leaving | Kilograms | Normal probability distribution |
| Average weight of replaced ewes "on-the-hoof" | Kilograms | Normal probability distribution |
| Average weight of replaced rams "on-the-hoof" | Kilograms | Normal probability distribution |
| Costs of maintenance supplies of machinery | Mexican pesos | Triangular probability distribution |
| Costs of maintenance supplies of the vehicle | Mexican pesos | Triangular probability distribution |
| Ewes "on-the-hoof" unit price to sell | Mexican pesos | Triangular probability distribution |
| Ewes in carcass unit price to sell | Mexican pesos | Triangular probability distribution |
| Ewes mortality | Percentage | Normal probability distribution |
| Ewes unit price to buy | Mexican pesos | Normal probability distribution |
| Lambing | Percentage | Normal probability distribution |
| Lambs "on-the-hoof" unit price to sell | Mexican pesos | Triangular probability distribution |
| Lambs in carcass unit price to sell | Mexican pesos | Triangular probability distribution |
| Number of lambs by every lambing | Heads | Normal probability distribution |
| Petrol | Liters | Triangular probability distribution |
| Preweaning lambs mortality | Percentage | Normal probability distribution |
| Rams "on-the-hoof" unit price to sell | Mexican pesos | Triangular probability distribution |
| Rams in carcass unit price to sell | Mexican pesos | Triangular probability distribution |
| Rams mortality | Percentage | Normal probability distribution |
| Rams unit price to buy | Mexican pesos | Normal probability distribution |
| Weaning animals mortality before to leave | Percentage | Normal probability distribution |

from the parents previously chosen in the selection step. To carry out the crossover between genetic information from both individuals, a pseudorandom crossover point is chosen to separate the bits of the parents and passing a complementary part of their genetic information each other. Mutation is responsible to change pseudorandomly and partially the genetic information of the individuals, creating new individuals who become part of the population to be able to try them. After this process is concluded, it must be subjected to a series of iterations to create new generations until the chosen criteria to stop the GA is reached. Once the GA is stopped, the best solution will be that individual with the best result according to the cost function; in this work case, it corresponds to the maximum value of NPV [14].

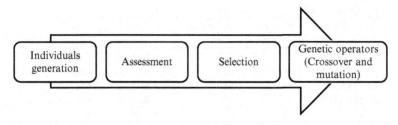

**Fig. 28.2**  Genetic algorithm general structure

The GA structure [15] is shown in Fig. 28.2.

The use of GA in optimization is focused in four basic choices: coefficients, rules, rule-base structure, and total optimization. This problem was raised with coefficients to establish the alternatives configuration.

- Stage 5: To calculate a value measure. Every time the pseudorandom samples were generated through the GA, the results statements were produced and the NPV was calculated with the MARR chosen.
- Stage 6: To describe a value measure. The results obtained from the simulations were analyzed.
- Stage 7: Conclusions. This step depended on the GA application in the economic evaluation process and the results obtained in such process, which determined the best alternative and variables coefficients.

## 3   Results

The NPV optimization results using a GA run with different crossover and mutation indexes are presented in Table 28.2, where the number of runs performed is shown in the first column. The second column indicates the best value found in the function which was optimized, for this study the maximum NPV. The next four columns correspond to the setting that produced the best values in each run. Finally, the last two columns show the crossover and mutation indexes used in each run.

According to the obtained results, the best one was identified with the setting of run no. 3, which corresponded to the variables 10-year planning horizon, 127 ewes as initial number, Suffolk breed, and a carcass type of sale.

From these results, the other variable values for the investment decision-making were able to be estimated: six rams as initial number and 40,863 m$^2$ of land area required.

Profitability speaking, if the initial investment is carried out as recommended, with 25% MARR and 10-year planning horizon, the NPV estimated would be MXN 22,533,867.59.

**Table 28.2** Results using genetic algorithms

| Run | Best NPV (MXN) | Planning horizon | Initial number of ewes | Type of breed | Type of sale | MI | CI |
|---|---|---|---|---|---|---|---|
| 1 | 22,222,168.21 | 9 | 127 | Suffolk | Carcass | 0.1 | 0.5 |
| 2 | 22,165,510.65 | 9 | 127 | Suffolk | Carcass | 0.1 | 0.7 |
| 3 | 22,533,867.59 | 10 | 127 | Suffolk | Carcass | 0.1 | 0.9 |
| 4 | 22,287,097.17 | 9 | 127 | Suffolk | Carcass | 0.3 | 0.5 |
| 5 | 22,152,817.68 | 9 | 127 | Suffolk | Carcass | 0.5 | 0.5 |

*MI* mutation index, *CI* crossover index

# 4 Conclusion

The assessment of an economic project such as the profitability of fattening-sheep farming involves a relatively complex structure for its optimization, when uncertain parameters are considered.

The objective of this work was based in the maximized NPV nonlinear function, and involved the farming process simulation to estimate cash flows. Using different scenarios where the initial number of sheep was changed (therefore, the space requirements also varied), the initial investment was estimated.

Using Monte Carlo simulation, the sheep population was estimated on a yearly basis, as well as the sales and the farming and trading costs, analyzing the uncertain factors through probability distributions.

Based on the results obtained, applying a GA the investment decision-making could increase the profitability of a sheep farming project.

The best NPV estimated for this analysis was obtained from the setting: 10-year planning horizon, 127 ewes as initial number, Suffolk breed, and a carcass type of sale.

These results have a coherent behavior in view of the fact that Suffolk is an expensive type of breed (observed in the initial investment); nevertheless, its high price is rewarded by its high carcass yield, having as a premise that the carcass sale is better paid in the market than "on-the-hoof" sale.

**Acknowledgments** This work was supported in part by the Division of Research and Postgraduate Studies of the Technological Institute of Orizaba.

Secretariat of Agriculture, Livestock, Rural Development, Fisheries and Food, State of Veracruz Delegation, Fortin District for facilitating all the information about sheep farming in the region.

All the expert people that helped to understand the fattening-sheep farming system, but their names are omitted to avoid forgetting someone's name.

# Appendix

## *Glossary*

- Sheep: Singular and plural general name to refer lambs, ewes, rams, and wethers as a whole.
- Lamb: A sheep either male or female younger than 1 year.
- Ewe: A female sheep older than 1 year.
- Ram (occasionally called tup): A male sheep older than 1 year.
- Wether: A castrated male sheep older than 1 year.
- "On-the-hoof" type of sale: Sheep are sold by weight without any further process.
- Carcass type of sale: Sheep are slaughtered to produce meat, obtaining about 50% of the live animal weight.
- Semiextensive system: Sheep are tended during the day and they get supplementary food in feeding troughs at the end of the afternoon.

# References

1. Angulo-Fernandez IP, Aguilar-Lasserre AA, Gonzalez-Huerta MA, Moras-Sanchez CG. (2011) Investment decision based on a genetic algorithm to optimize the profitability of fattening-sheep farming. In: Proceedings of the world congress on engineering 2011 (WCE 2011). Lecture Notes in Manufacturing Engineering and Engineering Management, London, 6–8 July 2011, pp 656–660
2. Anonymous (2008) Household running costs database. INEGI Web site (in Spanish). National Institute of Statistics and Geography. http://www.inegi.org.mx. Accessed 10 Jan 2011
3. Anonymous (2008) Countries by commodity (imports) database. FAOSTAT Web site. http://faostat.fao.org. Accessed 12 Jan 2011
4. Anonymous (2009) Livestock summary by state. SIAP Web site. Agrifood and Fishery Information Service, Mexico. http://www.siap.gob.mx
5. Anonymous (2008) Agri-food and fisheries data service. SAGARPA Web site. Secretariat of Agriculture, Livestock, Rural Development, Fisheries and Food, Mexico. http://www.sagarpa.gob.mx. Accessed 4 Jan 2011
6. Vilaboa Arroniz J (2006) Sheep meat trading. Engormix Web site (in Spanish). http://www.engormix.com. Accessed 16 Jul 2009
7. Rebollar-Rebollar S et al (2008) Economic optima in Pelibuey lambs fed in corrals, vol 24. Universidad y Ciencia, Tabasco, pp 67–73
8. Holland JH (1975) Adaptation in natural and artificial systems. University of Michigan Press, Ann Arbor
9. Bustos MJR (2005) Artificial intelligence in agricultural and livestock sectors (in Spanish). In: Research Seminar-I/2005-II/299622/Version 1.1. National University of Colombia, Bogota
10. Renkema TJW, Berghout EW (1997) Methodologies for information systems investment evaluation and the proposal stage: a comparative review. Inform Softw Technol 39(1):1–13
11. Anonymous. Sheep database, Fortin District, State of Veracruz Delegation. Secretariat of Agriculture, Livestock, Rural Development, Fisheries and Food

12. Blank LT, Tarquin AJ (2000) Engineering economy (in Spanish), 4th edn. McGraw-Hill, Colombia, pp 628–629
13. Koeslag JH, Kirchner Salinas F et al (2008) Sheep (in Spanish), 3rd edn. Trillas, Mexico
14. Haupt RL, Haupt SE (1998) Practical genetic algorithm, 2nd edn. Wiley, Hoboken, pp 22, 36–47
15. Martin del Brio B, Sanz Molina A (2007) Artificial neural networks and fuzzy systems (in Spanish), 3rd edn. Alfaomega Editor Group, Mexico, pp 289–292
16. Anonymous (2008) Livestock primary database. FAOSTAT Web site. Food and Agriculture Organization of the United Nations. http://faostat.fao.org. Accessed 15 Jan 2011

# Chapter 29
# High-Precision Machining by Measuring and Compensating the Error Motion of Spindle's Axis of Rotation in Radial Direction

Ahmed A.D. Sarhan and Atsushi Matsubara

## 1  Introduction

The present global market competition has attracted the manufacturer's attention on automation and high-precision machining via condition monitoring of machine tools and processes as a method of improving the quality of products, eliminating inspection, and enhancing manufacturing productivity [1–6]. The cutting forces are the most important indicator for that as they could tell limits of cutting conditions, accuracy of the workpiece, tool wear, and other process information, which are indispensable for process feedback control [7–14]. Hence, reliable cutting force measurement systems are investigated. The most common method to measure cutting forces in machining operations is through table dynamometers. Although table dynamometers provide accurate and effective force measurement, they are more suitable for laboratory or experimental use rather than for practical application on production machines, due to the limitation of workpiece size, mounting constraints, high sensitivity to overload, and high costs [15–17]. Furthermore, the dynamic characteristics of table dynamometers are strongly dependent on the workpiece mass, which may change during machine operation. To overcome limitations of mass and size of workpiece, a force sensor can be integrated into the spindle itself instead of installing it on the machine table, thus converting an ordinary spindle into a so-called monitoring spindle [18–20].

A.A.D. Sarhan (✉)
Center of Advanced Manufacturing and Material Processing, Department of Engineering Design and Manufacture, Faculty of Engineering, University of Malaya, Kuala Lumpur 50603, Malaysia
e-mail: ah_sarhan@um.edu.my

A. Matsubara
Department of Micro Engineering, Graduate School of Engineering, Kyoto University, Yoshida-honmachi, Sakyo-ku, Kyoto 606-8317, Japan
e-mail: matsubara@prec.Kyoto-u.ac.jp

S.-I. Ao and L. Gelman (eds.), *Electrical Engineering and Intelligent Systems*,
Lecture Notes in Electrical Engineering 130, DOI 10.1007/978-1-4614-2317-1_29,
© Springer Science+Business Media, LLC 2013

The authors have employed displacement sensors, as they are cheap and small enough to be built in the spindle structure. Displacement signals are translated into cutting force information by calibration. However, monitoring the quality is problematic, because sensors also detect the displacement caused by machine tool spindle error motions in the radial direction [21–26]. In this research, we develop a spindle with displacement sensors in $X$- and $Y$-axes directions near the front bearings of the spindle to monitor the spindle displacement. Cutting forces are estimated from the displacement signals by the simple signal processing technique. Monitoring tests are carried out under end-milling operations on cast iron workpieces. By comparing the estimate with the measured cutting force by using a dynamometer, the machine tool spindle error motions in sensor output are investigated, and its compensation scheme is proposed. With its compensation scheme implemented, the experimental result shows that the monitoring system is reliable for the adaptive control of machining accuracy for end-milling process.

## 2  Measurement of the Spindle Displacement to Investigate the Spindle's Axis Error Motions

### 2.1  Experimental Set-up

In order to measure the machine tool spindle displacement caused by cutting force during cutting to investigate the error motion of spindle's axis of rotation, displacement sensors are installed on the spindle unit of a high-precision machining center. The machine used in the study is a vertical-type machining center (GV503 made by Mori Seiki Co., Ltd.). The spindle has constant position preloaded bearings with oil–air lubrication, and the maximum rotational speed is 20,000 $\text{min}^{-1}$. Four eddy-current displacement sensors are installed on the housing in front of the bearings to detect the radial motion of the rotating spindle. Tables 29.1 and 29.2, respectively, show the specifications of the machining center and the displacement sensor used in this research. A thin collar with a fine cylindrical surface is attached to the spindle as a sensor target. Figure 29.1 shows the sensor locations. The two sensors, $S_1$ and $S_3$, are aligned opposite in the $X$ direction, and the other two sensors, $S_2$ and $S_4$, are aligned opposite in the $Y$ direction.

### 2.2  Experimental Procedure

A slot end-milling test is carried out to investigate the machine tool spindle error motion in radial direction. Table 29.3 shows the tool and cutting conditions. The cutting time for one operation should be short enough to avoid the thermal disturbances on displacement signals.

**Table 29.1**  Specifications of the machining center

| | | |
|---|---|---|
| Spindle | Spindle speed (min$^{-1}$) | 200–20,000 |
| | Power 15 min/cont. (kW) | 22/18.5 |
| | Tool interface | 7/24 taper no. 40 with nose face contact |
| Feed drive | Max. rapid traverse rate (mm/min) | 33,000 |
| | Max. feed rate (mm/min) | 10,000 |
| | Max. acceleration rate ($G$) | $X$-axis: 0.67, $Y$-axis: 0.64, and $Z$-axis: 0.56 |
| | Travel distance (mm) | $X$-axis: 630, $Y$-axis: 410, and $Z$-axis: 460 |
| Machine size | Width × length × height (mm) | 2,320 × 3,780 × 2,760 |
| | Mass (kg) | 5,500 |
| CNC servo system | 64-bit CPU (RISC processor) + high gain servo amplifier | |

**Table 29.2**  Specifications of the displacement sensor

| Detection principle | Eddy current |
|---|---|
| Measurement range (mm) | 0–1 |
| Output scale (V) | 0–5 |
| Diameter (mm) | 5.4 |
| Length (mm) | 18 |
| Sensitivity (mm/V) | 0.2 |
| Linearity (% of full scale) | ±1 |
| Dynamic range (kHz) | 1.3 (−3 dB) |

**Fig. 29.1**  Locations of the sensors

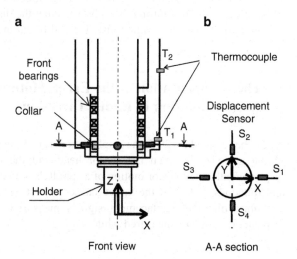

First, the spindle is rotated at 1,500 min$^{-1}$ without cutting for 30 min for the first warm-up. Then, the first cutting test is carried out at the same spindle speed. The spindle is warmed up again at the speed of 3,000 min$^{-1}$ for another 30 min without cutting followed by the second cut at the speed of 3,000 min$^{-1}$.

**Table 29.3** The tool and cutting conditions

| | |
|---|---|
| Spindle speed (min$^{-1}$) | 1,500, 3,000, 6,000, and 12,000 |
| Axial depth of cut (mm) | 10 |
| Feed mm/min | 600 |
| Cutting tool | Coated carbide end mill, diameter: 10 mm |
| Holder | BT40-C20 (collet)-20 (extension) |
| Cutting mode | Down cut |
| Coolant | No |
| Workpiece material | Carbon steel (S50C) |

Similarly, the third and forth warm-up and cutting tests at 6,000 and 12,000 min$^{-1}$ are carried out. The displacement signals are digitized with a 16-bit A/D board, and the dynamometer signals are digitized with a 12-bit A/D board. The sampling frequency is set so that 40 points per one spindle revolution can be obtained.

## 2.3 Experimental Results

Figure 29.2 shows an example of the output of the displacement sensor signals in $X$-axis direction before, during, and after cutting at spindle speed of 1,500 min$^{-1}$.

As can be seen in Fig. 29.2, before cutting, all the displacement profiles involve periodical type of fluctuation. These periodic fluctuations are related to the error motion of a spindle's axis of rotation in radial direction and roundness errors of the target surface. During cutting, it is observed that the displacement profile shifts but it returns to its original position after the end of cutting.

## 3 The Compensation of the Error Motion of Spindle's Axis of Rotation in Radial Direction

As can be seen in Fig. 29.2, the displacement sensor measures not only the variation of the gap size between the sensor head and the target surface but also the displacements due to error motion of a spindle's axis of rotation in radial direction and roundness errors of the target surface. These errors are simply compensated by subtracting the displacement signals measured while air cutting from the displacement signals measured while cutting.

## 3.1 The Concept of the Error Motion Compensation Scheme

Figure 29.3 shows the concept of the spindle displacement measurement. When the spindle axis shifts by $\Delta x$ (μm) in $X$ direction due to the cutting force, the displacement signals from $S_1$ and $S_3$ are as follows.

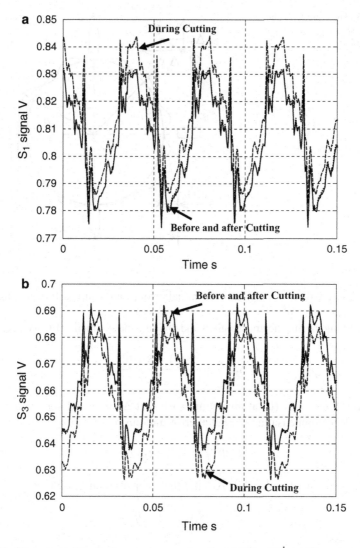

**Fig. 29.2** The output of the displacement sensor signals at 1,500 min$^{-1}$. (**a**) The output of the $S_1$ sensor signals. (**b**) The output of the $S_3$ sensor signals

$$S_1(\theta) = G[R_1 - r(\theta) - \Delta x] \qquad (29.1)$$

$$S_3(\theta) = G[R_3 - r(\theta + \pi) + \Delta x] \qquad (29.2)$$

where $G$: sensor sensitivity (mV/μm), $R_i$: the distance between the spindle center and detection surface of the sensor $S_i$ (μm) ($i = 1, ..., 4$), $\theta$: rotation angle of the

**Fig. 29.3** The concept of the
spindle displacement
measurement

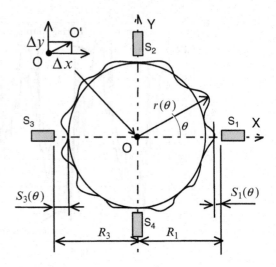

spindle (rad), and $r(\theta)$: the sum of the radial error motion and surface roughness of
the sensor target (μm).

Subtracting the displacement signals and dividing the subtraction by two, we obtain:

$$S_x(\theta) = \frac{[S_3(\theta) - S_1(\theta)]}{2} \tag{29.3}$$

Letting $S_x(\theta) = S_{xo}(\theta)$ at air cutting such that $\Delta x = 0$, and subtracting $S_{xo}(\theta)$
from $S_x(\theta)$ to compensating the radial error motions.

$$S_x(\theta) - S_{xo}(\theta) = G \cdot \Delta x \tag{29.4}$$

Then we obtain the axis shift $\Delta x$ after the compensation as follows:

$$\Delta x = \frac{[S_x(\theta) - S_{xo}(\theta)]}{G} \tag{29.5}$$

Similarly, the axis shift in $Y$ direction, $\Delta y$, is calculated from the displacement
signals from $S_2$ and $S_4$. The axis shift is called the spindle displacement hereafter.

## 3.2   Investigate the Accuracy of the Compensation

The accuracy of the compensation is investigated by comparing the compensated
spindle displacement with cutting forces measured by using a table type tool
dynamometer (Model 9257B made by Kistler). Figure 29.4 shows the compensated
spindle displacement and measured cutting force with the dynamometer at different

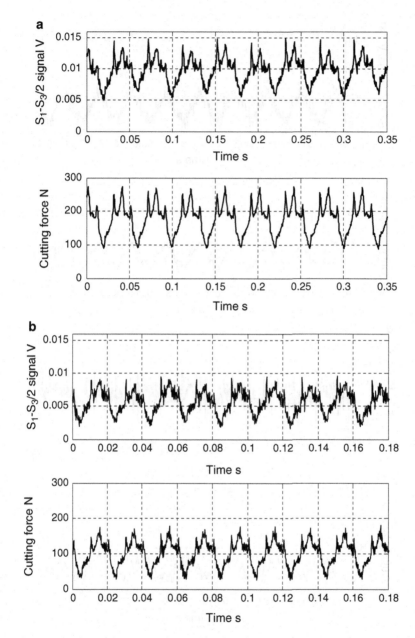

**Fig. 29.4** The compensated spindle displacement and measured cutting force. (**a**) Spindle speed: 1,500 min$^{-1}$. (**b**) Spindle speed: 3,000 min$^{-1}$. (**c**) Spindle speed: 6,000 min$^{-1}$. (**d**) Spindle speed: 12,000 min$^{-1}$

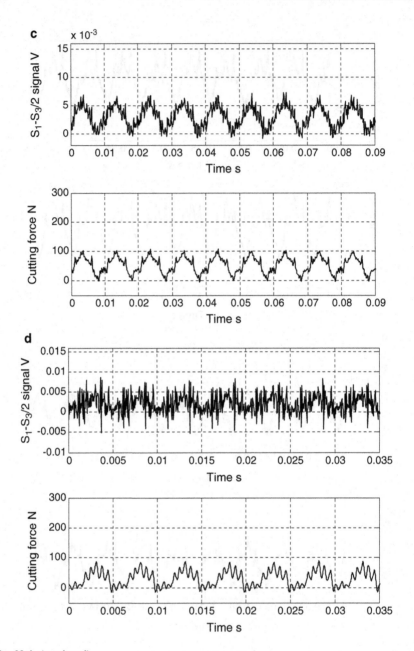

**Fig. 29.4** (continued)

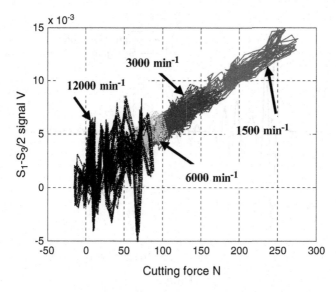

**Fig. 29.5** The relation between the spindle displacements and cutting forces at different spindle speeds

spindle speeds. As can be seen in Fig. 29.4, the displacement profiles looks similar to the force profiles.

Figure 29.5 shows the relation between the spindle displacements and cutting forces at different spindle speeds. The quantization error in digital measurement system is 0.1526 mV. As can be seen in Fig. 29.5, the spindle displacement has a linear relationship with the cutting force when the spindle speed is $<12{,}000 \ \mathrm{min}^{-1}$. In the case where the spindle speed is $12{,}000 \ \mathrm{min}^{-1}$, the relationship between the spindle displacement and the cutting force shows an unstable characteristic. This characteristic may come from the dynamics limitations of the measurement system.

## 4   The Dynamics Limitations of the Measurement System

To investigate the dynamic limitations of the measurement system, the spindle structure is excited on the tool tip and tool holder. The spindle is hammered by an impact force hammer while the spindle is stopping. Both impact force and displacement are recorded synchronously and processed using a Fourier analyzer system to identify the frequency bandwidth. The sampling frequency used is 5 kHz. Figure 29.6 shows examples of the displacement signal ($S_3$) and the force signal. The reason why the first peak of the force is opposite to the one of the displacements is that the displacement sensor detects the gap which is decreased in the hammered direction. Figure 29.7 shows the frequency response of displacement sensor $S_3$. The first major resonance peak is at approximately 570 Hz and the second one is at

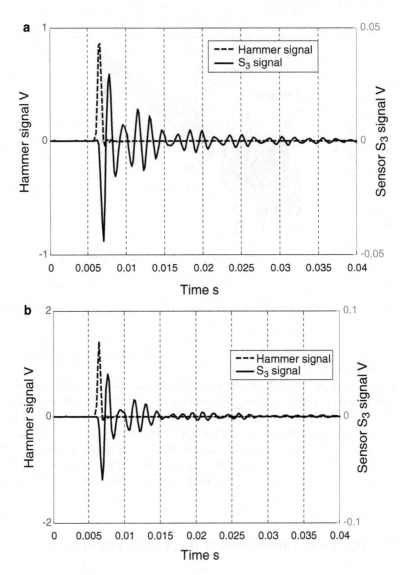

**Fig. 29.6** The displacement signals ($S_3$) and the force hammer signals. (**a**) Exited at the tool tip. (**b**) Exited at the tool holder

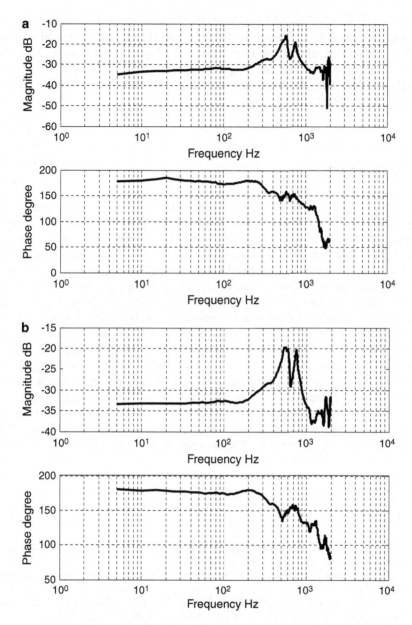

**Fig. 29.7** Frequency response of displacement sensor $S_3$. (**a**) Frequency response from tool tip to sensing point. (**b**) Frequency response from tool holder to sensing point

750 Hz. The transfer function measurements indicate that the sensor system can respond to cutting forces whose bandwidth is 200 Hz whereas the magnitude and phase shift remains almost invariant. Hence, the sensor system can measure the force components precisely at the spindle speeds up to 12,000 $min^{-1}$, which is totally matching with the results shown in Fig. 29.5. Similar results are obtained from the other displacement sensors.

# 5   Conclusion

Displacement sensors are installed in the spindle structure of a machining center for the monitoring of cutting forces. We present how to monitor the spindle displacement more precisely by using four eddy-current displacement sensors. The displacement sensor measures not only the variation of the gap size between the sensor head and the target surface but also the displacements due to error motion of a spindle's axis of rotation in radial direction and roundness errors of the target surface. The error motion of a spindle's axis of rotation in radial direction and roundness errors of the target surface are compensated by subtracting the displacement signals measured while air cutting from the displacement signals measured while cutting. The accuracy of the compensation is investigated by comparing the compensated spindle displacement with cutting forces measured by using a table type tool dynamometer. The test results show that the monitoring system is reliable for the adaptive control of machining accuracy for end-milling process. However, in the case where the spindle speed is 12,000 $min^{-1}$, the relationship between the spindle displacement and the cutting force shows an unstable characteristic. This characteristic may come from the dynamics limitations of the measurement system.

# References

1. Martin KF (1994) A review by discussion of condition monitoring and fault diagnosis in machine tools. Int J Mach Tools Manuf 34(4):527–551
2. Andrews GC, Tlusty J (1983) A critical review of sensors for unmanned machining. Ann CIRP 32–2:563–572
3. Byrne G, Dornfeld D, Inasaki I, Ketteler G, Konig W, Teti R (1995) Tool condition monitoring (tcm)—status of research and industrial application. Ann CIRP 44(2):541–567
4. Strafford KN, Audy J (1997) Indirect monitoring of machinability in carbon steels by measurement of cutting forces. J Mater Process Technol 67(1–3):150–156
5. DimlaSr DE, Lister PM (2000) On-line metal cutting tool condition monitoring: I: force and vibration analyses. Int J Mach Tools Manuf 40(5):739–768
6. Song D-Y, Otani N, Aoki T, Kamakoshi Y, Ohara Y, Tamaki H (2005) A new approach to cutting state monitoring in end-mill machining. Int J Mach Tools Manuf 45(7–8):909–921
7. Sarhan A, Sayed R, Nasr AA, El-Zahry RM (2001) Interrelation between cutting force variation and tool wear in end-milling. J Mater Process Technol 109(3):229–235

8. Lee P, Altintas Y (1996) Prediction of ball end milling forces from orthogonal cutting data. Int J Mach Tools Manuf 36:1059–1072
9. Feng HS, Menq CH (1994) The prediction of cutting forces in the ball end milling process—II. Cut geometry analysis and model verification. Int J Mach Tools Manuf 34:711–719
10. Sokolowski S, Kosmol J (1996) Intelligent monitoring system designer. In: Japan/USA symposium on flexible automation, vol 2. ASME, New York
11. Feng HS, Menq CH (1994) The prediction of cutting forces in the ball end milling process model-I. Formulation and model building procedure. Int J Mach Tools Manuf 34:697–710
12. Teti R, Jawahir IS, Jemielniak K, Segreto T, Chen S, Kossakowska J (2006) Chip form monitoring through advanced processing of cutting force sensor signals. CIRP Ann Manuf Technol 55(1):75–80
13. DimlaSnr DE (2000) Sensor signals for tool-wear monitoring in metal cutting operations— a review of methods. Int J Mach Tools Manuf 40(8):1073–1098
14. Huang SN, Tan KK, Wong YS, de Silva CW, Goh HL, Tan WW (2007) Tool wear detection and fault diagnosis based on cutting force monitoring. Int J Mach Tools Manuf 47 (3–4):444–451
15. Matsubara A, Kakino Y, Ogawa T, Nakagawa H, Sato T (2000) Monitoring of cutting forces in end-milling for intelligent machine tools. In: Proceedings of the 5th international conference on progress of machining technology, Beijing, p 615
16. Chung YL, Spiewak SA (1994) A model of high performance dynamometer. J Eng Ind (ASME) 16:279–288
17. Santochi M, Dini G, Tantussi G, Beghini M (1997) A sensor-integrated tool for cutting force monitoring. CIRP Ann Manuf Technol 46(1):49–52
18. Altintas Y (1992) Prediction of cutting forces and tool breakage in milling from feed drive current measurement. J Eng Ind (ASME) 114(4):386–391
19. Sarhan AAD, Sugihara M, Saraie H, Ibaraki S, Kakino Y (2006) Monitoring method of cutting force by using additional spindle sensors. JSME Int J Ser C 49(2):307–315
20. Donaldson RR (1972) A simple method for separating spindle error from test ball roundness. Ann CIRP 21:125–26
21. Tu JF, Bossmanns B, Hung Spring C C (1997) Modeling and error analysis for assessing spindle radial error motions. Precis Eng 21(2–3):90–101
22. Shinno H, Mitsui K, Tatsue Y, Tanaka N, Omino T, Tabata T, Nakayama K (1987) A new method for evaluating error motion of ultra precision spindle. CIRP Ann Manuf Technol 36(1):381–384
23. Fujimaki K, Mitsui K (2007) Radial error measuring device based on auto-collimation for miniature ultra-high-speed spindles. Int J Mach Tools Manuf 47(11):1677–1685
24. Castro HFF (2008) A method for evaluating spindle rotation errors of machine tools using a laser interferometer. Measurement 41(5):526–537
25. Noguchi S, Tsukada T, Sakamoto A (1995) Evaluation method to determine radial accuracy of high-precision rotating spindle units. Precis Eng 17(4):266–273
26. Sarhan AAD, Hassan MA, Matsubara A, Hamdi M (2011) Compensation of machine tool spindle error motions in the radial direction for accurate monitoring of cutting forces utilizing sensitive displacement sensors. In: Proceedings of the world congress on engineering 2011 (WCE 2011), vol I, London, 6–8 July 2011, pp 535–539

# Chapter 30
# Modeling a Complex Production Line Using Virtual Cells

Luís Pinto Ferreira, Enrique Ares Gómez, Gustavo Peláez Lourido, and Benny Tjahjono

## 1 Introduction

Simulation is considered one of the worthwhile tools that can be used to analyze the behavior of a production line especially when the application of analytical methods proves to be difficult [1, 2]. The complexity of production systems justifies the use of simulation techniques in the detection of critical problems during the design or redesign of new systems, or in the diagnosis of existing systems in order to improve their performance [3]. According to (Sadowski) in Torbörn Ilar [4], a successfully developed simulation project is one which produces useful information, within an adequate time-span, to support decision-making.

The work described in this chapter consists of the development of a simulation model based on a real case and is aimed at a very specific class of production lines, with a four closed loop network configuration, used commonly in the automotive sector. This study is an extension of the work of Resano and Pérez [5, 6], who designed one of the first analytical models for an assembly line in the vehicle sector as a network of four closed loops of machines, decoupled by intermediate buffers formed by conveyors. They consider that machines process pallets, which are not univocally related to each other. The use of simulation model presented in this

L.P. Ferreira (✉)
Escola Superior de Estudos Industriais e de Gestão (Technical Scientific Unity of Industrial Engineering and Production), Instituto Politécnico do Porto, Porto, Portugal
e-mail: Luispintoferreira@eu.ipp.pt

E.A. Gómez • G.P. Lourido
Área Ingeniería de los Procesos de Fabricación, Universidad de Vigo, Vigo, Spain
e-mail: enrares@uvigo.es; gupelaez@uvigo.es

B. Tjahjono
Manufacturing Department, Cranfield University, Cranfield, UK
e-mail: b.tjahjono@cranfield.ac.uk

S.-I. Ao and L. Gelman (eds.), *Electrical Engineering and Intelligent Systems*, Lecture Notes in Electrical Engineering 130, DOI 10.1007/978-1-4614-2317-1_30, © Springer Science+Business Media, LLC 2013

chapter, the blocking and starvation phenomena of a complex production line can be analyzed. These models also consider the proportion of four- and two-door cars between the door disassembly and assembly stations.


## 2  Conceptual Modeling Based on Virtual Cells

The concept of virtual cells (or also known as reconfigurable cells) can be conceptually used in the modeling of a production line. The concept of the virtual cell was first introduced by McLean et al. [7] in the context of the development of control software for an automated production line of small lots of parts. A virtual cell stems from the purely logical integration, and not a physical one, of the resources needed for the manufacture of a product [8]. A virtual cell is the logical grouping of resources within a controller. When an order is placed (which determines a set of work stations to manufacture a specific product), the virtual cell controller takes on the control of those stations and makes communication and iteration among them possible [7, 9–11]. The virtual cell differs from a traditional production cell in two aspects: the concept of *machine sharing* and the *physical layout of the machines*. In the traditional production system, a family of products is expected to be processed entirely within a dedicated cell of machines. Any movement of parts between the different cells is not advisable. Thus, common problems are the duplication of equipment and its low rate of usage. The deployment of the machinery influences all of the production activities. Therefore, any changes in the product to be manufactured with the purpose of meeting the market demand imply changing the line layout, which consequently impacts the time expenditure and its associated costs [12, 13]. In contrast, virtual cells are configured according to the dynamic and temporary allocation of the workstations so as to process a specific product [8]. In a virtual cell, the position of machinery is not critical [12]. Thus, according to Khilwani et al. [14], virtual cells have taken on a relevant role in cases of variable demand as well as in those where there are changes in the structure of manufactured products. According to Baykasoglu [15], the main difference between a virtual cell and the physical manufacturing cell resides in the dynamic nature of the former. In the case of physical cells, the location of work stations is fixed and can be perfectly identifiable. This does not happen with virtual cells since they will vary according to the requirements at hand. According to Ares et al. [16], a virtual cell is any production scenario, from the lowest levels in a manufacturing hierarchy (for example, a CNC machine) to the highest (for example, a factory). Its internal organization allows for the manufacture of different types of products, as well as that of establishing significant parameters in its production capacity.

Figure 30.1 shows an example of how three virtual cells can be grouped in a job-shop production, in which three types of products are processed and where each of the geometric features belonging to the system represent machines that enable operations to be undertaken according to the flow of each product [17].

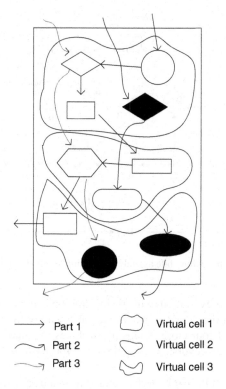

| | | | | |
|---|---|---|---|---|
| ⟶ | Part 1 | | Virtual cell 1 |
| ⌇ | Part 2 | | Virtual cell 2 |
| ⋯⟶ | Part 3 | | Virtual cell 3 |

**Fig. 30.1** Line of virtual cells based on job-shop production (adapted from [17])

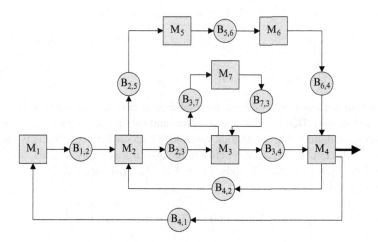

**Fig. 30.2** Main automobile assembly line and subassembly lines [5, 6]

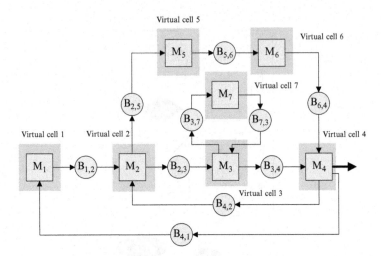

**Fig. 30.3** Modeling of the automobile assembly line using seven virtual cells [18]

Figure 30.2 is a representation of the main automobile production line used in this research, along with the door and front axle subassembly lines. The configuration exhibits a network of four closed loops of machines (M) and intermediate buffers (B) connected by conveyors. The cycle time for each machine is fixed and deterministic.

The total number of the car bodies, door and front axle assembly pallets stored in each of the intermediate buffers of the three first closed loops remain constant at any time and are defined by (30.1)–(30.3) [5, 6].

$$n_{12} + n_{23} + n_{34} + n_{41} = 237 \tag{30.1}$$

$$n_{25} + n_{56} + n_{64} + n_{42} = 450 \tag{30.2}$$

$$n_{37} + n_{73} = 138 \tag{30.3}$$

The fourth closed loop defines the relationship between the number of pallets with car doors in different subassembly states and the number of pallets of cars with disassembled doors using an external variable $(x)$, according to (30.4). This variable represents the four-door car ratio between the door disassembly stations, located at $M_2$ and the door assembly stations, located at $M_4$. This variable can have values between 0 and 1.

$$n_{25} + n_{56} + n_{64} + 60 = (216 + n_{23} + n_{34})(1 + x) \tag{30.4}$$

Figure 30.3 shows the automobile assembly line analyzed in this study. As with Resano and Pérez [5, 6], seven virtual cells were constructed so as to better understand the trajectories of the state of the intermediate buffers formed by

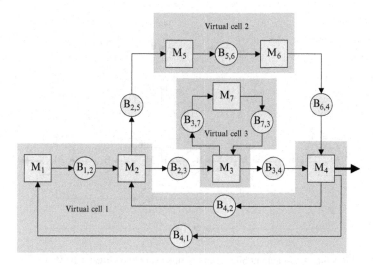

**Fig. 30.4** Modeling of the automobile assembly line based on three virtual cells [18]

conveyors, positioned between each virtual cell and thus ensuring supply to the cells and avoiding starvation. These virtual cells are represented by each of the machines (M) presented in Fig. 30.3. In the automobile production line being analyzed, each of these constitutes a set of different types of stations (e.g. manual stations, automatic stations, semiautomatic stations), which are logically grouped with the purpose of undertaking a specific set of tasks [18]. For this study, this modeling of line configuration can be used to develop a decision-support system based on a discrete event simulation model for a multistage multiproduct automobile production line.

Another line configuration modeling option based on three virtual cells could be used for the analysis and optimization of the automobile assembly line being studied, as is presented in Fig. 30.4.

Another modeling example based on the concept of virtual cells is presented in Fig. 30.5, in which the automobile assembly line is structured on the basis of two virtual cells.

The main advantage of virtual cells is that they facilitate modeling of a complex network system, in a system of multistage lines, where each stage (see Fig. 30.6) is a virtual cell $M_j$, its entry and exit buffers are $B_{ij}$ and $B_{i,j+1}$, and $Q_{ij}$ is the supply flow of parts, from buffer $B_{ij}$ to the virtual cell $M_j$ [16, 17, 19–21].

Virtual cell $M_j$ can process various types of parts. The entry of parts on cell $M_j$ is undertaken sequentially and its rate is determined by the production strategy being considered. The $Q_{ij}(t)$ represents the supply flow, which crosses virtual cell $M_j$, from the entry buffer $B_{ij}$, to the exit buffer $B_{i,j+1}$. If the line consists of m virtual cells and each cell processes $n$ parts, there will then be $n \times m$ flows in the entire system. The flow of parts $Q_{ij}(t)$ is expressed by (30.5), where index $i$ represents the

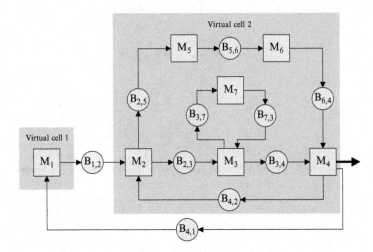

**Fig. 30.5** Modeling of the automobile assembly line using two virtual cells [18]

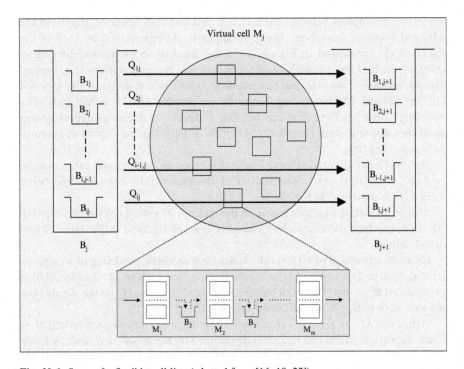

**Fig. 30.6** Stage of a flexible cell line (adapted from [16, 19, 22])

type of part, and index $j$ indicates the order that the virtual cell being analyzed occupies on the line [16, 21, 23, 24].

$$Q_{ij}(t) = S_{ij} \times V_{ij}(t) = (1/T_{ij}) \times V_{ij}(t), \quad i = 1, \ldots, n; \quad (30.5)$$

$$\text{subject to:} \quad \sum_{i=1}^{n} V_{ij}(t) \leq 1$$

where:

- $T_{ij}$ represents the processing time for each part ($T_{ij} \in$ IR, is known and fixed);
- $S_{ij}$ translates maximum production flow ($S_{ij} = 1/T_{ij}$);
- $V_{ij}(t)$ represents the dedicated coefficient for each virtual cell $M_j$ for each type of part $i$, in every moment in time $t$. The $V_{ij}(t)$ is obtained in accordance with the control strategy selected a priori ($V_{ij}(t) \in$ IR, $0 \leq V_{ij}(t) \leq 1$).

The space restrictions in each of the buffers $B_{ij}$ and $B_{i,j+1}$, represented in Fig. 30.6, require one to balance the flows in order to prevent line blockages. Considering that the variable $X_{ij}(t)$ indicates the instant value for the level of storage on each of the buffers, this variable will represent the state of the buffers at every moment. Thus, if $b_{ij}$ represents the maximum capacity for buffer $B_{ij}$, variable $X_{ij}(t)$ can be expressed by (30.6).

$$0 \leq X_{ij}(t) \leq b_{ij} \quad (30.6)$$

The evolution over time of the state variable $X_{ij}(t)$, which will be represented by $Y_{ij}(t)$, is provided by the difference between the supply flow of parts to $B_{ij}$, $Q_{ij-1}(t)$ and the supply flow of parts from buffer $B_{ij}$ to the $M_j$ cell, which is $Q_{ij}(t)$. Thus, $Y_{ij}(t)$ can be represented by (30.7). Therefore, if $Y_{ij}(t) > 0$, this means that the buffer is filling up; if $Y_{ij}(t) < 0$, this means that the buffer is becoming empty and if $Y_{ij}(t) = 0$, one can consider the flows to be balanced ($X_{ij}(t) = $ constant) [16, 17, 19, 21, 24, 25].

$$Y_{ij}(t) = Q_{ij-1}(t) - Q_{ij}(t) \quad (30.7)$$

The three running modes or buffer states for $M_j$ cells or for their $B_{ij}$ can be defined as:

- *Full output mode (on $M_{j-1}$)*. If buffer $B_{ij}$ is full, the parts being manufactured on cell $M_{j-1}$ cannot be placed here since this will cause the blockage of cell $M_{j-1}$. In this case, the instant value of the supply level can be expressed as $X_{ij}(t) = b_{ij}$.
- *Null input mode (on $M_j$)*. If buffer $B_{ij}$ is empty, there are no parts to supply cell $M_j$, causing a situation of starvation on $M_j$. In this case, the instant value for the storage level is provided by the condition: $X_{ij}(t) = 0$.
- *Normal mode*. Here, the buffer is neither full nor empty. In this running mode, the value of $X_{ij}(t)$ is provided by the following expression: $0 < X_{ij}(t) < b_{i,j+1}$.

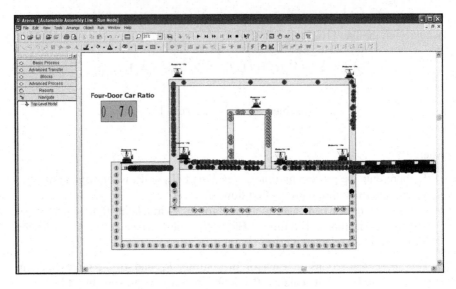

**Fig. 30.7** Simulation model for $x = 0.7$

## 3 Characteristics of the Decision-Making Support System Developed

In the context of this project, all the work was developed in an Arena® simulation environment [26]. Its simulation language constitutes a visual and flexible programming tool, directed at the object, since it simultaneously combines the construction of simulation models with the integration of different commonly used languages: Visual Basic, C, C++. This language is based on the Siman simulation language [18, 27, 29].

The purpose of the use of an Arena simulation environment was that of enabling the production engineer to evaluate the performance of the automobile assembly line, through the variation of different parameters, thus contributing to an improved specification, characterization and definition of the most efficient control system. With this objective in mind, a support system for decision-making was developed; this enables the automatic generation of different simulation models. In the initial stage of simulation, the user is able to interact with the system to be developed, through the introduction of various parameters such as [18, 28, 29]: the four-door car ratio $(x)$; the processing time for each machine; the production sequence in accordance with each type of car (two- or four-door); the speed and length of the intermediate buffers formed by conveyors; simulation time; the number of pallets circulating on the first three closed loops. Figure 30.7 shows an example of a screenshot of the simulation model developed in Arena® incorporating the abovementioned parameters.

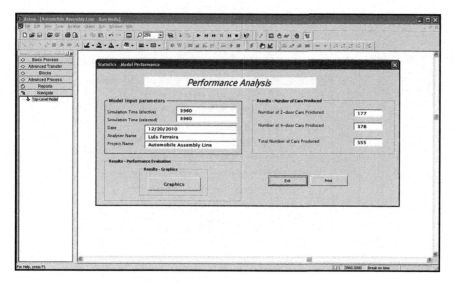

**Fig. 30.8**  Main selected input parameters

Another contribution made by this study is the automatic report generator in the simulation environment; this presents information on the main performance indexes of the simulated models, using graphic visualization, which required the integration of various applications in the Arena simulation environment, such as: Visual Basic for Applications (VBA), Visual Basic (VB), Microsoft Access (Database), and Crystal Report (Graphics Edition). So at the end of each simulation, a summary of the main selected input parameters is presented to the user (see Fig. 30.8). From this interface, the user can access to another module (see Fig. 30.9), where he or she has the opportunity to edit, in the form of graphics, information concerning the simulated model's performance. In this context, it is worth highlighting the visual aspects of the graphics produced (two examples are presented in Figs. 30.10 and 30.11), which allow the user to have a better perception of the performance of the simulated models. From among those considered, the following performance indicators are highlighted: machine usage levels; number of cars for each of the types produced; number of operations undertaken on each of the machines; production time/vehicle; relationship between the time cycle of each machine and the number of pallets on each upstream buffer.

## 4   Validation of the Simulation Model Proposed

In Resano and Pérez [5, 6], the incompatibility between (30.2) and (30.4) was demonstrated. These define the sum of the number of pallets on the intermediate buffers, which constitute the second and fourth closed loop, with some of the

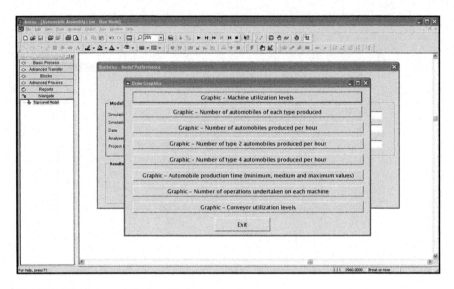

**Fig. 30.9** Simulation results (graphics)

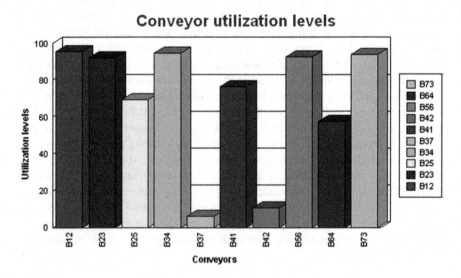

**Fig. 30.10** Conveyor utilization levels

maximum and minimum capacities of the corresponding intermediate buffers, for the specific values of variable $x$. This incompatibility reveals that the automobile assembly line cannot operate in practice for $x < 0.37$, in both the stationary and transitory regimes, as well as for $x > 0.97$, in the stationary regime. As with any simulation project, an important phase is the validation of the model constructed;

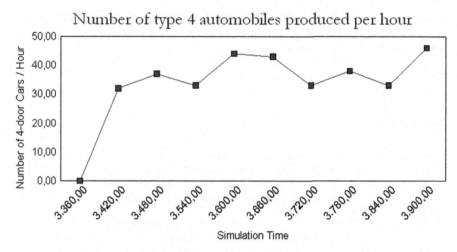

**Fig. 30.11**  Number of type 4 automobiles produced per hour

this ensures that it is a true representation of reality and can then be used for decision-making. Through the simulation model developed in the context of this study, and for the automobile production line being analyzed, one also verified that the system did not function for the values of $x < 0.37$ and $x > 0.97$, due to the phenomena of blockage and starvation occurring on the machines integrating the analyzed model. In the same way, one also noticed that, once the minimum buffer capacity mentioned in Resano and Pérez [5, 6] was reached, a continuous supply of transport and assembly pallets to machines was assured. In the context of this work, the validation of the simulation model proposed was confirmed, on the basis of these two presuppositions [18, 28, 29].

## 5   Conclusion

In this paper, an automobile assembly line and different preassembly lines were modeled as a network of four closed loop machines and intermediate buffers formed by conveyors, constituting a configuration which is widely used in these kinds of assembly line. A representative model of this line was developed with the purpose of providing the production engineers with a better understanding and assessment of its performance; it represents each moment in the state of the system and its evolution over time, thus allowing for improved communication between the model and the user. In fact, the visual components in the model allows for a clear understanding of the decision-making agents who will eventually use them, since they will be able to view the interaction that occurs between the entities which make up the model, allowing for a better understanding of the simulation results.

372                                                                    L.P. Ferreira et al.

This study has contributed to the importance of simulation in the development of a computer-based model and virtual environments, which is a reproduction of the behavior of many real systems.

# References

1. Benedettini O, Tjahjono B (2009) Towards an improved tool to facilitate simulation modelling of complex manufacturing systems. Int J Adv Manuf Technol 43:191–199
2. Papadopoulos CT, O'Kelly MEJ, Vidalis MJ, Spinellis D (2009) Analysis and design of discrete part production lines. In: Springer optimization and its applications, vol 31. Springer, New York. ISBN 978-0-387-89493-5
3. Ferreira JJP (1995) Suporte do Ciclo de Vida dos Sistemas Integrados de Fabrico através de Modelos Executáveis sobre Infra-estruturas de Integração. Tese de Doutoramento em Engenharia Electrotécnica e de Computadores, Faculdade de Engenharia da Universidade do Porto
4. Ilar T (2008) Production simulation as a management tool. Doctoral Thesis, University of Technology, Luleå, ISSN: 1402-1544
5. Resano Lázaro A (2007) Análisis Funcional y Optimización de la Distribuición en Planta de una Línea de Ensamblaje de Automóviles. PhD Thesis, Departamento de Ingeniería Mecánica
6. Resano Lázaro A, Luis Pérez CJ (2008) Analysis of an automobile assembly line as a network of closed loops working in both, stationary and transitory regimes. Int J Prod Res 46 (17):4803–4825, ISSN: 1366-588X
7. Mclean CR, Bloom HM, Hopp TH (1982) The virtual manufacturing cell. In: Proceedings of fourth ifAC/IFIP conference on information control problems in manufacturing technology, USA, pp 105–111
8. Alves AC (2007) Projecto Dinâmico de Sistemas de Produção Orientados ao Produto. Tese de Doutoramento em Engenharia de Produção e Sistemas, Universidade do Minho
9. Drolet J, Abdulnour G, Rheault M (1996) The cellular manufacturing evolution. Comp Ind Eng 31(1–2):139–142
10. Mak KL, Wang XX (2002) Production scheduling and cell formation for virtual cellular manufacturing systems. Int J Adv Manuf Technol 20:144–152
11. Nomden G, van der Zee D-J (2008) Virtual cellular manufacturing: configuring routing flexibility. Int J Prod Econ 112(1):439–451
12. Ko K-C, Egbelu PJ (2003) Virtual cell formation. Int J Prod Res 41(11):2365–2389
13. Fung RYK, Liang F, Jiang Z, Wong TN (2008) A multi-stage methodology for virtual cell formation oriented agile manufacturing. Int J Adv Manuf Technol 36:798–810
14. Khilwani N, Ulutas BH, Attila Islier A, Tiwari MK (2009) A methodology to design virtual cellular manufacturing systems. J Intell Manuf, Springer, 22(4): 533–544
15. Baykasoglu A (2003) Capability-based distributed layout approach for virtual manufacturing cells. Int J Prod Res 41(11):2597–2618
16. Ares Gómez E, Peláez Lourido G, Gómez Lourido R (1990) Modelo Dinamico para la Evaluacion de la Eficiencia de Estrategias de Produccion en Ambiente CIM. In: Anales de Ingeniería Mecánica, Revista de la Asociación Española de Ingeniería Mecánica, Diciembre 1990
17. Peláez Lourido G (1999) Arquitectura y Modelo integral de un Sistema de Fabricación multietapa/multiproducto. Thesis Doctoral, Universidade de Vigo
18. Ferreira LP, Ares Gómez E, Peláez Lourido G, Tjahjono B (2011) Optimization of a multiphase multiproduct production line based on virtual cells. In: Lecture notes in engineering and computer science: Proceedings of the world congress on engineering (WCE2011), London, 6–8 July 2011, pp 616–621

19. Ares Gómez JE (1986) Estructura Jerárquica de Metodologías para la Implantación y Gestión de Sistemas de Fabricación Flexible. Tesis Doctoral, Universidad de Santiago de Compostela

20. Villa A, Ares E (1988) In: Mital A (ed) Methodology to evaluate dispatching rules for a manufacturing line by discrete event dynamic model, vol 6. Elsevier, Amsterdam, pp 485–491. ISBN 0-4444-42929

21. Ares Gómez JE, Pérez Garcia JA, Peláez Lourido GC (1992) Diseño de Células en el Lay-Out de un Taller y Análisis del Flujo Multi-Etapa/Multi-Producto en Un Entorno CIM. In: X Congreso Nacional de Ingeniería Mecánica, Revista de la Asociación Española de Ingeniería Mecánica, Septiembre 1992

22. Ares E, Peláez G (1997) Integración de línea y célula flexibles en un modelo jerárquico para la planificación y programación de la producción. In: XII Congreso Nacional de Ingeniería Mecánica, Anales de Ingeniería Mecánica, Revista de la Asociación Española de Ingeniería Mecánica, Febrero 1997

23. Ares Gómez E, Ollero A, Arcostanzo M, Villa A (1987) Planificación de la producción en sistemas de fabricación en estaciones interconectadas en red. In: Anales de Ingeniería Mecánica, Revista de la Asociación Española de Ingeniería Mecánica, Diciembre 1987

24. Villa A, Fiorio G, Ares E (1987) In: Kusiak A (ed) Production planning and control in multi-stage multi-product systems. Elsevier, Amsterdam, pp 247–257. ISBN 0-444-70272-5

25. Villa A, Ares Gómez JE (1987) A methodology to analyze workshop lines by discrete event dynamic models. In: Mital A (ed), IXth International conference on production research, vol I, Cincinnati, pp 2072–2078

26. David Kelton W, Sadowski RP, Sturrock DT (2007) Simulation with arena, 4th edn. McGraw-Hill, New York. ISBN 978-0-07-110685-6

27. Takus DA, Profozich DM (1997) ARENA Software Tutorial. In: Proceedings of 1997 winter simulation conference, Atlanta, pp 541–544

28. Ferreira LP, Ares Gómez E, Peláez Lourido GC, Quintas JD, Tjahjono B (2011) Analysis and optimisation of a network of closed-loop automobile assembly line using simulation. Int J Adv Manuf Technol, Springer, 59:351–366

29. Ferreira LP, Ares Gómez E, Peláez Lourido G, Salgado M, Quintas JD (2011) Analysis on the influence of the number of pallets circulating on an automobile closed-loop assembly line. Int J Adv Eng Sci Technol 2(2):119–123

# Chapter 31
# Optimal Quantity Discount Strategy for an Inventory Model with Deteriorating Items

Hidefumi Kawakatsu

## 1 Introduction

This paper presents a model for determining optimal all-unit quantity discount strategies in a channel of one seller (wholesaler) and one buyer (retailer). Many researchers have considered the seller's quantity discount decision. By offering a discounted price to induce the buyer to order in larger quantities, the seller can increase her/his profit through reductions in her/his total transaction cost associated with ordering, shipment, and inventorying. Monahan [1] was one of the early authors who formulated the transaction between the seller and the buyer, and proposed a method for determining an optimal all-unit quantity discount policy with a fixed demand (see also Rosenblatt and Lee [2] and Data and Srikanth [3]). Lee and Rosenblatt [4] generalized Monahan's model to obtain the "exact" discount rate offered by the seller, and to relax the implicit assumption of a lot-for-lot policy adopted by the seller. Parlar and Wang [5] proposed a model using a game theoretical approach to analyze the quantity discount problem as a perfect information game. For more work, see also Sarmah et al. [6]. These models assumed that both the seller's and the buyer's inventory policies can be described by classical economic order quantity (EOQ) models. The classical EOQ model is a cost-minimization inventory model with a constant demand rate. It is one of the most successful models in all the inventory theories due to its simplicity and easiness.

In many real-life situations, retailers deal with perishable products such as fresh fruits, food-stuffs, and vegetables. The inventory of these products is depleted not only by demand but also by deterioration. Yang [7] has developed the model to determine an optimal pricing and an ordering policy for deteriorating items with quantity discount, which is offered by the vendor. His model assumed that the

H. Kawakatsu (✉)
Department of Economics and Information Science, Onomichi University,
1600 Hisayamada-cho, Onomichi 722-8506, Japan
e-mail: kawakatsu@onomichi-u.ac.jp

S.-I. Ao and L. Gelman (eds.), *Electrical Engineering and Intelligent Systems*,
Lecture Notes in Electrical Engineering 130, DOI 10.1007/978-1-4614-2317-1_31,
© Springer Science+Business Media, LLC 2013

deterioration rate of the vendor's inventory is equal to its rate of the retailer's inventory, and focused on the case where both the buyer's and the vendor's total profits can be approximated using Taylor series expansion.

In this study, we discuss a quantity discount problem between a seller (wholesaler) and a buyer (retailer) under circumstances where both the wholesaler's and the retailer's inventory levels of the product are depleted not only by demand but also by deterioration. The wholesaler purchases products from an upper-leveled supplier (manufacturer) and then sells them to the retailer who faces her/his customers' demand. The shipment cost is characterized by economies of density [8]. The wholesaler is interested in increasing her/his profit by controlling the retailer's order quantity through the quantity discount strategy. The retailer attempts to maximize her/his profit considering the wholesaler's proposal. Our previous work has formulated the above problem as a Stackelberg game between the wholesaler and the retailer to show the existence of the wholesaler's optimal quantity discount pricing policy, which maximizes her/his total profit per unit of time [9]. Our previous study showed how to determine the optimal quantity discount policy in the case of $N_2 = 1$, where $N_2$ will be defined in Sect. 2. In this study, we reveal additional properties associated with the optimal policy in the case of $N_2 \geq 2$. Numerical examples are presented to illustrate the theoretical underpinnings of the proposed model.

## 2  Notation and Assumptions

The wholesaler uses a quantity discount strategy in order to improve her/his profit. The wholesaler proposes, for the retailer, an order quantity per lot along with the corresponding discounted wholesale price, which induces the retailer to alter her/his replenishment policy.

We consider the two options throughout the present study as follows:

*Option $V_1$*: The retailer does not adopt the quantity discount proposed by the wholesaler. When the retailer chooses this option, she/he purchases the products from the wholesaler at an initial price in the absence of the discount, and she/he determines her/himself an optimal order quantity, which maximizes her/his own total profit per unit of time.

*Option $V_2$*: The retailer accepts the quantity discount proposed by the wholesaler.

The main notations used in this paper are listed below:

$Q_i$: the retailer's order quantity per lot under Option $V_i$ ($i = 1, 2$).
$S_i$: the wholesaler's order quantity per lot under Option $V_i$.
$T_i$: the length of the retailer's order cycle under Option $V_i$ ($T_1 \leq T_2$).
$h_s$, $h_b$: the wholesaler's and the retailer's inventory holding costs per item and unit of time, respectively.
$a_s$, $a_b$: the wholesaler's and the retailer's ordering costs per lot, respectively.
$\xi(T_i)$: the shipment cost per shipment from the wholesaler to the retailer.

$c_s$: the wholesaler's unit acquisition cost (unit purchasing cost from the upper-leveled manufacturer).

$p_s$: the wholesaler's initial unit selling price, i.e., the retailer's unit acquisition cost in the absence of the discount.

$p_b$: the retailer's unit selling price, i.e., unit purchasing price for her/his customers.

$y$: the discount rate for the wholesale price proposed by the wholesaler, i.e., the wholesaler offers a unit discounted price of $(1 - y)p_s$ $(0 \leq y < 1)$.

$\theta_s$, $\theta_b$: the deterioration rates of the wholesaler's inventory and of the retailer's inventory, respectively $(\theta_s < \theta_b)$.

$\mu$: the constant demand rate of the product.

The assumptions in this study are as follows:

(1) The retailer's inventory level is continuously depleted due to the combined effects of its demand and deterioration. In contrast, the wholesaler's inventory is depleted by deterioration during $[jT_i, (j + 1)T_i)$ $(j = 0, 1, 2, \ldots)$, but at time $jT_i$ her/his inventory level decreases by $Q_i$ because of shipment to the retailer.

The retailer's inventory level $I^{(b)}(t)$ and the wholesaler's inventory level $I^{(s)}(t)$ at time $t$ $(jT_i \leq t < (j+1)T_i)$ can, respectively, be expressed by the following differential equations [9]:

$$dI^{(b)}(t)/dt = -\theta_b I^{(b)}(t) - \mu, \tag{31.1}$$

$$dI^{(s)}(t)/dt = -\theta_s I^{(s)}(t). \tag{31.2}$$

(2) The rate of replenishment is infinite and the delivery is instantaneous.
(3) Backlogging and shortage are not allowed.
(4) The quantity of the item can be treated as continuous for simplicity.
(5) Both the wholesaler and the retailer are rational and use only pure strategies.
(6) The shipment cost is characterized by economies of density [8], i.e., the shipment cost per shipment decreases as the retailer's lot size increases. We assume, for simplicity, that $\xi(T_i) = \beta - \alpha Q(T_i) > 0$.
(7) The length of the wholesaler's order cycle is given by $N_i T_i$ $(N_i = 1, 2, \ldots)$ under Option $V_i$ $(i = 1, 2)$, where $N_i$ is a positive integer. This is because the wholesaler can possibly improve her/his total profit by increasing the length of her/his order cycle from $T_i$ to $N_i T_i$. In this case, the wholesaler's lot size can be obtained by the sum of $N_i$ times of the retailer's lot size and the cumulative quantity of the waste product to be discarded during $[0, N_i T_i)$.

## 3 Retailer's Total Profit and Optimal Response

In our previous work [9], we have formulated the retailer's total profits per unit of time under Options $V_1$ and $V_2$ available to the retailer, and then revealed the retailer's optimal response. For this reason, in the following we briefly summarize the results associated with the retailer's profits under these two options and her/his optimal response.

## 3.1   Retailer's Total Profit

If the retailer chooses Option $V_1$, her/his order quantity per lot and her/his unit acquisition cost are, respectively, given by $Q_1 = Q(T_1)$ and $p_s$, where $p_s$ is the unit initial price in the absence of the discount. In this case, she/he determines her/himself the optimal order quantity $Q_1 = Q_1^*$, which maximize her/his total profit per unit of time.

Under assumption (1), Kawakatsu [9] has formulated the retailer's total profit per unit of time, which is given by

$$\pi_1(T_1) = \frac{p_b \mu T_1 - p_s Q(T_1) - h_b \int_0^{T_1} I^{(b)}(t)dt - a_b}{T_1}$$
$$= \rho(p_b \theta_b + h_b) - \frac{(p_s + h_b/\theta_b)Q(T_1) + a_b}{T_1}, \tag{31.3}$$

where

$$\rho = \mu/\theta_b, \tag{31.4}$$

$$Q(T_1) = \rho(e^{\theta_b T_1} - 1). \tag{31.5}$$

We showed that there exists a unique finite $T_1 = T_1^*$ ($>0$), which maximizes $\pi_1(T_1)$ in (31.3) [9].

The optimal order quantity is therefore given by

$$Q(T_1) = \rho(e^{\theta_b T_1^*} - 1). \tag{31.6}$$

The total profit per unit of time becomes

$$\pi_1^* = \rho\theta_b[(p_b + h_b/\theta_b) - (p_s + h_b/\theta_b) e^{\theta_b T_1^*}]. \tag{31.7}$$

In contrast, if the retailer chooses Option $V_2$, the order quantity and unit discounted wholesale price are, respectively, given by $Q_2 = Q(T_2) = \rho(e^{\theta_b T_2} - 1)$ and $(1 - y)p_s$.

The retailer's total profit per unit of time can therefore be expressed by

$$\pi_2(T_2, y) = \rho(p_b \theta_b + h_b) - \frac{[(1-y)p_s + h_b/\theta_b]Q(T_2) + a_b}{T_2} \tag{31.8}$$

## 3.2   Retailer's Optimal Response

The retailer prefers Option $V_1$ over Option $V_2$ if $\pi_1^* > \pi_2(T_2, y)$, but when $\pi_1^* < \pi_2(T_2, y)$, she/he prefers $V_2$ to $V_1$. The retailer is indifferent between the two options if $\pi_1^* = \pi_2(T_2, y)$, which is equivalent to

$$y = \frac{1}{p_s} \cdot \frac{(p_s + h_b/\theta_b)[Q_2(T_2) - \rho\theta_b T_2\, e^{\theta_b T_1^*}] + a_b}{Q_2(T_2)}. \tag{31.9}$$

Let us denote, by $\Psi(T_2)$, the right-hand side of (31.9). It can easily be shown from (31.9) that $\Psi(T_2)$ is increasing in $T_2$ ( $\geq T_1^*$).

## 4   Wholesaler's Total Profit

This section formulates the wholesaler's total profit per unit of time, which depends on the retailer's decision. Figure 31.1 shows both the wholesaler's and the retailer's transitions of inventory level in the case of $N_i = 3$ under Option $V_i$ ($i = 1, 2$). The length of the wholesaler's order cycle is divided into $N_i$ shipment cycles as described in assumption (7), where $N_i$ is also a decision variable for the wholesaler. The wholesaler ships the goods to the retailer at time $jT_i$ ($j = 0, \ldots, N_i - 1$). The wholesaler's inventory is depleted only due to the deterioration during the interval $[jT_i, (j+1)\,jT_i)$, but at time $jT_i$ her/his inventory level drops from $z_j(jT_i)$ to $(z_j(jT_i) - Q_i)$ because of the shipment to the retailer, where $z_j(jT_i)$ denotes the wholesaler's inventory level at the end of $j$th shipment cycle. In contrast, the retailer's inventory is decreased due to the combined effects of its demand and deterioration. Obviously, we have $S_i = z_0(jT_i)$.

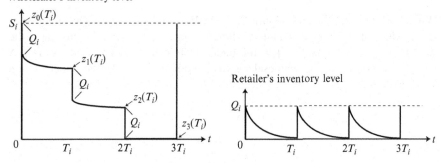

**Fig. 31.1**  Transition of inventory level ($N_i = 3$)

## 4.1   Total Profit Under Option $V_1$

If the retailer chooses Option $V_1$, her/his order quantity per lot and unit acquisition cost are given by $Q_1 = Q(T_1)$ and $p_s$, respectively.

The wholesaler's order quantity per lot can be expressed as $S_1 = S(N_1, T_1)$. Under assumption (1), for a given $N_1$, the wholesaler's total profit per unit of time under Option $V_1$ can be given by the following equation [9]:

$$
\begin{aligned}
P_1(N_1, T_1) &= \frac{1}{N_1 T_1} \cdot \left[ p_s N_1 Q(T_1) - N_1 \xi(T_1) - c_s S(N_1, T_1) \right. \\
&\quad \left. - h_s \sum_{j=1}^{N_1-1} \int_{(j-1)T_1}^{jT_1} I_j^{(s)}(t)\, dt - a_s \right] \\
&= \frac{\left(p_s + \frac{h_s}{\theta_s} + \alpha\right) Q(T_1) - \beta}{T_1} - \frac{\left(c_s + \frac{h_s}{\theta_s}\right) S(N_1, T_1) + a_s}{N_1 T_1},
\end{aligned}
\tag{31.10}
$$

where

$$
I_j^{(s)}(t) = z_j(T_1) e^{\theta_s (jT_1 - t)},
\tag{31.11}
$$

$$
z_j(T_1) = Q(T_1)[e^{(N_1-j)\theta_s T_1} - 1] / (e^{\theta_s T_1} - 1),
\tag{31.12}
$$

$$
S(N_1, T_1) = Q(T_1)(e^{N_1 \theta_s T_1} - 1) / (e^{\theta_s T_1} - 1).
\tag{31.13}
$$

## 4.2   Total Profit Under Option $V_2$

When the retailer chooses Option $V_2$, she/he purchases $Q_2 = Q(T_2)$ units of the product at the unit discounted wholesale price $(1 - y)p_s$. In this case, the wholesaler's order quantity per lot under Option $V_2$ is expressed as $S_2 = S(N_2, T_2)$, accordingly the wholesaler's total profit per unit of time under Option $V_2$ is given by

$$P_2(N_2, T_2, y) = \frac{1}{N_2 T_2} \cdot \left[ (1-y)p_s N_2 Q(T_2) - N_2 \xi(T_2) - c_s S(N_2, T_2) \right.$$

$$\left. - h_s \sum_{j=1}^{N_2-1} \int_{(j-1)T_2}^{jT_2} I_j^{(s)}(t) dt - a_s \right]$$

$$= \frac{\left[ (1-y)p_s + \frac{h_s}{\theta_s} + \alpha \right] Q(T_2) - \beta}{T_2} - \frac{\left( c_s + \frac{h_s}{\theta_s} \right) S(N_2, T_2) + a_s}{N_2 T_2},$$

$$(31.14)$$

where

$$Q(T_2) = \rho(e^{\theta_b T_1} - 1), \tag{31.15}$$

$$S(N_2, T_2) = Q(T_2)(e^{N_2 \theta_s T_2} - 1)/(e^{\theta_s T_2} - 1). \tag{31.16}$$

## 5   Wholesaler's Optimal Policy

The wholesaler's optimal values for $T_2$ and $y$ can be obtained by maximizing her/his total profit per unit of time considering the retailer's optimal response, which was discussed in Sect. 3.2.

Henceforth, let $\Omega_i$ ($i = 1, 2$) be defined by

$$\Omega_1 = \{(T_2, y) | y \leq \Psi(T_2)\},$$

$$\Omega_2 = \{(T_2, y) | y \geq \Psi(T_2)\}.$$

Figure 31.2 depicts the region of $\Omega_i$ ($i = 1, 2$) on the $(T_2, y)$ plane.

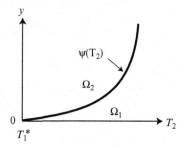

**Fig. 31.2** Characterization of retailer's optimal responses

## 5.1 Under Option $V_1$

If $(T_2, y) \in \Omega_1 \backslash \Omega_2$ in Fig. 31.2, the retailer will naturally select Option $V_1$. In this case, the wholesaler can maximize her/his total profit per unit of time independently of $T_2$ and $y$ on the condition of $(T_2, y) \in \Omega_1 \backslash \Omega_2$. Hence, for a given $N_1$, the wholesaler's locally maximum total profit per unit of time in $\Omega_1 \backslash \Omega_2$ becomes

$$P_1^* = \max_{N_1 \in N} P_1(N_1, T_1^*). \tag{31.17}$$

where $N$ signifies the set of positive integers.

## 5.2 Under Option $V_2$

On the other hand, if $(T_2, y) \in \Omega_2 \backslash \Omega_1$, the retailer's optimal response is to choose Option $V_2$. Then the wholesaler's locally maximum total profit per unit of time in $\Omega_2 \backslash \Omega_1$ is given by

$$P_2^* = \max_{N_2 \in N} \hat{P}_2(N_2), \tag{31.18}$$

where

$$\hat{P}_2(N_2) = \max_{(T_2, y) \in \Omega_2 \backslash \Omega_1} P_2(N_2, T_2, y). \tag{31.19}$$

More precisely, we should use "sup" instead of "max" in (31.19).

For a given $N_2$, we show below the existence of the wholesaler's optimal quantity discount pricing policy $(T_2, y) = (T_2^*, y^*)$, which attains (31.19). It is easily proven that $P_2(N_2, T_2, y)$ in (31.14) is strictly decreasing in $y$, and consequently the wholesaler can attain $\hat{P}_2(N_2)$ in (31.19) by letting $y \to \Psi(T_2) + 0$. By letting $y = \Psi(T_2)$ in (31.14), the total profit per unit of time on $y = \Psi(T_2)$ becomes

$$P_2(N_2, T_2) = (p_s + h_b/\theta_b)\rho\theta_b e^{\theta_b T_1^*}$$
$$- \frac{C \cdot S(N_2, T_2) + HN_2 Q(T_2) + (a_b + \beta)N_2 + a_s}{N_2 T_2}, \tag{31.20}$$

where

$$C = (c_s + h_s/\theta_s), \tag{31.21}$$

$$H = (h_b/\theta_b + h_s/\theta_s - \alpha). \tag{31.22}$$

By differentiating $P_2(N_2, T_2)$ in (31.20) with respect to $T_2$, we have

$$\frac{\partial}{\partial T_2} P_2(N_2, T_2) = -\frac{\left\{ \begin{array}{c} [\rho\theta_b T_2 \, e^{\theta_b T_2} - Q_2(T_2)] \\ \times \left( C \frac{e^{N_2 \theta_s T_2} - 1}{e^{\theta_s T_2} - 1} - HN_2 \right) \\ + C\theta_s T_2 Q(T_2) \\ \times \frac{N_2 \, e^{N_2 \theta_s T_2} (e^{\theta_s T_2} - 1) - e^{\theta_s T_2} (e^{N_2 \theta_s T_2} - 1)}{(e^{\theta_s T_2} - 1)^2} \end{array} \right\} - [(a_b + \beta)N_2 + a_s]}{N_2 T_2}.$$

$$(31.23)$$

Let $L(T_2)$ express the terms enclosed in braces { } in the right-hand side of (31.23).

We here summarize the results of analysis in relation with the optimal quantity discount policy, which attains $\hat{P}_2(N_2)$ in (31.19) when $N_2$ is fixed to a suitable value. The proofs are given in Appendix A.

1. $N_2 = 1$:

   - $(c_s + h_b/\theta_b - \alpha) > 0$ :
     In this subcase, there exists a unique finite $\tilde{T}_2$ $(>T_1^*)$, which maximizes $P_2(N_2, T_2)$ in (31.20), and therefore $(T_2^*, y^*)$ is given by

     $$(T_2^*, y^*) \rightarrow (\tilde{T}_2, \Psi(\tilde{T}_2)). \qquad (31.24)$$

     The wholesaler's total profit then becomes

     $$\hat{P}_2(N_2) = \rho\theta_b[(p_s + h_b/\theta_b) \, e^{\theta_b T_1^*} - (c_s + h_b/\theta_b - \alpha) \, e^{\theta_b T_2^*}]. \qquad (31.25)$$

   - $(c_s + h_b/\theta_b - \alpha) \leq 0$ :
     In this subcase, the optimal policy can be expressed by

     $$(T_2^*, y^*) \rightarrow (\hat{T}_2, 1). \qquad (31.26)$$

   where $\hat{T}_2(>T_1^*)$ is the unique finite positive solution to $\psi(T_2) = 1$.
   The wholesaler's total profit is therefore given by

   $$\hat{P}_2(N_2) = -\frac{(c_s - \alpha)Q(T_2^*)}{T_2^*} - \beta - a_s. \qquad (31.27)$$

2. $N_2 \geq 2$:
   In this case, $\tilde{T}_2$ is the unique solution (if it exists) to

   $$L(N_2, T_2) = (a_b + \beta)N_2 + a_s. \qquad (31.28)$$

The optimal quantity discount pricing policy is given by (31.24).

In the following, the mathematical results are briefly summarized in the special case where $\theta_s = \theta_b = \theta$ and $\alpha \leq (h_b - h_s)/\theta$.

- $C[(N_2 \theta T_1^* - 1)e^{N_2 \theta T_1^*} + 1] + HN_2[(\theta T_1^* - 1)e^{\theta T_1^*} + 1] < [(a_b + \beta)N_2 + a_s]/\rho$ :

  In this subcase, there exists a unique finite $\hat{T}_2(>T_1^*)$, which maximizes $P_2(N_2, T_2)$ in (31.20), and therefore $(T_2^*, y^*)$ is given by (31.24).

  The wholesaler's total profit is given by

  $$\hat{P}_2(N_2) = \rho \theta [(p_s + h_b/\theta_b)e^{\theta T_1^*} - C e^{N_2 \theta T_2^*} - H e^{\theta T_2^*}]. \tag{31.29}$$

- $C[(N_2 \theta T_1^* - 1)e^{N_2 \theta T_1^*} + 1] + HN_2[(\theta T_1^* - 1)e^{\theta T_1^*} + 1] \geq [(a_b + \beta)N_2 + a_s]/\rho$ :

  In this subcase, $P_2(N_2, T_2)$ is nonincreasing in $T_2$, and consequently $T_2 = T_1^*$. The wholesaler's total profit is expressed by

  $$(T_2^*, y^*) \to (T_1^*, \Psi(T_1^*)). \tag{31.30}$$

### 5.3   Under Options $V_1$ and $V_2$

In the case of $(T_2, y) \in \Omega_1 \cap \Omega_2$, the retailer is indifferent between Options $V_1$ and $V_2$. For this reason, this study confines itself to a situation where the wholesaler does not use a discount pricing policy $(T_2, y) \in \Omega_1 \cap \Omega_2$.

## 6   Numerical Examples

Table 31.1 reveals the results of sensitivity analysis in reference to $Q_1^*, p_1 (=p_s), S_1^*, N_1^*, P_1^*, Q_2^*, p_2^* (=(1 - y^*)p_s), S_2^*, N_2^*, P_2^*$ for $(c_s, p_s, p_b, a_b, h_s, h_b, \theta_s, \theta_b, \mu, \alpha, \beta) = (100, 300, 600, 1200, 1, 1.1, 0.01, 0.015, 5, 2, 100)$ when $a_s = 500, 1000, 2000,$ and $3000$.

In Table 31.1a, we can observe that both $S_1^*$ and $N_1^*$ are nondecreasing in $a_s$. As mentioned in the above section, under Option $V_1$, the retailer does not adopt the quantity discount offered by the wholesaler, which signifies that the wholesaler cannot control the retailer's ordering schedule. In this case, the wholesaler's cost associated with ordering should be reduced by increasing her/his own length of order cycle and lot size by means of increasing $N_1$.

Table 31.1b shows that, under Option $V_2$, $S_2^*$ increases with $a_s$, in contrast, $N_2^*$ takes a constant value, i.e., we have $N_2^* = 1$. Under Option $V_2$, the retailer accepts the quantity discount proposed by the wholesaler. The wholesaler's lot size can therefore be increased by stimulating the retailer to alter her/his order quantity per

**Table 31.1** Sensitivity analysis

(a) Under option $V_1$

| $a_s$ | $Q_1^*$ | $p_1$ | $S_1^*\ (N_1^*)$ | $P_1^*$ |
|---|---|---|---|---|
| 500 | 47.35 | 300 | 47.35 (1) | 1012.99 |
| 1,000 | 47.35 | 300 | 99.09 (2) | 963.48 |
| 2,000 | 47.35 | 300 | 99.09 (2) | 907.01 |
| 3,000 | 47.35 | 300 | 155.61 (3) | 854.45 |

(b) Under option $V_2$

| $a_s$ | $Q_2^*$ | $p_2^*$ | $S_2^*\ (N_2^*)$ | $P_2^*$ |
|---|---|---|---|---|
| 500 | 107.63 | 285.01 | 107.63(1) | 998.56 |
| 1,000 | 117.65 | 281.63 | 117.65(1) | 972.80 |
| 2,000 | 135.77 | 275.49 | 135.77(1) | 926.23 |
| 3,000 | 152.04 | 270.04 | 152.04(1) | 884.43 |

lot through the quantity discount strategy. If the wholesaler increases $N_2$ one step, her/his lot size also significantly jumps up since $N_2$ takes a positive integer. Under this option, the wholesaler should increase her/his lot size using the quantity discount rather than increasing $N_2$ when $a_s$ takes larger values.

We can also notice in Table 31.1 that we have $P_1^* < P_2^*$. This indicates that using the quantity discount strategy can increase the wholesaler's total profit per unit of time.

# 7 Conclusion and Future Work

In this study, we have discussed a quantity discount problem between a wholesaler and a retailer under circumstances where both the wholesaler's and the retailer's inventory levels of the product are depleted not only by demand but also by deterioration. The wholesaler is interested in increasing her/his profit by controlling the retailer's order quantity through the quantity discount strategy. The retailer attempts to maximize her/his profit considering the wholesaler's proposal. We have formulated the above problem as a Stackelberg game between the wholesaler and the retailer to show the existence of the wholesaler's optimal quantity discount policy that maximizes her/his total profit per unit of time in the same manner as our previous work [9]. Our previous study mainly showed that there exists a unique finite quantity discount pricing policy in the case of $N_2 = 1$. In this study, we have derived the additional properties associated with the optimal policy in the case of $N_2 > 1$. It should be pointed out that our results are obtained under the situation where the inventory holding cost is independent of the value of the item. The relaxation of such a restriction is an interesting extension.

## Appendix A

In this appendix, we discuss the existence of the optimal quantity discount pricing policy, which attains $\hat{P}_2(N_2)$ in (31.19) when $N_2$ is fixed to a suitable value.

1. $N_2 = 1$ :
By differentiating $P_2(N_2, T_2)$ in (31.20) with respect to $T_2$, we have

$$\frac{\partial}{\partial T_2} P_2(N_2, T_2) = -\frac{[\rho\theta_b\,e^{\theta_b T_2} - Q(T_2)](c_s + h_b/\theta_b - \alpha) - (a_b + a_s + \beta)}{T_2^2}.$$

(31.31)

It can easily be shown from (31.31) that the sign of $\partial P_2(N_2, T_2)/\partial T_2$ is positive when $(c_s + h_b/\theta_b - \alpha) = 0$. In contrast, in the case of $(c_s + h_b/\theta_b - \alpha) \gtreqqless 0$, $\partial P_2(N_2, T_2)/\partial T_2 \geq 0$ agrees with

$$\theta_b T_2\,e^{\theta_b T_2} - (\theta_b\,e^{\theta_b T_2} - 1) \lesseqqgtr \frac{a_b + a_s + \beta}{\rho(c_s + h_b/\theta_b - \alpha)}.$$

(31.32)

Let $L_1(T_2)$ express the left-hand side of Inequality (31.32), we have

$$L_1'(T_2) = \theta_b^2 T_2\,e^{\theta_b T_2} \quad (>0),$$

(31.33)

$$L_1(T_1^*) = \frac{a_b}{\rho(p_s + h_b/\theta_b)} \quad (>0),$$

(31.34)

$$\lim_{T_2 \to +\infty} L_1(T_2) = +\infty.$$

(31.35)

From (31.33)–(31.35), the existence of an optimal quantity discount pricing policy can be discussed for the following two subcases:

- $(c_s + h_b/\theta_b - \alpha) > 0$ :
Equation (31.34) yields

$$L_1(T_1^*) < \frac{a_b + a_s + \beta}{\rho(c_s + h_b/\theta_b - \alpha)} \quad (>0).$$

(31.36)

Equations (31.33), (31.35), and (31.36) indicate that the sign of $\partial P_2(N_2, T_2)/\partial T_2$ changes from positive to negative only once. This signifies that $P_2(N_2, T_2)$ first increases and then decreases as $T_2$ increases, and thus there exists a unique finite $\tilde{T}_2$ $(>T_1^*)$, which maximizes $P_2(N_2, T_2)$ in (31.20). Hence, $(T_2^*, y^*)$ is given by (31.24).

- $(c_s + h_b/\theta_b - \alpha) \leq 0$ :

In this subcase, we have

$$L_1(T_1^*) > \frac{a_b + a_s + \beta}{\rho(c_s + h_b/\theta_b - \alpha)} \quad (<0). \tag{31.37}$$

Equations (31.33), (31.35), and (31.37) signify that the sign of $\partial P_2(N_2, T_2)/\partial T_2$ is positive, and consequently the optimal policy can be expressed by (31.26).

2. $N_2 \geq 2$ :
   By differentiating $P_2(N_2, T_2)$ in (31.20) with respect to $T_2$, we have

$$\frac{\partial}{\partial T_2} P_2(N_2, T_2) = -\frac{L(T_2) - (a_b + \beta)N_2 - a_s}{N_2 T_2^2}. \tag{31.38}$$

Then $\partial P_2(N_2, T_2)/\partial T_2 \geq 0$ agrees with

$$L(N_2, T_2) \leq (a_b + \beta)N_2 + a_s. \tag{31.39}$$

If we assume that there exists a unique solution to (31.28), the optimal quantity discount pricing policy can be given by (31.24).

In the special case where $\theta_s = \theta_b = \theta$ and $\alpha \leq (h_b - h_s)/\theta$, by differentiating $P_2(N_2, T_2)$ in (31.20) with respect to $T_2$, we have

$$\frac{\partial}{\partial T_2} P_2(N_2, T_2) = -\frac{1}{N_2 T_2^2} \{ C\rho [(N_2 \theta T_2 - 1)e^{N_2 \theta T_2} + 1] - (a_b + \beta)N_2 \\ -a_s + HN_2\rho [(\theta T_2 - 1) e^{\theta T_2} + 1] \}. \tag{31.40}$$

Then $\partial P_2(N_2, T_2)/\partial T_2 \geq 0$ agrees with

$$C[(N_2 \theta T_2 - 1)e^{N_2 \theta T_2} + 1] + HN_2[(\theta T_2 - 1) e^{\theta T_2} + 1] \\ \leq \frac{(a_b + \beta)N_2 + a_s}{\rho}. \tag{31.41}$$

Let us denote, by $L_a(T_2)$, the left-hand side of Inequality (31.41), and we have

$$L_a'(T_2) = \theta^2 N_2 T_2 (CN_2 e^{N_2 \theta T_2} + He^{\theta T_2}) \quad (>0), \tag{31.42}$$

$$L_a(T_1^*) = C\left[(N_2\theta T_1^* - 1)e^{N_2\theta T_1^*} + 1\right]$$
$$+HN_2\left[(\theta T_1^* - 1)e^{\theta T_1^*} + 1\right], \tag{31.43}$$

$$\lim_{T_2 \to +\infty} L_a(T_2) = +\infty. \tag{31.44}$$

On the basis of the above results, we show below that an optimal quantity discount pricing strategy exits.

- $C[(N_2\theta T_1^* - 1)e^{N_2\theta T_1^*} + 1] + HN_2[(\theta T_1^* - 1)e^{\theta T_1^*} + 1] < [(a_b + \beta)N_2 + a_s]/\rho$ :
  In this subcase, the sign of $\partial P_2(N_2, T_2)/\partial T_2$ varies from positive to negative only once, and consequently there exists a unique finite $\tilde{T}_2$ ($>T_1^*$), which maximizes $P_2(N_2, T_2)$ in (31.20). Hence, $(T_2^*, y^*)$ is given by (31.24).
- $C[(N_2\theta T_1^* - 1)e^{N_2\theta T_1^*} + 1] + HN_2[(\theta T_1^* - 1)e^{\theta T_1^*} + 1] \geq [(a_b + \beta)N_2 + a_s]/\rho$ :
  This subcase provides $\partial P_2(N_2, T_2)/\partial T_2 \leq 0$, and therefore the optimal policy can be expressed by (31.30).

# References

1. Monahan JP (1984) A quantity discount pricing model to increase vendor's profit. Manag Sci 30 (6):720–726
2. Rosenblatt MJ, Lee HL (1985) Improving pricing profitability with quantity discounts under fixed demand. IIE Trans 17(4):338–395
3. Data M, Srikanth KN (1987) A generalized quantity discount pricing model to increase vendor's profit. Manag Sci 33(10):1247–1252
4. Lee HL, Rosenblatt MJ (1986) A generalized quantity discount pricing model to increase vendor's profit. Manag Sci 32(9):1177–1185
5. Parlar M, Wang Q (1995) A game theoretical analysis of the quantity discount problem with perfect and incomplete information about the buyer's cost structure. RAIRO/Oper Res 29 (4):415–439
6. Sarmah SP, Acharya D, Goyal SK (2006) Buyer vendor coordination models in supply chain management. Eur J Oper Res 175(1):1–15
7. Yang PC (2004) Pricing strategy for deteriorating items using quantity discount when demand is price sensitive. Eur J Oper Res 157(2):389–397
8. Behrens K, Gaigne C, Ottaviano GIP, Thisse JF (2006) How density economies in international transportation link the internal geography of trading partners. J Urban Econ 60(2):248–263
9. Kawakatsu H (2011) A wholesaler's optimal quantity discount policy for deteriorating items. Lecture notes in engineering and computer science. In: Proceedings of the world congress on engineering 2011 (WCE 2011), London, 6–8 July 2011, pp 540–544

# Chapter 32
# Effect of Cutting Parameters on Surface Finish for Three Different Materials in Dry Turning

Mohammad Nazrul Islam and Brian Boswell

## 1 Introduction

Surface finish of the machined parts is one of the important criteria by which the success of a machining operation is judged [1]. It is also an important quality characteristic that may dominate the functional requirements of many component parts. For example, good surface finish is necessary to prevent premature fatigue failure; to improve corrosion resistance; to reduce friction, wear, and noise; and finally to improve product life. Therefore, achieving the required surface finish is critical to the success of many machining operations.

Over the years, cutting fluids have been applied extensively in machining operations for various reasons, such as to reduce friction and wear, hence improving tool life and surface finish; to reduce force and energy consumption; and to cool the cutting zone, thus reducing thermal distortion of the workpiece and improving tool life, and facilitating chip disposal. However, the application of cutting fluid poses serious health and environmental hazards. Operators exposed to cutting fluids may have various health problems. If not disposed of properly, cutting fluids may adversely affect the environment and carry economic consequences.

To overcome these problems, a number of techniques, such as dry turning, turning with minimum quantity lubrication (MQL), and cryogenic turning, have been proposed. Dry turning is characterized by the absence of any cutting fluid, and unlike MQL and cryogenic turning, it does not require any additional delivery system. Hence, from the environmental perspective, dry turning is ecologically desirable; and from an economic perspective, it decreases manufacturing costs by 16–20% [2]. Nevertheless, in spite of all economic and environmental benefits, the quality of the component parts produced by dry turning should not be sacrificed.

M.N. Islam (✉) • B. Boswell
Department of Mechanical Engineering, Curtin University,
GPO Box U1987, Perth, WA 6845, Australia
e-mail: M.N.Islam@curtin.edu.au; B.Boswell@curtin.edu.au

S.-I. Ao and L. Gelman (eds.), *Electrical Engineering and Intelligent Systems*,
Lecture Notes in Electrical Engineering 130, DOI 10.1007/978-1-4614-2317-1_32,
© Springer Science+Business Media, LLC 2013

The two major functions of cutting fluids are (1) to increase tool life and (2) to improve the surface finish of manufactured parts. However, with the advent of various new tool materials and their deposition techniques, the tool lives of modern tools have increased significantly. At present, dry machining is possible without considerable tool wear; as such, research work has been focused on the surface finish aspect of dry turning.

Investigations of the surface finish of turned parts have received notable attention in the literature, but most of the reported studies concentrate on a single work material such as free machining steel [3], composite material [4], bearing steel [5], SCM 400 steel [6], tool steel [7], MDN250 steel [8], and alloy steel [9]. However, the work material has significant effects on the results of machining operations. Therefore, any study on machining operations would not be complete unless it covered a wide range of materials. Consequently, three work materials encompassing diverse machinability ratings were selected for this study, which is an extended and revised version of our previous work [10].

## 2  Scope

Several factors directly or indirectly influence the surface finish of machined parts, such as cutting conditions, tool geometry, work material, machine accuracy, chatter or vibration of the machine tool, cutting fluid, and chip formation. The objective of this research was to investigate the effects of major input parameters on the surface finish of parts produced by dry turning and to optimize the input parameters. From a user's point of view, cutting parameters—cutting speed, feed rate, and depth of cut—are the three major controllable variables; as such they were selected as input parameters.

*Surface roughness* represents the random and repetitive deviations of a surface profile from the nominal surface. It can be expressed by a number of parameters such as arithmetic average, peak-to-valley height, and ten-point height. Yet, no single parameter appears to be capable of describing the surface quality adequately. In this study, *arithmetic average* has been adopted to represent surface roughness, as it is the most frequently used and internationally accepted parameter. The arithmetic average represents the average of the absolute deviations from the mean surface level, which can be calculated using the following formula:

$$R_a = \frac{1}{L} \int_0^L |Y(x)| \, dx \qquad (32.1)$$

where $R_a$ is the arithmetic average roughness, $Y$ is the vertical deviation from the nominal surface, and $L$ is the specified distanced over which the surface roughness is measured [11]. For this research, a surface finish analyzer capable of measuring multiple surface finish parameters was employed. The results were then analyzed by three techniques—traditional analysis, Pareto ANOVA analysis, and Taguchi's signal-to-noise ratio (S/N) analysis.

In the traditional analysis, the average values of the measured variables were used. This tool is particularly suitable for monitoring a trend of change in the relationship of variables.

Pareto ANOVA analysis is an excellent tool for determining the contribution of each input parameter and their interactions with the output parameters (surface roughness). It is a simplified ANOVA analysis method that does not require an ANOVA table. Further details on Pareto ANOVA can be found in Park [12].

The Taguchi method applies the *signal-to-noise ratio* to optimize the outcome of a manufacturing process. The signal-to-noise ratio can be calculated using the following formula:

$$S/N = -10 \log \frac{1}{n} \left( \sum_{i=1}^{n} y_i^2 \right) \qquad (32.2)$$

where S/N is the signal-to-noise ratio (in dB), $n$ is the number of observations, and $y$ is the observed data.

The above formula is suitable for quality characteristics in which the adage "the smaller the better" holds true, which is the case for surface roughness. The higher the value of the S/N ratio, the better the result is because it guarantees optimum quality with minimum variance. A thorough treatment of the Taguchi method can be found in Ross [13].

Finally, regression analysis technique was applied to obtain prediction models for estimating the surface roughness of each selected material.

## 3   Experimental Work

The experiments were planned using Taguchi's orthogonal array methodology [13], and a two-level L$_8$ orthogonal array was selected for our experiments. Three parts were produced using three materials with varying machinability properties: aluminum (AISI 6061), mild steel (AISI 1030), and alloy steel (AISI 4340). Each part was divided into eight segments. Some important properties and chemical compositions of the work materials compiled from [14] are listed in Tables 32.1 and 32.2, respectively.

**Table 32.1**  Properties of work materials [14]

| Properties | Unit | AISI 6061 | AISI 1030 | AISI 4340 |
|---|---|---|---|---|
| Machinability | % | 190 | 71 | 50 |
| Hardness | BH | 95 | 149 | 217 |
| Modulus of elasticity | GPa | 68.9 | 205 | 205 |
| Specific heat capacity | J/g °C | 0.896 | 0.486 | 0.475 |

**Table 32.2** Properties of work materials [14]

| AISI 6061 | | AISI 1030 | | AISI 4340 | |
|---|---|---|---|---|---|
| Aluminum | 95.8–98.6% | Carbon | 0.27–0.34% | Carbon | 0.370–0.430% |
| Chromium | 0.040–0.35% | Iron | 98.67–99.13% | Chromium | 0.700–0.900% |
| Copper | 0.15–0.40% | Manganese | 0.60–0.90% | Iron | 95.195–96.33% |
| Iron | ≤0.70% | Phosphorous | ≤0.040% | Manganese | 0.600–0.800% |
| Magnesium | 0.80–1.20% | Sulfur | ≤0.050% | Molybdenum | 0.200–0.300% |
| Manganese | ≤0.15% | | | Nickel | 1.65–2.00% |
| Other, each | ≤0.050% | | | Phosphorous | ≤0.0350% |
| Other, total | ≤0.15% | | | Silicon | 0.150–0.300% |
| Silicon | 0.40–0.80% | | | Sulfur | ≤0.0400% |
| Titanium | ≤0.15% | | | | |
| Zinc | ≤0.25% | | | | |

**Table 32.3** Input variables

| Input parameters | Unit | Symbol | Levels | |
|---|---|---|---|---|
| | | | Level 0 | Level 1 |
| Cutting speed | m/min | $A$ | 54 | 212 |
| Feed rate | mm/rev | $B$ | 0.11 | 0.22 |
| Depth of cut | mm | $C$ | 0.5 | 1.5 |

The nominal size of each part was 160 mm length and 40 mm diameter. The experiment was carried out on a Harrison conventional lathe with 330 mm swing. For holding the workpiece, a three-jaw chuck supported at dead center was employed. Square-shaped inserts with enriched cobalt coating (CVD TiN–TiCN–$Al_2O_3$–TiN) manufactured by Stellram, USA, were used as the cutting tools. The inserts were mounted on a standard PSDNN M12 tool holder. A new cutting tip was used for machining each part to avoid any tool wear effect. Details of cutting conditions used—cutting speed, feed rate, and depth of cut—are given in Table 32.3. The range of depth of cut was chosen taking into consideration the finishing operation for which surface finish is more relevant.

# 4    Results and Analysis

## 4.1    Pareto ANOVA Analysis

The Pareto ANOVA analysis for aluminum (AISI 6061) is given in Table 32.4. It shows that feed rate ($B$) has the most significant effect on surface roughness with a contribution ratio ($P \cong 65\%$), followed by cutting speed ($A$) ($P \cong 9\%$), and depth of cut ($C$) ($P \cong 3\%$). The interactions between cutting speed and feed rate ($A \times B$) and feed rate and depth of cut ($B \times C$) also played roles with contributing ratios ($P \cong 13\%$) and ($P \cong 9\%$), respectively. It is worth pointing out that the total

**Table 32.4** Pareto ANOVA analysis for aluminum (AISI 6061)

| Sum at factor level | Factor and interaction | | | | | |
|---|---|---|---|---|---|---|
| | A | B | AxB | C | AxC | BxC |
| 0 | -7.51 | -1.62 | -7.84 | -6.88 | -5.76 | -7.52 |
| 1 | -4.30 | -10.20 | -3.98 | -4.94 | -6.05 | -4.29 |
| Sum of squares of difference (S) | 10.31 | 73.69 | 14.88 | 3.75 | 0.08 | 10.43 |
| Contribution ratio (%) | 9.11 | 65.13 | 13.15 | 3.32 | 0.07 | 9.22 |
| Pareto diagram | | | | | | |
| Cumulative contribution | 65.13 | 78.28 | 87.50 | 96.61 | 99.93 | 100.00 |
| Check on significant interaction | AxB two-way table | | | | | |
| Optimum combination of significant factor level | A1B0C1 | | | | | |

**Table 32.5** Pareto ANOVA analysis for mild steel (AISI 1030)

| Sum at factor level | Factor and interaction | | | | | |
|---|---|---|---|---|---|---|
| | A | B | AxB | C | AxC | BxC |
| 0 | -57.84 | -31.29 | -52.68 | -42.42 | -47.39 | -42.21 |
| 1 | -28.97 | -55.52 | -34.14 | -44.39 | -39.43 | -44.60 |
| Sum of squares of difference (S) | 833.31 | 587.22 | 343.75 | 3.90 | 63.40 | 5.71 |
| Contribution ratio (%) | 45.36 | 31.96 | 18.71 | 0.21 | 3.45 | 0.31 |
| Pareto diagram | | | | | | |
| Cumulative contribution | 45.36 | 77.32 | 96.03 | 99.48 | 99.79 | 100.00 |
| Check on significant interaction | AxB two-way table | | | | | |
| Optimum combination of significant factor level | A1B0C0 | | | | | |

contribution of the main effects is about 77% compared to the total contribution of the interaction effects of 23%. Therefore, it will be moderately difficult to optimize the diameter error by selection of input parameters.

The Pareto ANOVA analysis for mild steel (AISI 1030) is given in Table 32.5. It illustrates that cutting speed (A) has the most significant effect on surface roughness with a contribution ratio ($P \cong 45\%$), followed by feed rate (B) ($P \cong 31\%$), and depth of cut (C) ($P \cong 0.2\%$). The interactions between cutting speed and feed rate (A × B) and cutting and depth of cut (A × C) also played roles

**Table 32.6** Pareto ANOVA analysis for alloy steel (AISI 4340)

| Sum at factor level | Factor and interaction | | | | | |
|---|---|---|---|---|---|---|
| | A | B | AxB | C | AxC | BxC |
| 0 | -7.89 | -0.13 | -8.04 | -6.93 | -6.94 | -6.94 |
| 1 | -6.76 | -14.53 | -6.62 | -7.72 | -7.72 | -7.71 |
| Sum of squares of difference (S) | 1.28 | 207.42 | 2.00 | 0.62 | 0.61 | 0.59 |
| Contribution ratio (%) | 0.60 | 97.60 | 0.94 | 0.29 | 0.29 | 0.28 |
| Pareto diagram | 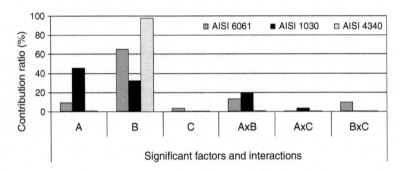 | | | | | |
| Cumulative contribution | 97.60 | 98.54 | 99.14 | 99.44 | 99.72 | 100.00 |
| Check on significant interaction | AxB two-way table | | | | | |
| Optimum combination of significant factor level | A1B0C0 | | | | | |

**Fig. 32.1** Comparison of Pareto diagrams for different materials

with contributing ratios ($P \cong 18\%$) and ($P \cong 3\%$), respectively. It is worth noting that the total contribution of the main effects remains roughly the same (about 78%), although in this case the contribution of cutting speed ($A$) is increased notably with expense of feed rate. As the total contribution of the interaction effects remains high (22%), it will be moderately difficult to optimize surface roughness by selection of input parameters.

The Pareto ANOVA analysis for alloy steel (AISI 4340) is given in Table 32.6. It shows that feed rate ($B$) has the most significant effect on surface roughness with a contribution ratio ($P \cong 98\%$). All other effects, both main and interaction effects, were almost negligible. Therefore, it will be relatively easy to optimize the surface roughness by selection of proper feed rate.

A comparison of Pareto diagrams for different materials is illustrated in Fig. 32.1, in which the dominant effect of feed rate on surface finish is evident. The effect of feed rate on surface roughness is well known, and most of the widely applied geometric models for surface roughness include feed rate and tool nose

**Table 32.7** Response table for mean S/N ratio for aluminum (AISI 6061)

| Cutting parameters | Symbol | Mean S/N ratio | | |
|---|---|---|---|---|
| | | Level 0 | Level 1 | Max–Min |
| Cutting speed | $A$ | −7.515 | −4.304 | 3.211 |
| Feed rate | $B$ | −1.617 | −10.201 | 8.584 |
| Depth of cut | $C$ | −6.878 | −4.940 | 1.937 |
| Interaction $A \times B$ | $A \times B$ | −7.838 | −3.981 | 3.857 |

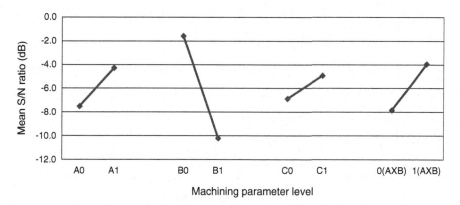

**Fig. 32.2** Response graphs of S/N ratio for aluminum (AISI 6061)

radius. However, results showed considerable interaction effects between cutting speed and feed rate, which influenced the surface finish. It appears that with the increase of material hardness the interaction effect diminishes.

Within the selected range of variation, depth of cut showed negligible effect on surface roughness (Fig. 32.1). Similar results have been reported by previous studies [9, 15, 16]. The Pareto ANOVA analyses (Tables 32.4–32.6) showed that in all cases high cutting speed ($A_1$) and low feed rate ($B_0$) produced the best surface roughness, which is in line with conventional machining wisdom. The cutting speed must be selected high enough to avoid formation of a built up edge (BUE).

## 4.2 Response Tables and Graphs

The response table and response graph for aluminum are illustrated in Table 32.7 and Fig. 32.2, respectively. As the slopes of the response graphs represent the strength of contribution, the response graphs confirm the findings of the Pareto ANOVA analysis given in Table 32.4. Figure 32.2 shows that high level of depth of ($C_1$) was the best depth of cut. Because the interaction $A \times B$ was significant, an $A \times B$ two-way table was applied to select their levels. The two-way table is not

**Table 32.8** Response table for mean S/N ratio for mild steel (AISI 1030)

| Cutting parameters | Symbol | Mean S/N ratio | | |
| --- | --- | --- | --- | --- |
| | | Level 0 | Level 1 | Max–Min |
| Cutting speed | $A$ | −14.460 | −7.243 | 7.217 |
| Feed rate | $B$ | −7.823 | −13.881 | 6.058 |
| Depth of cut | $C$ | −10.605 | −11.099 | 0.494 |
| Interaction $A \times B$ | $A \times B$ | −13.169 | −8.534 | 4.635 |

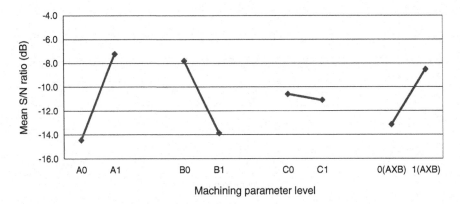

**Fig. 32.3** Response graphs of S/N ratio for mild steel (AISI 1030)

included in this work due to space constraints. From the $A \times B$ two-way table, the optimum combination of factors $A$ and $B$ in order to achieve a lowest surface finish was determined as $A_1B_0$. Therefore, the best combination of input variables for minimizing surface roughness was $A_1B_0C_1$; i.e., high level of cutting speed, low level of feed rate, and high level of depth of cut.

The response table and response graph for mild steel are illustrated in Table 32.8 and Fig. 32.3, respectively. The response graph confirms the findings of the Pareto ANOVA analysis given in Table 32.5. Low level of depth of ($C_0$) was the best depth of cut (Fig. 32.3). From the $A \times B$ two-way table, the optimum combination of factors $A$ and $B$ in order to achieve a lowest surface finish was determined as $A_1B_0$. Therefore, the best combination of input variables for minimizing surface roughness was $A_1B_0C_0$; i.e., high level of cutting speed, low level of feed rate, and low level of depth of cut.

The response table and response graph for alloy steel are shown in Table 32.9 and Fig. 32.4, respectively. The response graph confirms the findings of the Pareto ANOVA analysis given in Table 32.6. Low level of depth of ($C_0$) was the best depth of cut (Fig. 32.4). From the $A \times B$ two-way table, the optimum combination of factors $A$ and $B$ was set to $A_1B_0$. Therefore, the best combination of input variables for minimizing surface roughness was $A_1B_0C_0$; high level of cutting speed, low level of feed rate, and low level of depth of cut.

**Table 32.9**  Response table for mean S/N ratio for alloy (AISI 4340)

| Cutting parameters | Symbol | Mean S/N ratio | | |
|---|---|---|---|---|
| | | Level 0 | Level 1 | Max–Min |
| Cutting speed | A | −7.895 | −6.764 | 1.131 |
| Feed rate | B | −0.128 | −14.530 | 14.402 |
| Depth of cut | C | −6.934 | −7.724 | 0.790 |
| Interaction $A \times B$ | $A \times B$ | −8.037 | −6.622 | 1.416 |

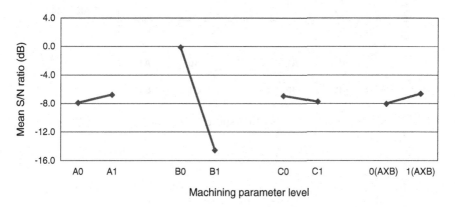

**Fig. 32.4**  Response graphs of S/N ratio for alloy steel (AISI 4340)

## 4.3  Traditional Analysis

Variations in surface roughness for input parameters cutting speed and feed rate are illustrated in Fig. 32.5. Effect of depth of cut is omitted because it demonstrated negligible percent contribution in Pareto ANOVA analyses (Tables 32.4–32.6).

The graph shows that for aluminum and mild steel, with the increase of cutting speed surface roughness more or less remained steady or even deteriorated, whereas for alloy steel it improved considerably (Fig. 32.5a). The graph also shows that with the increase of feed rate the surface roughness values also increase, although this increase is higher at low cutting speed (Fig. 32.5b).

Figure 32.5 demonstrates that, contrary to traditional machining wisdom, the material with the higher machinability rating did not always produce a better surface finish. Therefore, surface roughness by itself is not a reliable indicator of machinability. The reason behind this is that the optimum cutting conditions for different materials are different; whereas in our experiment we selected the same cutting parameters for all the materials selected, conservatively based on the optimum cutting condition suitable for the material most difficult to machine, alloy steel AISI 4340, primarily to protect the tool. Furthermore, due to the interaction effects between cutting speed and feed rate, surface roughness is not always related to machinability rating.

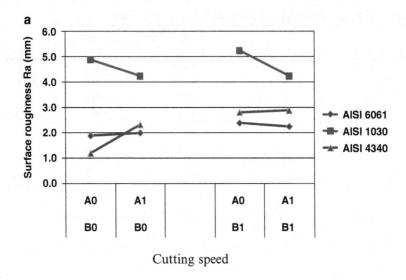

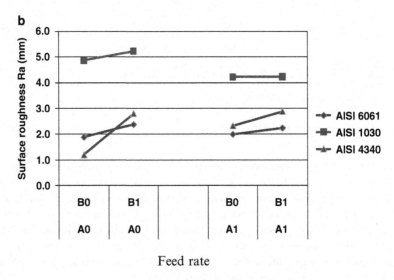

**Fig. 32.5** Variation of surface roughness for input parameters

## 4.4 Regression Analysis

To establish the prediction model, the software package *XLSTAT* was applied to perform the regression analysis using the experimental data. The regression analysis results and prediction model for aluminum are given in Table 32.10. The coefficient of determination ($R^2$) is 0.815, which indicates that the prediction model has a satisfactory "goodness of fit".

**Table 32.10** Results of regression analysis for aluminum (AISI 6061)

Summary of statistics

| Variable | Observations | Minimum | Maximum | Mean | Std. deviation |
|---|---|---|---|---|---|
| $R_a$ | 8 | 0.590 | 3.533 | 2.305 | 1.146 |
| A | 8 | 54.000 | 212.000 | 133.000 | 84.455 |
| B | 8 | 0.110 | 0.330 | 0.220 | 0.118 |
| C | 8 | 0.500 | 1.500 | 1.000 | 0.535 |

Goodness-of-fit statistics

| | |
|---|---|
| Observations | 8.000 |
| Sum of weights | 8.000 |
| DF | 4.000 |
| $R^2$ | 0.815 |
| Adjusted $R^2$ | 0.677 |
| MSE | 0.424 |
| RMSE | 0.651 |
| MAPE | 22.951 |

Analysis of variance

| Source | DF | Sum of sq. | Mean sq. | F | Pr > F |
|---|---|---|---|---|---|
| Model | 3 | 7.490 | 2.497 | 5.887 | 0.060 |
| Error | 4 | 1.696 | 0.424 | | |
| Corrected total | 7 | 9.187 | | | |

Equation of the model

$R_a = 0.913 - 2.579 \times 10^{-3} \times A + 8.572 \times B - 0.151 \times C$

The regression analysis results and prediction model for mild steel are given in Table 32.11. In this case, the coefficient of determination ($R^2$) is 0.843, which is slightly better compared to aluminum.

The regression analysis results and prediction model for alloy steel are given in Table 32.12. In this case, the coefficient of determination ($R^2$) is 0.993, which shows the best "goodness of fit" for the three materials selected. It is worth noting that "goodness of fit" improved with material hardness (see hardness data given in Table 32.2) due to reduction of interaction effects.

Tables 32.10–32.12 show that the model equation for each material is different. It is worth pointing out that currently available geometric models do not work in practice because they do not consider the formation of BUE and change of tool profile caused by tool wear. Therefore, any future analytical model should include material characteristics to make it meaningful.

**Table 32.11** Results of regression analysis for mild steel (AISI 1030)

Summary of statistics

| Variable | Observations | Minimum | Maximum | Mean | Std. deviation |
|---|---|---|---|---|---|
| $R_a$ | 8 | 1.043 | 6.480 | 4.051 | 1.875 |
| A | 8 | 54.000 | 212.000 | 133.000 | 84.455 |
| B | 8 | 0.110 | 0.330 | 0.220 | 0.118 |
| C | 8 | 0.500 | 1.500 | 1.000 | 0.535 |

Goodness-of-fit statistics

| | |
|---|---|
| Observations | 8.000 |
| Sum of weights | 8.000 |
| DF | 4.000 |
| $R^2$ | 0.843 |
| Adjusted $R^2$ | 0.726 |
| MSE | 0.964 |
| RMSE | 0.982 |
| MAPE | 20.396 |

Analysis of variance

| Source | DF | Sum of sq. | Mean sq. | F | Pr > F |
|---|---|---|---|---|---|
| Model | 3 | 20.764 | 6.921 | 7.177 | 0.044 |
| Error | 4 | 3.858 | 0.964 | | |
| Corrected total | 7 | 24.622 | | | |

Equation of the model
$R_a = 4.363 - 1.615 \times 10^{-2} \times A + 8.924 \times B - 0.128 \times C$

## 5 Concluding Remarks

From the experimental work conducted and the subsequent analysis, the following conclusions can be drawn:

- Feed rate has a dominant effect on surface finish; the interaction between cutting speed and feed rate also plays a major role, which is influenced by the properties of work material. With the increase of material hardness, the interaction effect diminishes.
- Within the selected range, depth of cut showed negligible effect on surface roughness.
- Surface roughness by itself is not a reliable indicator of machinability, due to nonoptimal cutting conditions and interaction effects of additional factors.
- The model equations resulting from regression analysis for different materials show significant differences. Therefore, any future analytical model should include material aspect to make it meaningful.

**Table 32.12** Results of regression analysis for alloy steel (AISI 4340)

Summary of statistics

| Variable | Observations | Minimum | Maximum | Mean | Std. deviation |
|---|---|---|---|---|---|
| $R_a$ | 8 | 0.860 | 5.527 | 3.183 | 2.303 |
| A | 8 | 54.000 | 212.000 | 133.000 | 84.455 |
| B | 8 | 0.110 | 0.330 | 0.220 | 0.118 |
| C | 8 | 0.500 | 1.500 | 1.000 | 0.535 |

Goodness-of-fit statistics

| | |
|---|---|
| Observations | 8.000 |
| Sum of weights | 8.000 |
| DF | 4.000 |
| $R^2$ | 0.993 |
| Adjusted $R^2$ | 0.988 |
| MSE | 0.062 |
| RMSE | 0.250 |
| MAPE | 8.198 |

Analysis of variance

| Source | DF | Sum of sq. | Mean sq. | F | Pr > F |
|---|---|---|---|---|---|
| Model | 3 | 36.875 | 12.292 | 197.161 | <0.0001 |
| Error | 4 | 0.249 | 0.062 | | |
| Corrected total | 7 | 37.124 | | | |

Equation of the model

$R_a = -1.163 - 4.641 \times 10^{-4} \times A + 19.508 \times B + 0.117 \times C$

# References

1. Droozda TJ, Wick C (eds) (1983) Tool and manufacturing engineers handbook, vol 1: machining. SME, Dearborn
2. Sreejith PS, Ngoi BRA (2000) Dry machining: machining of the future. J Mater Process Technol 101:287–291
3. Davim JP (2001) A note on the determination of optimal cutting conditions for surface finish obtained in turning using design of experiments. J Mater Process Technol 116:305–308
4. Manna A, Bhattacharyya B (2002) A study on different tooling system during machining of Al/SiC-MMC. J Mater Process Technol 123:476–481
5. Dilbag S, Rao PV (2006) A surface roughness prediction model for hard turning process. Int J Adv Manuf Technol 32:1115–1124
6. Thamizhmanil S, Saparudin S, Hasan S (2007) Analysis of surface roughness by turning process using Taguchi method. J Achiev Mater Manuf Eng 20:503–506
7. Isik Y (2007) Investigating the machinability of tool steels in turning operations. Mater Des 28:1417–1424

8. Lalwani DI, Mehta NK, Jain PK (2008) Experimental investigations of cutting parameters influence on cutting forces and surface roughness in finish hard turning of MDN250 steel. J Mater Process Technol 206:167–179
9. Rafi NH, Islam MN (2009) An investigation into dimensional accuracy and surface finish achievable in dry turning. Mach Sci Technol 13(4):571–589
10. Islam MN, Boswell B (2011) An investigation of surface finish in dry turning. In: Lecture notes in engineering and computer science: proceedings of the world congress on engineering 2011 (WCE2011), London, 6–8 July 2011, pp 895–900
11. Australian Standard (1982) AS2536 surface texture. Standards Association of Australia
12. Park SH (1996) Robust design and analysis for quality engineering. Chapman & Hall, London
13. Ross PJ (1988) Taguchi techniques for quality engineering. McGraw-Hill, New York
14. MatWeb (2011) Material property data. http://www.matweb.com/. Accessed 19 Feb 2011
15. Dhar NR, Kamruzzaman M, Ahmed M (2006) Effect of minimum quantity lubrication (MQL) on too wear and surface roughness in turning AISI-4340 steel. J Mater Process Technol 172:299–304
16. Lima JG, Avila RF, Abrao AM, Faustino M, Davim JP (2005) Hard turning: AISI 4340 high strength low alloy steel and AISI D2 cold work tool steel. J Mater Process Technol 169:388–395

# Chapter 33
# Impact of Radio Frequency Identification on Life Cycle Engineering

**Can Saygin and Burcu Guleryuz**

## 1  Introduction

Life cycle engineering is a decision-making methodology that considers product performance, functionality, environmental impact, and cost requirements for the duration of a product. Various concepts and methodologies, such as concurrent engineering (CE), design for environment [1], and environmentally conscious design and manufacturing (ECD&M) [2], are considered within the umbrella of life cycle engineering. Life cycle engineering, similar to an engineering design activity, involves iterative phases of design, planning, control, and assessment of all product life cycle related operations, as shown in Fig. 33.1.

A product life cycle includes a sequence of activities from design to material acquisition to disposal, all influenced by the designer [3–5]. A generic life cycle template, with various end-of-life options, is shown in Fig. 33.2.

Life cycle engineering is a complicated process and it is not possible to obtain an optimum, yet realistic, solution that satisfies all stakeholders. Therefore, sustainability principles and environmental responsibility concepts, guidelines, policies, and regulations must be introduced into the design of products and related processes and systems [6–8]. A comprehensive assessment of the product life cycle is essential to understand the impact of each decision variable on the environment and if possible to prevent or minimize the shifting of environmental impact from one stage to another in the life cycle. For a comprehensive review of environmentally conscious manufacturing and end-of-life (i.e., product recovery) issues in life cycle engineering, refer to [9, 10].

C. Saygin (✉) • B. Guleryuz
Mechanical Engineering Department, The University of Texas at San Antonio,
San Antonio, TX 78249-0670, USA
e-mail: can.saygin@utsa.edu

S.-I. Ao and L. Gelman (eds.), *Electrical Engineering and Intelligent Systems*,           403
Lecture Notes in Electrical Engineering 130, DOI 10.1007/978-1-4614-2317-1_33,
© Springer Science+Business Media, LLC 2013

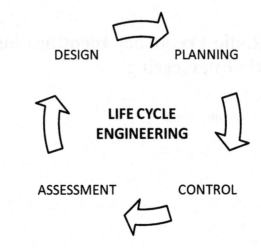

DESIGN      PLANNING

**LIFE CYCLE ENGINEERING**

ASSESSMENT      CONTROL

**Fig. 33.1** Life cycle engineering

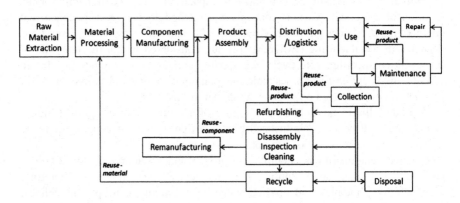

**Fig. 33.2** A generic life cycle model with end-of-life options

Each function in the life cycle of a product is typically dictated by the product design stage, which in turn has significant impact on life cycle performance metrics. Life cycle engineering goes beyond traditional supply chain management due to its holistic approach: (1) it captures not only forward logistics but also reverse logistics; (2) it focuses on a broader range of performance metrics, such as energy consumption and environmental impact, in addition to traditional production-related metrics.

Reverse logistics encompasses operations due to material reuse through recycling, component reuse through remanufacturing, and product reuse through maintenance, repair, collection, and refurbishing. Each activity shown in Fig. 33.2 utilizes energy and generates waste and various by-products that are not shown in the figure to keep it simple.

With such a broad perspective on products, processes, and systems, life cycle engineering can greatly benefit from technologies that provide operational visibility and facilitate effective design, planning, control, and assessment. Radio frequency identification (RFID) is such a technology that can provide stakeholders involved in the life cycle of a product (or a service) with the identity of each tagged entity in an automated and timely manner. The "digital visibility" of tagged entities can then trigger other information residing in corporate information systems to be retrieved in order to understand the status of these entities and to take necessary actions. In general, visibility can be defined as the extent to which the location of an entity is known in a system. For instance, not knowing where a certain inventory item is (i.e., lack of visibility), a manufacturing supervisor has to take into consideration "inventory search time" into overall manufacturing lead time approximation to quote on a job. In this case, lack of visibility is compensated by additional time. In other instances, additional capacity or additional inventory could be used to compensate for operational uncertainty due to lack of visibility. While these solutions can help with decision-making, they add cost associated with time, inventory, and capacity to the overall operation in the long run. RFID technologies can be deployed to avoid such ineffective solutions.

This chapter is organized as follows. An overview on RFID basics is given is Sect. 2. Problem definition is given in Sect. 3, which also discusses the case study at supply chain, facility, and station levels. Finally, Sect. 4 presents the conclusions.

## 2  Radio Frequency Identification

RFID has received a great deal of attention for its potential ability to perform noncontact object identification and to provide visibility at the point of use in a variety of different industries [11]. Although RFID is not a new technology as it dates back to the techniques developed to differentiate "friendly" aircraft from enemy warplanes in World War II, recent developments in computer technology and electronics have combined to make the RFID technology potentially viable for commercial purposes [12].

A typical RFID system consists of three components: (1) an electronic data carrying device, called a transponder or tag, (2) antennas and readers that facilitate tag interrogation, and (3) software, called middleware, that controls the RFID equipment, manages the RFID data, and distributes information to other remote data processing systems by interfacing with enterprise applications. An RFID system can be considered a wireless communication system since the reader communicates with the tags by using electromagnetic waves at radio frequencies [13]. RFID systems can be categorized as active and passive systems. In an active system, the tag (i.e., active RFID tag) has its own power source, which is a battery, enclosed in the transponder housing. In a passive system, the tag does not have its own power source; instead, it draws power from the reader's radio signals. Passive tags are inexpensive compared to active tags.

Information is the fuel that drives the economy and the society today [14]. As manufacturing operations go increasingly global, proper coordination among business and manufacturing units can be provided by sharing information in a timely manner [11]. Similarly, market and other uncertainties can be reduced and better managed by sharing information instead of building up inventories [15]. From supply chain level operations to shop floor level manufacturing execution, deploying RFID technologies can help facilitate information sharing and provide visibility in processes [16]. Further, with the existence of proper infrastructure, RFID can improve real-time exchange of data between locations and entities in a logistics network, facilitating better and more accurate information flow.

## 3  Problem Definition

Aerospace manufacturing and maintenance operations are very complex and involve the use of perishable materials packaged in a variety of containers, such as time and temperature-sensitive chemicals (TTSC) presented in this paper. Due to their only a few weeks long shelf life and $-100°F$ $(-73°C)$ storage requirement, TTSC require efficient means of tracking in terms of their consumption and storage.

In this study, an industrial RFID application for effective management of TTSC at three levels of deployment, as shown in Fig. 33.3, is presented as a case study. The problem was initiated at the manufacturing facility level. The major concern was to ensure that the required TTSC are available at all times to the operators to maximize the service level. First, the lack of TTSC results in loss of production and, in turn, loss of profits. Second, the chemicals that are not used in production within their lives expire and become another cost factor. Third, disposal of time-sensitive materials, once they reach their shelf life, to prevent their usage on a product is also major concern. In order to ensure the first objective (i.e., maximize service level), the manufacturer prefers to order higher quantities of variety of chemicals.

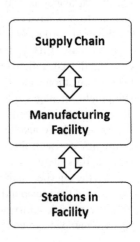

**Fig. 33.3** Levels of RFID
deployment

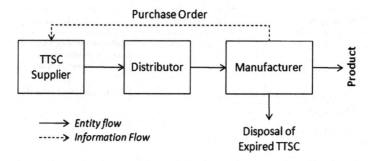

**Fig. 33.4** Supply chain level operation "as-is"

However, a large number of chemicals on the shop floor become an inventory management problem, as well as larger quantities expire, simply because they were ordered more than needed, becoming a disposal problem. They become a cost issue, as well as an environmental impact factor.

At the supply chain level, TTSC supplier receives orders from the manufacturer, as shown in Fig. 33.4, typically in large quantities. Detailed business processes are depicted in Fig. 33.5. Large volume is an advantage for the supplier and they operate as a mass production facility. The large quantity and high variety of TTSC requires a distributor, as the middle man, to manage the logistics of TTSC. The distributor allocates them in temperature-controlled buffers in the manufacturer's facility. Due to $-100°F$ ($-73°C$) storage requirement, the TTSC containers are typically covered with ice, which makes it very difficult to visually tell which type of chemical is in which container. The operators at the station level go through several containers in order to find the one they need for production.

## 3.1 Facility Level Inventory Management

The manufacturing environment presented in this case study involves tracking of approximately 5,000 TTSC items, which contain over 150 varieties, stored in approximately 100 buffers (i.e., each buffer holds 500 inventory items) that are used for assembly of a high-value product. The current practice is to order a higher quantity of materials than necessary determined by the baseline inventory of each buffer in order to attain a high service level, which is defined as the percentage of shop floor orders met on time. Inventory in each buffer is tracked manually. However, due to the pressures to complete the orders on time, operators who do not have the right inventory item in their designated buffers "borrow" inventory from other buffers. This undesired borrowing of items lead to discrepancy between the manually tracked inventory data and the actual inventory in each buffer, which leads to a lower service level and higher amount of expired materials. Overall, lack of real-time visibility on inventory levels leads to wasteful activities that add cost and increase lead time.

408 C. Saygin and B. Guleryuz

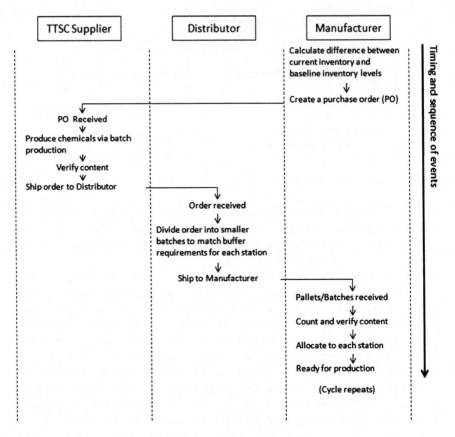

**Fig. 33.5** Business processes (simplified out of 54 steps) "as-is"

The RFID-based model to overcome the aforementioned deficiencies involves implementing a facility-wide decision-making model that utilizes RFID data coming from tagged inventory items stored in the buffers. The location of each buffer combined with time-stamped tag ID can be used to capture the actual status of each buffer and incoming and outgoing materials, thus inventory replenishment can be carried out in a more efficient manner.

The decision-making model uses a trend-adjusted exponential smoothing algorithm [11]. It uses two smoothing parameters, $0 < \alpha < 1$ and $0 < \beta < 1$, as coefficients for the average production demand and its trend, respectively. The adaptive inventory scheme looks at the difference between the current inventory level of a particular inventory item at a buffer and the associated forecast (i.e., predicted demand) in order to determine the amount of material that needs to be ordered. Total amount of materials needed is calculated by simply adding the required amount at each buffer. The purchase order is generated automatically due to the availability of RFID-based data and is shared in real-time with the materials distributor. The "operational visibility" gives the materials distributor enough time to plan and replenish buffers more effectively.

**Table 33.1** Smoothing parameters and performance measures

| $\alpha$ | $\beta$ | Expired | Normal use | Ordered | Stock-out | Service level | Reallocated items |
|---|---|---|---|---|---|---|---|
| 0.2 | 0.2 | 8[a] | 1 | 8 | 1 | 1 (96%) | 7,185 |
| 0.2 | 0.5 | 2 | 4 | 1 | 4 | 3 | 14,508 |
| 0.2 | 0.8 | 5 | 5 | 2 | 5 | 5 | 16,700 |
| 0.5 | 0.2 | 3 | 3 | 4 | 3 | 4 | 15,044 |
| 0.5 | 0.5 | 7 | 9 | 3 | 9 | 9 (91.5%) | 18,175 |
| 0.5 | 0.8 | 4 | 7 | 5 | 7 | 7 | 19,257 |
| 0.8 | 0.2 | 1 | 8 | 6 | 8 | 8 | 18,077 |
| 0.8 | 0.5 | 6 | 6 | 7 | 6 | 6 | 19,903 |
| 0.8 | 0.8 | 9 | 2 | 9 | 2 | 2 | 18,930 |
| Difference[b] | | 150 | 2,900 | 2,100 | 2,850 | 4.5% | 13,000 |

[a]The lower the relative weight the better (i.e., a lower relative weight is desired)
[b]Difference (i.e., variation of results) between the best (1) and the worst (9) results under each performance measure

In order to demonstrate the validity of the proposed model, a simulation study was carried out, which benchmarks the current practice against the new model. The simulation model is built around a simplified version of the actual manufacturing environment.

It includes 18 buffers, 23 different types of chemicals, a total inventory level of approximately 5,000 items, and 10 replications of a production period of 7 months. The simulation study uses six performance measures to evaluate the overall performance of the two inventory models. These are defined as expired, normal use, ordered, stock-out, reallocated, and service level. Expired refers to the total number of TTSC that expire, which needs to be disposed of in an environmentally friendly manner. Normal use is the total number of time-sensitive materials that are available in designated buffers and are used in production. Ordered is the total number of time-sensitive materials ordered. Stock-out is the total number of TTSC requested by an operator but was not available in the buffer at the time of the demand. Reallocated is the total number of TTSC an operator obtains from an adjacent station.

The proposed model was tested via simulation using different values for the two smoothing parameters $\alpha$ and $\beta$ in order to determine their most effective values. As shown in Table 33.1, each level-combination of the smoothing parameters, $\alpha$ and $\beta$ (i.e., sub-models), yields a different performance for different performance measures. For instance, on the one hand, the sub-model, where $\alpha = 0.8$ and $\beta = 0.2$, yields the best result for Expired. On the other hand, the same sub-model ranks number 8 for normal use, stock-out, and service-level metrics. Similar comparisons can be made for the other sub-models.

The relationship among the performance measures needs to be taken into consideration when analyzing the results. For instance, the higher the number of chemicals ordered, the higher the number of inventory items likely to expire. However, in this case, it is more likely that operators will not run out of stock due to higher level of inventory items available for production. Overall, since $\alpha = 0.2$ and $\beta = 0.2$ yield the highest service level, which is the primary

**Table 33.2** Proposed model ($\alpha = 0.2$ and $\beta = 0.2$) against current practice

| Comparison | Service level | Expired | Normal use | Stock-out | Reallocated | Ordered |
|---|---|---|---|---|---|---|
| Current practice | 91% | 9,310 | 59,897 | 6,179 | 7,046 | 74,363 |
| Proposed model | 95% | 1,957 | 62,999 | 3,265 | 7,185 | 66,437 |

performance metric identified by the team at the manufacturing site, with the lowest number of reallocations, $\alpha = 0.2$ and $\beta = 0.2$ combination is selected to be the best combination of all nine sub-models.

After the most effective settings for the smoothing parameters are identified, the proposed model using these settings is benchmarked against the current practice. As shown in Table 33.2, the proposed model outperforms the current practice with higher service level, fewer amount of expired materials, more items withdrawn from designated buffers (i.e., normal use), fewer stock-outs, and much less ordered. Although there are relatively more reallocated items in the proposed model, it is due to the visibility provided by RFID that facilitates "controlled" inventory sharing; items unavailable at designated buffers are tracked easily and reallocated from other buffers in order to maximize service level.

## 3.2  Supply Chain Level Inventory Management

After the validation of the RFID-based inventory management model, the presence of RFID at facility level is investigated in terms of its impact at the supply chain level. There are three business partners involved (Fig. 33.4): supplier, distributor, and manufacturer. The supplier produces TTSC using mass production strategy; the larger the order quantities, the better for the supplier. The distributor tracks the orders at the manufacturing site and replenishes them twice a week. The process flow of the current practice at the supply chain level includes 54 business processes, as shown in Fig. 33.5 as a simplified process. These processes involve a variety of nonvalue added activities, such as coding forms by hand, data entry using computer keyboard, regrouping inventory several times, sorting, error checking, data verification, and reconciliation.

The current process flow is analyzed to determine nonvalue added activities. Then, a second sub-group out of the nonvalue added group is selected by considering "what if RFID was deployed." A thorough investigation of each step, along with interviews with experts and additional data collection, the RFID-enhanced value stream map is reduced to 18 steps, as opposed to 54, which represents 66.7% improvement in number of steps. Such an improvement not only makes the supply chain more reliable but also reduces the overall lead time drastically, which is estimated to be a reduction of 60%.

The supply chain level study revealed that technologies, such as RFID, allow supply chain partners to redefine the rules of business and roles of participants within the life cycle of a product (or a service). For instance, the RFID-enhanced supply chain operations lead to elimination of cross-docking and staging, which are carried out by the distributor in the as-is (non-RFID) practice. Elimination of such nonvalue added activities reduces the involvement of the distributor, thus diminishes its business potential. If the further deployment of RFID technologies is considered, then the supplier can actually deliver items directly to the manufacturing site and the reason for existence of the materials distributor becomes questionable. Such an approach removes the middle man and leads to the "vendor managed inventory" concept where the materials manufacturer is authorized to manage inventories at the customer site and to make decisions, such as when and how much inventory to ship.

### 3.3 Station Level Inventory Management

In passive RFID systems, information from tags is modulated (i.e., backscatter process) onto the reader carrier signal and reflected back to the reader. Read-rate is defined as the ratio of the number of tags that are read over the total number of tags in the interrogation zone of the reader. Due to various reasons, there could be unreadable tags in the interrogation zone of the reader, and its normal operation is disrupted leading to read-rates lower than 100%, which is contradictory to RFID's premise about "total visibility." The reasons include radio frequency (RF) interference, RF absorbing materials, and environmental conditions [17, 18]. In this study, $-100°F$ storage requirement causes excessive ice buildup, rendering RFID technology unreliable. In addition, RF absorbing chemicals also adversely affect read-rates and lead to longer reading times.

In order to overcome such technological deficiencies, an RFID-based smart freezer is developed and implemented [19]. The freezer utilizes systematic selection of antenna configuration and antenna power control to maximize read-rates. The RFID antenna configuration design methodology ensures a 99% read-rate of items while minimizing the required number of RFID antennas in the confined cold chain environments with non-RF friendly materials. The RFID-based smart freezer performance is verified through lab testing and on-site prototyping on an industrial freezer operating at $-100°F$.

## 4  Conclusion and Future Work

RFID is an enabling technology to increase operational efficiency when combined with effective decision-making models that utilize RFID data [20]. Without coupling it with decision-making, it does not by itself bring solutions to life cycle engineering issues. RFID technology simply facilitates visibility in a process. Tools, standards, and roadmaps that lead to effective utilization of such visibility to improve performance, reduce cost, and expedite decision-making are crucial for a successful RFID deployment.

In this paper, the case study clearly shows that a successful RFID project should have an answer to each of the following in an integrated manner: (1) RFID-specific technological constraints (read-rate issues at freezer level) at the point of use; (2) typically operational issues at facility level (such as operators "borrowing" chemicals from other buffers) that can be controlled by introducing new policies; and (3) beyond the four walls of a manufacturer, supply chain issues that need to be resolved.

Due to variety of applications, functions, and information requirements of decisions, it is not possible to simply rely on RFID tags over the whole product life cycle. During the life cycle of a product, voluminous data and information, such as product design data, process plans, engineering documents, etc., are created, revised, exchanged, transferred, stored, merged, and converted into different forms. Some applications need more information than just product ID; specific sensors can be used to gather such data. Others may need a combination of RFID data and more advanced processing and data storage ability for decision-making. Therefore, modeling of product life cycle information must be the first step in order to eventually determine to what extend RFID can be a solution.

Within the scope of this paper, RFID provides several advantages for making effective end-of-life decisions. First, identifying the chemicals automatically removes all the manual identification processes that were in place before deploying RFID. Second, tracking such chemicals and documenting their disposal process are required by federal laws; manual documentation methods were replaced by RFID-driven automatic documentation procedures. Last but not least, RFID facilitates a closed-loop control mechanism that ensures "material balance" among purchasing, usage, and disposal of chemicals, which leads to reduction in consumption without losing productivity.

# References

1. Sun J, Han B, Ekwaro-Osire S, Zhang H-C (2003) Design for environment: methodologies, tools, and implementation. Trans Soc Des Process Sci 7(1):59–75
2. Wanyama W, Ertas A, Zhang H-C, Ekwaro-Osire S (2003) Life-cycle engineering: issues, tools, and research. Int J Comp Integr Manuf 16(4–5):307–316
3. Alves C, Ferrao PMC, Preitas M, Silva AJ, Luz SM, Alves DE (2009) Sustainable design procedure: the role of composite materials to combine mechanical and environmental features for agricultural machines. Mater Des 30:4060–4068
4. McAloone TC, Andreasen MM (2004) Design for utility, sustainability, and societal virtues: developing product service systems. In: International design conference—design 2004, May 18–21, Dubrovnik, Croatia
5. Borg JC, Yan X-T, Juster NP (2000) Exploring decisions' influence on life-cycle performance to aid design for multi-X. Artif Intell Anal Manuf 14:91–113
6. Kobayashi H (2006) A systematic approach to eco-innovative product design based on life cycle planning. Adv Eng Inform 20:113–125
7. Kobayashi H (2005) Strategic evolution of eco-products: a product life cycle planning methodology. Res Eng Des 16:1–16

8. Rosa C, Stevels A, Ishii K (2000) A new approach to end of life design advisor. In: Proceedings of the 2000 IEEE international symposium on electronics and the environment, Oak Brook, pp 99–104
9. Gungor A, Gupta SM (1999) Issues in environmentally conscious manufacturing and product recovery: a survey. Comp Ind Eng 36:811–853
10. Ilgin MA, Gupta SM (2010) Environmentally conscious manufacturing and product recovery (ECMPRO): a review of the state of the art. J Environ Manag 91:563–591
11. Mills-Harris MD, Soylemezoglu A, Saygin C (2007) Adaptive inventory management using RFID data. Int J Adv Manuf Technol 32:1045–1051
12. Soylemezoglu A, Zawodniok MJ, Sarangapani J (2009) RFID-based smart freezer. IEEE Trans Ind Electron 56(7):2347–2356
13. Keskilammi M, Sydanheimo L, Kivikoski M (2003) Radio frequency technology for automated manufacturing and logistics. Part 1: passive RFID systems and the effects of antenna parameters on operational distance. Int Adv Manuf Technol 21:769–774
14. Wyld DC (2005) RFID: the right frequency for government. IBM Center for Business of Government (E-Government series), Washington
15. Shaw MJ (2000) Information-based manufacturing with the web. Int J Flex Manuf Syst 12(2/3):115–129
16. Saygin C, Sarangapani J, Grasman S (2007) A systems approach to viable RFID implementation in the supply chain. In: Jung H, Chen F, Jeong B (eds) Trends in supply chain design and management: technologies and methodologies. Springer, London, pp 3–27. ISBN 978-1-84628-606-3
17. Clarke RH, Twede D, Tazelaar JR, Boyer KK (2006) Radio frequency identification (RFID) performance: the effect of tag orientation and package contents. Packaging Technol Sci 19 (1):45–54
18. Cha K, Zawodniok MJ, Ramachandran A, Sarangapani J, Saygin C (2006) Interference mitigation and read-rate improvement in RFID-based network-centric environments. J Sens Rev 26(4):318–325
19. Soylemezoglu A, Zawodniok MJ, Cha K, Hall D, Birt J, Saygin C, Sarangapani J (2006) A testbed architecture for auto-ID technologies. Assemb Autom 26(2):127–136
20. Saygin C, Guleryuz B (2011) Impact of radio frequency identification (RFID) on life cycle engineering. In: Lecture notes in engineering and computer science of the world congress on engineering 2011 (WCE 2011), London, 6–8 July 2011, pp 855–860

# Index

**A**

Adaptive algorithm, 35, 36
Architecture, 32, 40, 43–46, 49, 51, 84, 85, 136–140, 172, 184, 196, 198, 200, 203
Assembly, 40, 43, 45, 46, 48–50, 65, 66, 109–114, 117, 118, 286, 293, 307, 361–366, 368, 371, 407
Asset maintenance, 263, 264, 268
Asymmetry, 231, 236
Auction, 82–92
Authentication, 122, 127, 131–133, 172, 202, 204
Automotive components manufacturer, 327–334

**B**

Bayesian networks, 110, 115–118
Bid adjustment, 81–92
Binding, 220, 223, 225
Black–Litterman approach, 249–260

**C**

Closed loops, 81–92, 314, 361, 364, 368, 371, 412
Complementary n-tuples, 23, 24, 26
Complex stochastic Boolean systems, 15–26
Contemporary, 118, 263–270, 328
Context, 16, 109–113, 118, 148, 149, 196, 199, 202, 203, 205, 207–209, 211, 266, 328, 362, 368, 370, 371
Control, 54, 55, 58, 81, 88, 109–118, 137, 140, 171, 173, 184, 195–198, 201, 202, 205, 210, 285, 292, 293, 305, 306, 309, 328, 330, 331, 337, 347, 348, 358, 362, 367, 368, 384, 403, 405, 411, 412
Critical success factors (CSFs), 286, 288, 289, 297, 327
Cryptography, 39
CSFs. *See* Critical success factors (CSFs)
Custom instruction, 39–51
Cutting force, 347, 348, 350, 352–355, 358
Cutting parameters, 398–401

**D**

Data prefetch, 29, 37
Design capabilities (DC), 285–297
Deteriorating items, 375–385
Displacement sensor, 348–351, 355, 357, 358
Duality, 15–26
Dynamic environments, 81, 83

**E**

E-commerce, 207–211, 216, 217
Economic analysis, 337–339, 343
Efficient market hypothesis (EMH), 251, 260
Energy and entropy features, 122, 124, 127, 128, 133
Exhaust, 293, 301–310

**F**

Financial crisis, 231, 240, 247, 259
Fingerprinting, 155–168
Fitness, 69, 72, 75–78, 227–228
Futures, 232, 236
Fuzzy logic, 53–62, 136, 155–168

S.-I. Ao and L. Gelman (eds.), *Electrical Engineering and Intelligent Systems*,
Lecture Notes in Electrical Engineering 130, DOI 10.1007/978-1-4614-2317-1,
© Springer Science+Business Media, LLC 2013

**G**
Genetic algorithms, 53, 65–79, 337–344
Granular computing, 135–146
Granular neural network, 135, 136

**H**
Hash tables, 34, 35
Heart
    murmurs detection, 147–153
    sounds analysis, 149
High precision machining, 347–358
Hypothesis test, 1–12

**I**
IEEE 802.11b, 174, 186
IEEE 802.11g, 174, 186
Image segmentation, 135, 141, 142
Individualised, 263–270
Indoor location, 155–168
Intention to purchase (ITP), 210, 213,
    215, 216
Intrinsic order, 16–26
Intrinsic order graph, 16, 19–26
Inventory management, 407–410
Investment decision making, 338,
    343, 344
ITP. *See* Intention to purchase (ITP)

**L**
Lean manufacturing, 327–334
Life cycle engineering, 403–412

**M**
Malaysia, 285, 288, 290, 294, 297,
    328–331
Malliavin calculus, 240, 241, 247
Metalogic, 95–108
Meta-reasoning, 96, 104, 105, 107
Microarchitecture, 43–45, 51
Mobile computing, 109
Modeling, traffic, 304
Monte Carlo simulation, 338, 339, 344
Multi-robot, 81, 82, 84, 88, 89
Multistage multiproduct production
    lines, 365

**N**
Number of failures, 273–283

**O**
Objective criteria, 148, 150
Optimal hedge ratio (OHR), 231–236
Optimisation, 32, 34, 53, 57, 58, 65–79,
    249–260, 316, 319–322, 324, 337, 343,
    344, 365
Options, 239, 241, 243, 287, 376, 377, 379,
    384, 403, 404
OWL 2, 96–103, 105, 107

**P**
Palmprint, 121–133
Pareto ANOVA analysis, 391–396
Partial exploration, 1–12
Pattern search, 159–163, 169
PD. *See* Product development (PD)
Performance, 29, 40, 53, 81, 128, 141, 148,
    156, 171, 183, 217, 251, 266, 275, 286,
    301, 331, 361, 403
Permutations algorithms, 39, 46, 48, 49
Petri nets, 1–5, 10
Phishing, 207–217
3PL. *See* Third party logistics service
    provider (3PL)
Portfolio optimization, 249–254, 257–259
Post-sales, 313–324
Precedence-constrained sequencing and
    scheduling, 65–79
Prediction limits, 273–283
Preexecution, 29–37
Price sensitivities, 239–247
Product development (PD), 285–290, 292,
    294–297, 404, 412
Proton, 285–297, 330, 331

**Q**
Quality-of-service (QoS), 195–205
Quantity discounts, 375–385
Quota sampling, 2, 5–9, 11

**R**
Radio frequency identification (RFID), 155,
    157, 403–412
Reliability centred maintenance (RCM), 263,
    265–269
Reverse logistics, 313–324, 404
RFID. *See* Radio frequency identification
    (RFID)
RISC architecture, 49, 51
Robotics, 109, 113, 116

Robust optimization, 316, 319, 320, 324
Rules, 53, 56, 57, 60, 95–108, 160, 162, 254, 266, 290, 341, 343, 410

**S**
Security, 41, 118, 171–181, 184, 192, 195–205, 207–210, 212, 214, 216, 249, 250, 253
Semantic web ontologies, 95
Sigmoid, 122, 124, 125, 127–129, 131–133
Simulated annealing, 53–62
Simulation, modeling, 338, 361, 365, 367–369, 371, 409
SIPD. *See* Supplier involvement in product development (SIPD)
Spindle error motion, 348
Stack architecture, 49, 51
Stackelberg game, 376, 385
State space, 1–12
Supplier involvement in product development (SIPD), 286, 287, 290, 292, 294, 296
Support vector machine (SVM), 126–130, 133
Surface roughness, 352, 390–398, 400
SVM. *See* Support vector machine (SVM)

**T**
Taguchi method, 391
Task allocation, 67, 81–92
Taxonomy, 81, 113, 211
Third party logistics service provider (3PL), 314, 315, 323
Total productive maintenance (TPM), 263, 265–269
Total profit, 66, 69, 75, 376–385
Toyota production system (TPS), 327, 329, 331–333

TPM. *See* Total productive maintenance (TPM)
TPS. *See* Toyota production system (TPS)
Tracking, 406, 407, 412
Transformation, 12, 30, 34, 219, 220, 223–227, 229, 275, 328
TRP ratio, 249–260
Type-2 fuzzy systems, 54–57

**U**
Uncertainties, 54, 55, 83, 84, 87, 89, 91, 92, 118, 207–217, 406
US equity market, 232, 236

**V**
Vertical handover, 195, 196, 199–201, 204–205
Video segmentation, 135–146
Viterbi decoder, 40–42

**W**
Weibull distribution, 273–280, 282, 283
WEP point-to-point links, 184, 192
Wi-Fi, 171–181, 183–192
Willingness to pay (WTP), 209, 213, 215, 216
Wireless network laboratory performance measurements, 172, 176, 184, 192
WLAN, 171, 183, 184, 195
WPA2 point-to-point links, 172
WTP. *See* Willingness to pay (WTP)

**Y**
Y-Comm, 196, 199–205